Plastics Application Technology for Safe and Lightweight Automobiles

Other SAE books of interest:

Composite Materials Handbook (CMH-17), Polymer Matrix Composites, Volumes 1–3
(Product Code: R-422.SET3)

Care and Repair of Advanced Composites, Second Edition
By Keith B. Armstrong
(Product Code: R-336)

Engineering Plastics and Plastic Composites in Automotive Applications
By Kalyan Sehanobish
(Product Code: T-122)

Plastics Application Technology for Safe and Lightweight Automobiles

Edited by Sudhakar R. Marur

INTERNATIONAL.

Warrendale, Pennsylvania, USA

INTERNATIONAL

400 Commonwealth Drive
Warrendale, PA 15096-0001 USA

E-mail: CustomerService@sae.org
Phone: 877-606-7323 (inside USA and Canada)
724-776-4970 (outside USA)
Fax: 724-776-0790

ISBN 978-0-7680-7640-0
SAE Order Number R-415
DOI 10.4271/R-415

Library of Congress Cataloging-in-Publication Data
Marur, Sudhakar R.
Plastics application technology for safe and lightweight automobiles / by Sudhakar R. Marur.
pages cm
"SAE order number R-415"—Title page verso.
Includes bibliographical references.
ISBN 978-0-7680-7640-0
1. Plastics in automobiles. 2. Automobiles—Safety measures. 3. Lightweight materials. 4. Reinforced plastics. I. Title.
TL240.5.P42S27 2013
629.2′32—dc23
2013016403

To purchase bulk quantities, please contact:
SAE Customer Service
Email: CustomerService@sae.org
Phone: 877-606-7323 (*inside USA and Canada*)
724-776-4970 (*outside USA*)
Fax: 724-776-0790

Visit the SAE International Bookstore at
BOOKS.SAE.ORG

Illustrations credit: P. Arunachala, B. Deshmukh

Dedication

The late Dr. Wim Bruijs
Director, Application Technology–Europe
SABIC Innovative Plastics
Bergen op Zoom
The Netherlands

Table of Contents

Preface

This book is a compilation of the collective experience and expertise of a team in building successful plastics applications for the automotive market for well over a decade. The genesis was an experimental business model, during the peak of the globalization era, to have a global center of excellence in Bangalore, India, for driving greater market penetration through application technology, with a special emphasis on computational modeling. The business model became successful within the very first year of its inception, and during the years that followed, thousands of plastics applications were designed and successfully launched for customers around the world. In addition, innovations during the course of these projects spawned many publications at the annual SAE International Congress and in peer-reviewed scientific journals, as well as several patents.

With the accumulation of experience on the development and deployment of various technologies, the need was felt to have the quintessence of this experience captured in a form that would benefit a wide spectrum of people—students of plastics and engineering, researchers, practicing professionals, academicians, and anyone interested in automotive technologies and thermoplastic material applications. The result is this volume, wherein each chapter focuses on an automotive segment covering various aspects of application technology—*from concept generation to part completion*—through relevant case studies.

Many senior leaders of SABIC Innovative Plastics have substantially contributed to the conceptualization and successful implementation of the global center of excellence model for building plastic applications, through an internal initiative known as *Factory Direct*, enabling direct contact of end customers with the design and engineering team. Notable among them are Charlie Crew, Greg Adams, Wim Bruijs, Jeroen Verhoeven, and Rick Pontillo. In addition, many leaders of the automotive business unit such as Stephen Shuler, Jim Wilson, Derek Buckmaster, Frank Mooijman, Geert-Jan Doggen, and UV have significantly supported and nurtured the growth of the team. Some of the finest talent in mechanical engineering and plastics engineering have contributed to the development of technology as well as to various chapters of this book. I wish to express my heartfelt thanks to the leadership and the wonderful team of brilliant colleagues at SABIC Innovative Plastics.

Finally, a special thanks to SAE International for publishing this.

Sudhakar R. Marur
Bangalore, October 24, 2012

Chapter 1
Introduction to Plastics Application Technology

Sudhakar R. Marur

1.1 Introduction

Plastics application technology deals with the paradigm of replacing any material such as metal, glass, and ceramics with plastics in a new or existing application. Application technology helps to position a new plastic material, as it is developed, for a new application; conversely, it directs new material development from the application requirements.

Traditionally, metal had been the primary choice for any auto build because it carried a rich legacy of knowledge, expertise, and infrastructure not only for its manufacturing but also for its target applications. In addition, metal offered a variety of benefits such as mechanical strength, stiffness, high-temperature performance, and better dimensional stability. It possessed resistance to ultraviolet rays and, more importantly, carried "perceived quality" in the eyes of the end customer. And last but not least, it has proven itself as a reliable material over time.

However, with auto design trends getting influenced by factors such as competitive cost model, styling differentiation, and driving comfort in general, and weight reduction, regulation-based safety compliance, and eco-friendliness [1-1] in particular, thermoplastics emerged as the candidate material for automotive applications.

When the cost and complexity of plastic and metal parts are compared, the cost of the latter increases dramatically as the part complexity increases [1-2]. When the forming complexity increases, the cost and performance advantages are arguably in favor of plastics.

In general, plastics can offer various advantages on the performance front, including weight reduction, parts consolidation, design freedom, ease of incorporation of new functionality, and resistance to corrosion and dents. On the cost side, this material can result in system-level cost reduction, elimination of secondary operations, ease of fabrication, shortened production cycle time, and higher tool life (for more details, see Section 1.4).

1.2 Application Development Cycle

An application to be built with plastics would typically have to go through some or all the phases of the application development cycle described in the following sections.

1.2.1 Voice of the Customer

Understanding the stated and unstated needs of the customer and the market is essential for building an application with plastics. Once the voice of the customer (VoC) is clearly understood, the right mix of material, design, manufacturing, and cost models can be used to arrive at the best application development framework.

1.2.2 Benchmarking

Benchmarking is a key phase, wherein various aspects of the proposed plastics application are compared and contrasted with other available applications. Also, it is quite common to adopt the process of teardown, by which an assembly is systematically dismantled into its constituent parts and subparts, so that the complete details of the entire assembly are clearly understood. The clarity thus obtained from benchmarking forms the basis for generating new concepts for functionality, material positioning, design, manufacturing, and even assembly.

1.2.3 Material Selection

One of the critical elements of application development is the identification of right material. To choose the right material, one has to have proper *a priori* understanding of end-use requirements of the application (elaborated in Section 1.3).

1.2.4 Styling and Industrial Design

Fusing aesthetics with functionality, styling brings in greater market differentiation to any product by appealing to the emotional needs of the end customer. Industrial design plays a very significant role in automotive applications, in both the exterior and interior segments of automotive.

1.2.5 Computer-Aided Design

Computer-aided design (CAD) brings the new concepts generated at styling and industrial design stage into 3-D engineering frame work, using computational geometry and graphics.

1.2.6 Computer-Aided Engineering

Any part modeled through CAD is analyzed, using computer-aided engineering (CAE) simulation tools that involve computational mechanics algorithms, to predict its performance under loads such as structural, impact, thermomechanical, and thermal loads. Essentially, the design is carried out for stiffness; strength; long-term behavior such as creep, fatigue, and wear; and impact performance. In the case of thermal and heat-transfer problems, computational fluid dynamics (CFD) tools are used to compute the temperature distribution and hot spots within a part subjected to a heat source. The analysis output gives insight into the performance of the part, leading to the effective design of the same.

1.2.7 Process Modeling

Plastic products or parts can be manufactured through various manufacturing processes such as injection molding, thermoforming, blow molding, extrusion, rotational molding, and calendaring. Numerical modeling techniques simulate these manufacturing processes in order to obtain the ideal manufacturing process parameters to be set in the actual machinery. Also, design for appearance (by eliminating surface defects, sink marks, weld lines, etc.), precision (e.g., dimensional stability), and moldability are carried out in this phase.

1.2.8 Tooling

During the tooling phase of the application development cycle, the tool maker would make decisions about tool material, robust tool-building mechanism, design for extended tool life, better cycle time, and so forth, based on the criteria laid out for material choice, processing conditions, and production volume.

1.2.9 Prototyping

Prototypes can be made through either soft tooling or rapid prototyping. Prototyping tests whether the material selection is appropriate and ensures that the component or part requirements do not exceed the design limits. Potential problem areas in performance, manufacturing, and assembly can be identified through prototyping. Also, prototyping yields preliminary product performance information and enables prelaunch assessment of the component or product.

1.2.10 Secondary Operations

Secondary operations are those that are carried out during the post-molding period, such as machining, welding, adhesion, assembly, printing, laser marking, and in-mold decoration.

1.2.11 Part Testing

Product performance tests can be conducted on functional prototypes or production parts. During these tests, any part or prototype is subjected to various loads, including impact, structural, thermal, and weathering, to evaluate its performance under real-life conditions. The design can then be refined, if necessary. Typically, strain gauge, photoelasticity, and thermal emission–based stress analyses are employed for assessing structural performance; environmental chambers are used for thermal cycling and accelerated aging testing; and holography is used for the life assessment of a component. However, care must be taken in interpreting the results of functional prototypes, as they may have been produced using nonproduction tooling or part modeling.

1.3 Material Selection Methodology

A critical aspect of the material selection process [1-3] is understanding the end-use requirements of an application, which can be broadly classified into the categories described in the following sections.

1.3.1 Screening of Material Properties

Various material properties, relating to mechanical, thermal, rheological, electrical, chemical resistance, weatherability, and so forth, need to be carefully studied and analyzed before identifying a suitable candidate material for a given application.

1.3.2 Conversion Processes

The type of conversion process that will be employed (e.g., injection molding, thermoforming, blow molding, or extrusion) to make the end product is an important factor in the material selection.

1.3.3 Structural Requirements

Assessment of the type of loading (static, quasi-static, dynamic, impact, vibration, and thermomechanical) and rate of loading to which the final part would be subjected is a critical aspect in identifying the suitable material.

1.3.4 Environmental Conditions

When a material's service temperature exceeds its safe operational temperature range, the material's properties may degrade, and parts made of the material may not perform their intended functions. In addition, knowledge of the service environment of the component is critical, especially the potential for interaction with chemical agents and outdoor exposure for long periods leading to material degradation.

1.3.5 Assembly and Secondary Operations

Possible methods of assembly need to be identified, such as mechanical fastening, welding, and adhesive bonding. Also, information on the secondary operations, such as the type of painting, printing or hot stamping, is also required at this stage.

1.3.6 Cost Factors

Various factors that would significantly influence the cost of the end product—prospective material cost, appropriate processing method based on cycle time, tooling cost for the chosen processing method, service life of the component, and contribution margin from the sale of the product—need careful consideration at this phase.

1.3.7 Regulations and Standards Compliance

Information about the standards or regulations that are applicable to the end component or product in the marketplace is essential in choosing the appropriate material. Here are some of the well-known organizations whose standards are widely adopted for material characterization and testing:

- ISO: International Standards Organization [1-4]
- ASTM International: Standards Worldwide [1-5]
- UL: Underwriters Laboratories [1-6]

The following are organizations that formulate standards and regulations related to automotive crash and safety:

- SAE International [1-7]
- NHTSA: National Highway Traffic Safety Administration [1-8]
- Euro NCAP: European New Car Assessment Program [1-9]
- IIHS: Insurance Institute for Highway Safety [1-10]
- RCAR: Research Council for Automobile Repairs [1-11]

1.4 Advantages of Plastics

Some of the key advantages that the end user would benefit from, when an application is built or a part thereof is replaced with plastics include the following:

1.4.1 Styling Freedom

Plastics offers greater styling freedom for various automotive segments. Some of the styling-dependent parts are glazing for roof, tailgate, windows, and the like; body panels in general and fenders in particular; air intake grilles; headlamp assembly; and instrument panels.

1.4.2 Material Property

Alloys and blends of plastics can easily be produced for the given application requirements. Some examples are low-temperature ductility for impact applications [1-12], conductivity in materials [1-13, 1-14], and additive-based properties as in laser direct structuring (LDS) process [1-15], wherein organometallic additives are selectively activated by a laser beam for subsequent copper plating of electronic circuits on the surface of a molded part. Also, the material properties can be tailored to have better corrosion and chemical resistance.

1.4.3 Performance

Increased levels of performance can be achieved through the use of plastics in applications such as energy absorbers with very low packaging space at very low temperatures [1-16], dimensional stability of fenders

[1-17], dent performance of fenders at low speeds, better thermal performance of glazed panels [1-18, 1-19], and corrosion resistance of underbody parts and body panels.

1.4.4 Part Integration

Many parts assembled together through traditional methods of manufacturing can be integrated with plastics in a single phase, resulting in the elimination of secondary operations and reduction in time and cost associated with assembly. Some of the key examples are the integration of elements of inner panel of the tailgate, various parts of a front-end module [1-20], instrument panel with an integrated chute and seamless airbag door [1-21], energy-absorbing features in fenders, and in-mold decorated bezels.

1.4.5 Weight Reduction

Because of its low density and higher design flexibility compared to traditional materials, plastic parts could be made that are lightweight but have the required stiffness. Design features that are difficult to achieve in metals can be done quite easily with plastics, such as ribs, closed profiles, hollow structures, and beads, which help in achieving similar stiffness while being lightweight. In any application, the required performance can be obtained with a lighter weight part, subsystem, or system, resulting in lower fuel consumption and emissions.

1.4.6 System-Level Cost Reduction

Generally, the system-level cost of an assembly made with plastic parts is less than those made with other materials. Because of the feasibility of part integration and elimination of the secondary process, plastics processing requires smaller assembly lines, which results in a major reduction in assembly time and energy requirements, as well as lower initial investments, thus reducing the overall system cost.

1.5 Key Automotive Plastics Applications

1.5.1 Safety and Energy Management

Engineering thermoplastics can offer ideal solutions to automotive safety through effective crash energy management. With increased styling needs and consequently the gradual reduction of packaging space for the energy-absorber design, plastics have provided feasible options through both material properties (withstanding impact loads at temperatures varying from −40 °C [−40 °F] to 80 °C [176 °F]) and geometric designs for effective crushing. The plastic energy absorber, as shown in Fig. 1.1, is typically positioned between the fascia and the back end beam.

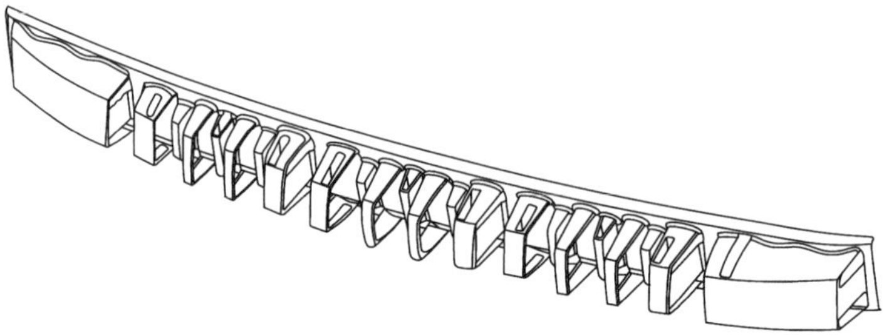

Fig. 1.1 Energy absorber for safety.

Energy absorbers can crush and absorb the impact energy and thereby offer protection to pedestrians [1-22], as well as to the vehicle from damage, during low-speed impacts usually with solitary bumper beams either with plastics [1-23] or plastic-metal hybrids (PMHs) [1-24]. The dual-stage energy-absorber concept has been developed to enable the use of the same energy absorber to increase pedestrian safety and reduce vehicle damage from low-speed impacts [1-25].

In the case of high-speed front, side, rear, and rollover crash scenarios, the vehicle crash worthiness or structural integrity emerges as the key factor for occupant safety. This is achieved by the energy absorption through plastic deformation of various body-in-white (BIW) components such as A, B, C, D pillars; crush box; front rails; roof rail; and floor rockers. Plastics could play a critical role in the lightweight design of such components. The classical example is the plastic-metal rail extension or hybrid crush box, which is mounted on rails and supports the bumper beam. These rail extensions deform in a progressive manner and absorb energy during a frontal crash and thereby offer increased safety [1-26]. Another example is the PMH reinforcement for BIW components, which are designed to offer safety during rollover crashes and side impact conditions [1-27], as elaborated in the second chapter.

1.5.2 Interiors and Occupant Safety

Several of interior parts, as depicted in Fig. 1.2a, function as human-machine interface, ensure occupant safety, and play a significant role in the aesthetics of the cabin.

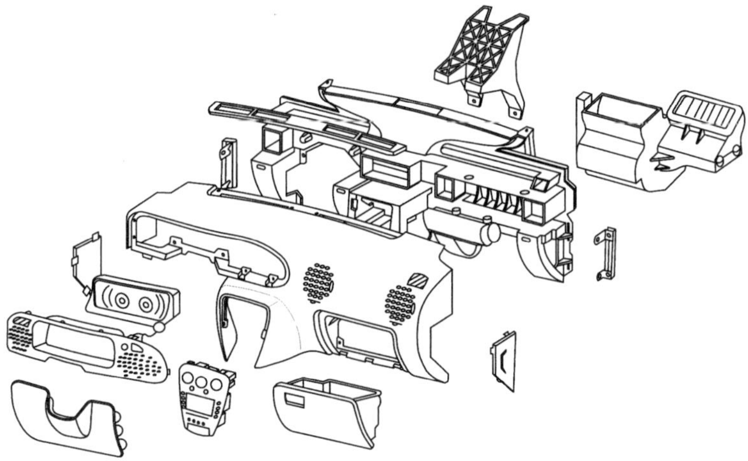

Fig. 1.2a Instrument panel assembly.

One of the key interior applications with plastics is the instrument panel retainer, which offers safety to the occupant during head impact [1-28] and knee impact [1-29], as well as providing greater part integration, cost savings, and ease of assembly. The airbag cover and assembly constitute an important interior application. A typical airbag cover is shown in Fig. 1.2b.

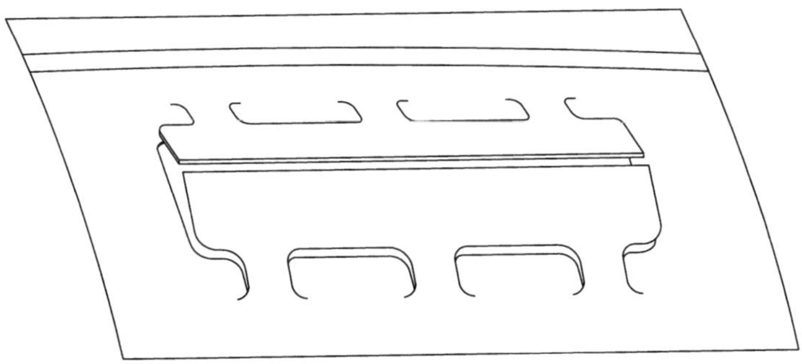

Fig. 1.2b Airbag cover.

Instrument panels with integrated chute and seamless airbag systems have been made possible with thermoplastics. Here, the chute containing the airbag assembly gets molded as part of the instrument panel retainer, as a single piece, eliminating the secondary operation of having to weld the chute to the retainer top [1-21]. The single-piece airbag door with laser-scored seams enables safe airbag deployment without any fragmentation [1-30].

Another key application with plastics is the steering wheel, in which all the structural and performance requirements can be met with lighter weight using open-section [1-31, 1-32] or closed-section components [1-33, 1-34]. The conceptual models are shown in Fig. 1.3.

Fig. 1.3 Plastic single- and double-piece steering wheel.

Energy-absorbing elements in the passenger compartment offer cushioning to occupants during crashes and thus are an important part of safety systems. Plastic energy absorbers mounted on the roof and under the liner offer protection to the head during vehicle rollover [1-35]. Strategically mounted, plastic energy-absorber elements in the door help in reducing abdominal injuries during the side crashes. The plastic console box, located between the driver and front passenger seat, helps to absorb the impact energy, besides offering adequate stiffness to retain survival space between the driver and passenger.

Similarly, door modules [1-36] and seating systems [1-37] can be built with plastics. Various other interior applications with plastics include the knee bolster, central console, door handle, and speaker grille cover (see Chap. 3).

1.5.3 Glazing

Usage of plastics, instead of glass, in some applications offers many advantages such as greater styling freedom, weight reduction, and moldability of complex curvature. Plastics can be used for components that traditionally used glass, such as the roof panel, tailgate, fixed quarter windows, and regular windows. A typical plastic roof panel and tailgate configuration is shown in Fig. 1.4.

Fig. 1.4 Plastic glazed panels.

However, several challenges have to be overcome to make such replacements a success by maintaining structural stiffness under wind load, induced during the high-speed vehicular movement; dimensional stability of large roof panels under temperature variation and sun load [1-38]; comparable or better acoustic performance inside the cabin compared to glass [1-1]; and thermal management within the cabin and load on the heating, ventilation, and air conditioning (HVAC) system [1-19]. Also, manufacturing of large transparent panels would require injection compression molding (ICM) or two-shot injection compression molding (2K-ICM), rather than the standard injection molding process, to overcome high residual stresses. Details of how these issues are addressed in building polycarbonate glazed parts are discussed in Chapter 4.

1.5.4 Plastic-Metal Hybrid Structures

The combination of plastics and metal in an application to form a plastic-metal hybrid (PMH) structure offers a great advantage of bearing the load through the metal while integrating parts and transferring of load through plastics. However, the key challenge is to arrive at a suitable method to join the metal and plastic together. Some of the patented technologies with closed [1-39] and open sections [1-40] make this a reality. Hybrid structures would be valuable in applications such as integrated tailgates (Fig. 1.5a), front-end modules [1-41], a well-known load-bearing structure in front of the automobile (Fig. 1.5b), the door module, cross car beams, and seating systems.

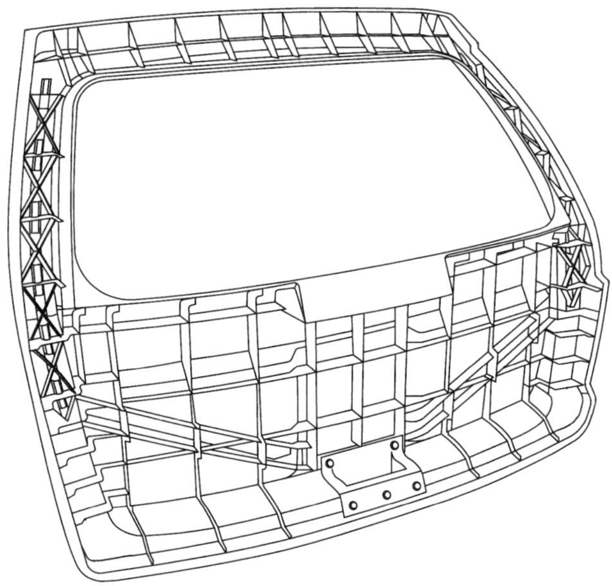

Fig. 1.5a Metal-plastic hybrid structures: tailgate.

Details on design, processing, and performance evaluation of open-channel and closed-channel hybrid structures, through a front-end module case study, are systematically explored in Chapter 5.

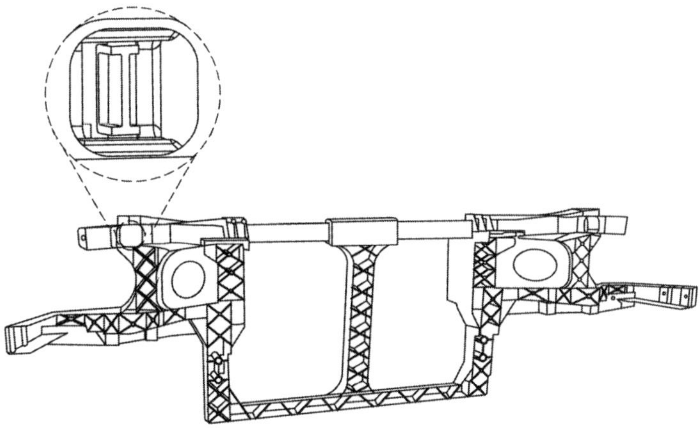

Fig. 1.5b Metal-plastic hybrid structures: front-end module.

1.5.5 Headlamps

In headlamp applications, plastics play a significant role in replacing bulk molding compound (BMC) or metal reflectors, glass lens, and bezels. Various parts of a conceptual headlamp assembly are shown in Fig. 1.6.

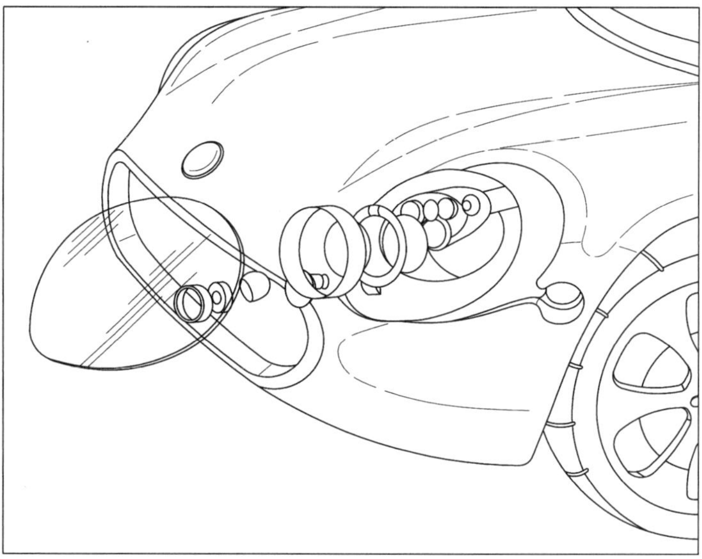

Fig. 1.6 Headlamp assembly.

To design a reflector with plastics, detailed thermal modeling through CFD [1-42], stress analysis, optical beam pattern studies, and tooling and injection molding process modeling must be done. In the case of bezel and lens, temperature would play the key role in material selection and processing.

Pedestrian safety features can be incorporated into the headlamp design, through plastic energy-absorber rings [1-43]. Various aspects of headlamp design are covered in Chapter 6. Heat sink with plastics is an emerging area of application with the next generation of LED (light-emitting diode) headlamps.

1.5.6 Body Panels

Key body panel parts include the fascia, wheel arch, fenders, exterior door panels, B-pillar claddings, tailgate, and fuel lid. Two major advantages of having large body panels such as fenders and tailgates with plastics are weight reduction and increased styling freedom. However, design of any plastic body panel has to address various issues during the manufacturing phase (surface finish, warpage, clamp tonnage, etc.), paint cycle phase (dimensional stability at higher temperatures), and its service life (capability to withstand mechanical, thermal, and impact loads). A typical fender with connection locations is shown in Fig. 1.7.

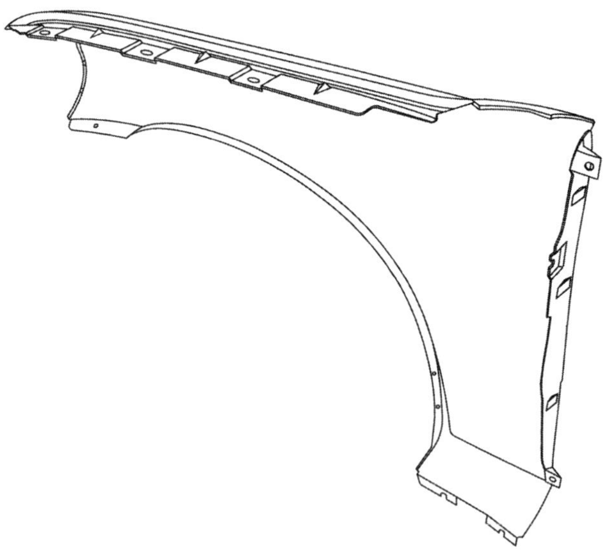

Fig. 1.7 Plastic body panels: fender.

Hence, the plastic fender design has to focus on the required stiffness and dimensional stability at high temperatures during the e-coat and paint cycles [1-44, 1-45] and due to that of sun load during its service life; on the rigidity to withstand impact loads; and on the surface finish through appropriate material, tooling, and processing conditions. In the case of a plastic tailgate, dimensional stability through flush and gap management is a critical challenge; in addition, warpage control of inner panel, structural rigidity to withstand loads, and multiple opening and closing cycles during its service life are additional challenges that need to be addressed through appropriate design and process modeling. See Chapter 7 for a detailed exposition of materials selection; the design and manufacturing process for fenders; and tailgate concepts including hybrids, tank flaps, and spoilers.

1.5.7 Under-the-Hood Components

Many parts that are located under the hood (UTH) and amenable for replacement with plastics include air intake manifolds, turbo air ducts, engine covers, oil pan, throttle body, pump housing, radiator end caps, brake booster valve, fuel lines, and wire coating. However, there are many challenges in this segment for plastics applications. The following paragraphs outline some of the critical challenges.

High temperature: As the UTH area gets packed and the space between the hood and components becomes reduced, the service temperature of these components soars. Hence, the ability to withstand such temperatures is a key requirement for both polymer material development and design of plastic parts.

Chemical resistance: Parts located UTH are exposed to fluids such as engine oil, brake fluid, transmission fluid, antifreeze in coolant, and gasoline. This physical contact with chemicals as well as compounds such as road salt warrants adequate resistance to be built into the material characteristics, so that these plastics applications serve their full design life term.

High heat tolerance and dimensional stability: Parts such as transmission body, fuel rails, and throttle body [1-46] are involved with fluid flow, which is critical for the vehicle performance. A model of a throttle body is shown in Fig. 1.8.

When metals are used to manufacture such parts, high tolerance is achieved through precision machining. If molded plastic parts are to replace the metal, then their dimensional performance becomes a critical design and processing challenge. In addition, if these parts are reinforced

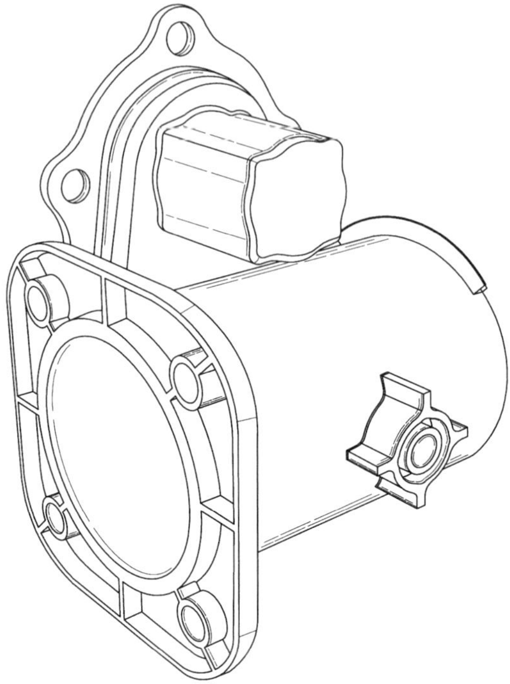

Fig. 1.8 Plastic UTH component: throttle body.

with fibers to be able to withstand structural loads, then maintaining dimensional stability becomes an even bigger challenge. Suitable material and multiphysics-based analysis are required for such part designs.

Heat aging of plastics: Temperature plays a significant role in the selection of materials for UTH parts. Prolonged exposure of UTH parts to elevated temperatures leads to the degradation of physical properties of plastics. Therefore, it is critical to account a priori for the effect of degradation over a period of a decade or the life of the vehicle, in the material design as well as the part design.

Noise reduction: Another important functional role of the materials and parts that go into the UTH segment is to effectively manage noise or demonstrate better acoustic performance, compared to those of metals. Such acoustic performance plays a significant role in determining candidate materials and the design of parts such as the engine cover and oil pan.

Different types of engineering thermoplastics employed for various UTH applications and two detailed case studies on metal replacement for turbo air duct and electronic throttle body are presented in Chapter 8, in addition to the recent advancements made in materials, design, processing, and secondary operations, for this unique and critical automotive segment.

1.6 Summary

Building automotive applications with plastics, either for a given design space or for specific metal replacement requirements, typically follows some or all the phases of the application development cycle, of which material selection is the critical phase. Subsequent chapters present an overview of the application development cycle for different automotive segments using engineering thermoplastics and the overall value proposition that they bring to the table for the end customer.

While light weighting has been the focus of plastic applications to reduce carbon emissions, the emerging automotive industry trends on sustainability—including life cycle assessment for products and processes, design for sustainability, bio-based or renewably sourced materials for plastics, and *end-of-life* options of a product, (e.g., recycling and reuse)—are essential in reducing greenhouse gas emissions and the resulting damage to environment. These topics are covered in detail in Chapter 9.

The objective of this book is to present an overview of plastics application technologies, through engineering principles and case studies, to build safe, lightweight, and environmentally friendly automotive segments or subsystems.

1.7 References

1-1. Stauber, R. and Vollrath, L. 2007. *Plastics in Automotive Engineering: Exterior Applications*. Munich: Carl Hanser.

1-2. Rosato, D.V., Rosato, D.V. and Rosato, M.G. 2001. *Plastics Design Handbook*. Berlin: Springer.

1-3. Strong, A.B. 2006. *Plastics: Materials and Processing, 3rd Ed.* Upper Saddle River, NJ: Prentice Hall.

1-4. International Standards Organization, home page, http://www.iso.org/iso/home.html. Accessed January 1, 2012.

1-5. “Plastic Standards,” ASTM International: Standards Worldwide, http://www.astm.org/Standards/plastics-standards.html. Accessed January 1, 2012.

1-6. “Plastics,” Underwriters Laboratories, http://www.ul.com/global/eng/pages/offerings/industries/chemicals/plastics/. Accessed January 1, 2012.

1-7. “SAE Standards,” SAE International, http://www.sae.org/standards/. Accessed January 1, 2012.

1-8. “NHTSA Regulations,” NHTSA, http://www.nhtsa.gov/Laws-Regs. Accessed January 1, 2012.

1-9. “Our Tests,” Euro NCAP, http://www.euroncap.com/tests.aspx. Accessed January 1, 2012.

1-10. Insurance Institute for Highway Safety, Highway Loss Data Institute, http://www.iihs.org/. Accessed January 1, 2012.

1-11. RCAR: Research Council for Automobile Repairs, home page, http://www.rcar.org/. Accessed January 1, 2012.

1-12. “Xenoy* Resin—Aesthetic Appeal, Impact Resistance,” Sabic, http://www.sabic-ip.com/gep/Plastics/en/ProductsAndServices/ProductLine/xenoy.html. Accessed January 1, 2012.

1-13. Weber, E.H. 2001. *Development and Modeling of Thermally Conductive Polymer/Carbon Composites.* Doctoral dissertation, Michigan Technological University, Houghton, MI. http://www.chem.mtu.edu/org/ctc/pdf/ehw%20dissertation.pdf. Accessed September 1, 2012.

1-14. “Thermally Conducting Plastics,” Ensinger, http://www.ensinger-online.com/en/compounds/products/thermally-conducting-plastics/. Accessed September 1, 2012.

1-15. “Laser Direct Structuring,” LPKF Laser & Electronics, http://www.lpkf.com/applications/mid/index.htm. Accessed January 1, 2012.

1-16. Sofi, F., Kulkarni, S., Haarda, M. and Takaaki, N. 2008. “A Novel Energy Absorber Design Technique for an Idealized Force-Deformation Performance.” SAE Paper No. 2008-01-0184. SAE International, Warrendale, PA.

1-17. Zuber, P.J. 1992. “CAE Processing Analysis of Plastic Fenders.” SAE Paper No. 922116. SAE International, Warrendale, PA.

1-18. Anand, E.S., Marur, S.R., Wittbrodt, J. and King, R. 2007. “Role of Adhesives in the Dimensional Stability of Polycarbonate Structural Panels.” *International Journal of Adhesion and Adhesives* 27(4): 338–350.

1-19. Gasworth, S. and Tankala, T. 2011. "Reduced Steady State Heating and Air Conditioning Loads via Reduced Glazing Thermal Conductivity." SAE Paper No. 2011-01-0126. SAE International, Warrendale, PA.

1-20. Maity, R.K., Prasad, P.S. and Goral, T. 2006. "Material Modeling and Finite Element Analysis of Hydroform: Short Glass Fiber Filled Thermoplastic Front-End Structures." SAE Paper No. 2006-01-0824. SAE International, Warrendale, PA.

1-21. Jaarda, E., Chaturvedi, M., Wieczorek, T., Meyers, A. and Chitteti, R. 2008. "Development of an Instrument Panel with an Integrated Chute and Seamless Airbag Door." SAE Paper No. 2008-01-1339. SAE International, Warrendale, PA.

1-22. Nagwanshi, D.K., Mana, D., Allen, K. and Bobba, S. 2010. "Diagnosing Vehicle Aggressiveness for Pedestrian Leg Impact and Development of Efficient Front End Energy Management Systems." SAE Paper No. 2010-01-1168. SAE International, Warrendale, PA.

1-23. Mana, D., Nagwanshi, D., Marks, M. and Arunachala, P. 2011. "Thermoplastic Rear Bumper Beams for Automobile Low-Speed Rear Impact." SAE Paper No. 2011-01-0544. SAE International, Warrendale, PA.

1-24. Nagwanshi, D., Chaturvedi, M., Marur, S. and Allen, K. 2010. "Hybrid 'Thermoplastics and Steel' Bumper Beam Solution to Protect the Vehicle in Low Speed Crashes." SAE Paper No. 2010-01-1009. SAE International, Warrendale, PA.

1-25. Jaarda, E.J., Mahfet, M.R., Mohapatra, S., Nagwanshi, D.K. and Sreeram, T.R. 2007. "Dual stage energy absorber." Patent No. US 7,568,746 B2.

1-26. Nagwanshi, D.K., Allen, K., Nemoto, T., Imai, S., Arunachala, P. 2012. "Energy absorbing device and methods of making and using the same." Patent Publication No. US 2012/0112479 A1.

1-27. Nagwanshi, D.K., Marur, S.R. and Marks, M.D. 2012. "Reinforced body in white and method of making and using the same." Patent Publication No. US 2012/0153669 A1.

1-28. "Standard No. 201: Occupant Protection in Interior Impact," Federal Motor Vehicle Safety Standards and Regulations, http://www.nhtsa.gov/cars/rules/import/fmvss/index.html#SN201. Accessed January 1, 2012.

1-29. "Standard No. 208: Occupant Crash Protection," Federal Motor Vehicle Safety Standards and Regulations, http://www.nhtsa.gov/

cars/rules/import/fmvss/index.html#SN208. Accessed January 1, 2012.

1-30. Wieczorek, T., Trappe, A., Meyers, A., Goyette, J., Lanning, V. and Chaturvedi, M. 2005. "An Efficient Solution for Hard or Soft Seamless Airbag Systems." SAE Paper No: 2005-01-1223. SAE International, Warrendale, PA.

1-31. Nayak, S., Garg, A., Kumar, S.M., Krishnamoorthy, N., Chaturvedi, M. and Marks, M. 2011. "Virtual Design Optimization of Thermoplastic Steering Wheel." SAE Paper No. 2011-01-0023. SAE International, Warrendale, PA.

1-32. Nayak, S., Garg, A., Kumar, S.M. and Chaturvedi, M. 2011. "Robust Thermoplastic Steering Wheel Design." SAE Paper No. 2011-26-0116. SAE International, Warrendale, PA. Paper presented at Symposium on International Automotive Technology (SIAT), Automotive Research Association of India (ARAI), Pune, India.

1-33. Garg, A., Surisetty, G.K., Chaturvedi, M. and Jaarda, E. 2009. "High Performance Thermoplastic Steering Wheel." SAE Paper No. 2009-26-0074. SAE International, Warrendale, PA.

1-34. Nayak, S., Garg, A., Chaturvedi, M., Wieczorek, T. and Marks, M. 2010. "Performance Evaluation of PU Over-molded Thermoplastic Steering Wheel." SAE Paper No. 2010-01-0916. SAE International, Warrendale, PA.

1-35. Marur, S.R., Garg, A.K., Nayak, S., Tankala, T.C. and Wilson, J.R. 2011. "Energy Absorber Elements and Vehicle Systems." Patent Application No. US 20110210579 A1.

1-36. Chaturvedi, M., Schijve, W. and Marks, M. 2009. "Advanced Thermoplastic Composites for Automotive Semi-Structural Applications." SAE Paper No. 2009-26-0086. SAE International, Warrendale, PA.

1-37. Naughton, P., Shembekar, P., Lokhande, A., Kauffman, K., Rathod, S. and Malunjkar, G. 2009. "Eco-Friendly Automotive Plastic Seat Design." SAE Paper No. 2009-26-0087. SAE International, Warrendale, PA.

1-38. Anand, E.S., Marur, S.R., Wittbrodt, J. and King, R. 2006. "Thermal Deformation of Polycarbonate Glazing Panels: Role of Connecting Mechanisms." SAE Paper No. 2006-01-0974. SAE International, Warrendale, PA.

1-39. Mooijman, F.R. and Prasad, P.S. 2008. "Closed hybrid structure and method." Patent Application No. US 2008/ 0317988 A1.

1-40. Mooijman, F.R., Prasad, P.S. and Poovanna, T.K. 2008. "Hybrid structure and method." Patent Publication No. US 2008/0138586 A1.

1-41. Goral, T., Prasad, P., Brown, M., Panter, K., Klages, J. and Longhouse, B. 2005. "Automotive Front End Structures Constructed by Over Molding Hydroform Metal Tubes to Engineering Thermoplastic Structures." SAE Paper No. 2005-01-1680. SAE International, Warrendale, PA.

1-42. Tankala, T., Marur, S.R. and Wilson, J. 2006. "Rapid Thermal Predictions for Automotive Headlamp Reflectors." SAE Paper No. 2006-01-1189. SAE International, Warrendale, PA.

1-43. Garg, A.K., Marur, S.R., Tankala, T. and Wilson, J. 2010. "Pedestrian Safe and Impact Resistant Headlamp Design Through a Novel Energy Absorber Ring Concept." SAE Paper No. 2010-01-0293. SAE International, Warrendale, PA.

1-44. Hardikar, N., Khandelwal, R., Venkatesha, N. and Doggen, G.J. 2010. "Prediction of Thermoplastic Fender Behavior During E-coat Bake Cycle—Part 1: FEA Methodology and Problem Formulation." SAE Paper No. 2010-01-0232. SAE International, Warrendale, PA.

1-45. Kancharla, A.K. Hardikar, N., Tankala, T. and Doggen, G.J. 2010. "Prediction of Thermoplastic Fender Behavior During E-coat Bake Cycle—Part 2: Influence of Temperature Distribution." SAE Paper No. 2010-01-0231. SAE International, Warrendale, PA.

1-46. Hardikar, N., Chunduru, S., Rexius, K., Tankala, T., Janardhan, Y. and Withey, C. 2005. "Six Sigma Study on the Effect of Geometric Tolerances at Low Airflow Rates in a Progressive Bore Throttle Body Using CFD." SAE Paper No. 2005-01-1916. SAE International, Warrendale, PA.

Chapter 2
Crash and Energy Management Systems

Dhanendra K. Nagwanshi, Somasekhar Bobba, Dinesh Mana, and Sudhakar R. Marur

2.1 Introduction

The worldwide efforts to save lives, prevent injuries, and reduce costs caused by road crashes demonstrate the increasing importance of automotive safety as a global concern. Improvement of various aspects of vehicle safety has emerged from legislation, consumer information, initiatives of individual manufacturers, and product liability considerations. The legislations aim for a minimum but high level of protection across the automotive product lines; consumer information aims to encourage the highest possible levels of safety; and auto industry policies increasingly promote safety as an indispensable product feature, enabling market differentiation. Improving vehicle safety is a key strategy to meet international and national road casualty reduction targets and to achieve a safer traffic system. Effective vehicle safety design depends primarily on continuing research and development, understanding of the source and mechanism of injury protection in a range of crash conditions, regular monitoring of performance in real world conditions, and confirmation that new technologies are used and accepted.

Automobile safety is the study and practice of vehicle design, construction, and equipment to minimize the occurrence and consequences of automobile accidents. Automobile safety systems can be categorized as active systems and passive systems. An active safety system comprises measures to help avoid a crash (crash avoidance), and a passive safety system tries to reduce injury in the event of a crash (crash protection). Injury is broadly related to the amount of kinetic energy applied to the human body frame. The tolerance of the human body to kinetic forces released in road traffic crashes is limited. Biomechanical research results reported over the years through international scientific conferences (International Research Council on Biomechanics of Injury (IRCOBI) [2-1], Stapp Car Crash Conference [2-2], etc.) indicate that the relationship between crash forces and resulting injury is now clearly understood for various parts of the body, types of injury for different categories of road users, and age groups. Various crash scenarios emerge from different impact energy levels, and appropriate crash countermeasures can be designed to address them.

The energy management systems and devices prevent or reduce the severity of injuries when a crash is imminent or actually happens. As the energy of a crash is related to the square of the velocity, any small increase in speed can increase tremendously the probability of the risk of injury. An effective energy management system absorbs maximum energy during the crash and limits the amount of energy to be transferred to the occupant or the vulnerable road users. Typically, energy management systems are designed to meet original equipment manufacturer (OEM) requirements for cost, performance, and weight. System cost is a critical requirement and is determined by the technology selected to meet performance and weight targets. The configuration (geometry), material, and process selected determine the performance of an energy management system. To cite an example, generic expanded polypropylene (EPP) foam is less expensive and significantly lighter than engineering plastics or metals. However, because of the inherent nature of EPP foam, that is, its slow response to impact, these solutions are less efficient than those of thermoplastics or metals. Metallic solutions are heavier than engineering plastic solutions and may also involve secondary operations like stamping, welding, and so forth. Deformable energy-absorbing elements made of plastics may help in designing the structural systems to be stiff enough to gain quick response during the initial stage of impact and absorb significant amount of kinetic energy through plastic deformation during the impact. A tuned stiffness, strength, and deformation may enhance pedestrian safety during crashes and reduce the cost associated with damage at low-speed vehicle crashes and occupant protection at high-speed impacts. Thermoplastic solutions may further offer various advantages such as manufacturability of complex shapes; ability to increase the energy-absorption efficiency; higher elongation to absorb more energy during plastic deformation; and reduction in the number of components needed to help minimize the processing, tooling, and assembly cost at the system level.

2.2 Safety as an Emerging Global Concern

Automobile safety had been a concern almost from the beginning of mechanized road vehicle development. Motor vehicle crashes killed more than 33,000 people and injured over 2.2 million others in 2009 in the United States alone. In addition to the terrible personal losses to people, these crashes make a huge economic impact on our society, with an estimated annual cost of $230 billion, representing an average of $750 for every person in the United States [2-3]. In Germany, 4152 fatalities have been recorded, and the calculated total accident costs amounted to approximately 30.52 billion euros in 2009 [2-4]. The number of fatalities (those who died within 24 hours) resulting from traffic accidents in 2010 was 4863 in Japan [2-5]. In Korea, 5838 fatalities in road traffic accidents have been reported for the year 2009 [2-6].

Motor vehicle crashes are one of the leading causes of fatalities across the globe. All the above statistics unequivocally prove that there is a lot of work to be done in the area of increasing safety of all road users: drivers, passengers, and vulnerable road users [2-7] such as motorcyclists and pedestrians. The regulatory bodies in various countries remain fully committed to improve motor vehicle safety through a coordinated effort involving research, education, enactment of legislation, and enforcement.

2.3 Regulatory and New Car Assessment Program Crash Test Requirements

Regulatory car crash tests and pedestrian subsystem tests have been developed by legislative bodies. Such tests incorporate various types and speeds of impact of the most common types of crashes and are incorporated in legislation. Each country has different legislative crash test requirements. In 1958, the United Nations established the World Forum for Harmonization of Vehicle Regulations, an international standards body advancing auto safety [2-8].

The New Car Assessment Program (NCAP) provides consumer information to prospective car buyers about the safety performance of cars during crashes and encourages manufacturers to introduce evidence-based safety designs beyond legislative norms.

Vehicle crash tests can be broadly categorized as

- Pedestrian impact tests
- Low-speed vehicle damageability or bumper structural tests
- High-speed crashes for occupant protection

2.3.1 Pedestrian Impact Tests

In highly urbanized regions such as Europe and Japan, legislators have already introduced vehicle-testing rules intended to reduce the risk of serious injuries if a pedestrian comes into contact with the front end of a moving vehicle. Studies show that pedestrians have a 90% chance of surviving a car crash at 30 km/h or below, but less than a 50% chance of surviving an impact at 45 km/h [2-9]. In 1991, the European Enhanced Vehicle-Safety Committee (EEVC) proposed a set of component tests representing the three most important mechanisms of injury: head, upper legs, and lower legs [2-10]. Fig. 2.1 represents the pedestrian impact test on a vehicle front bumper.

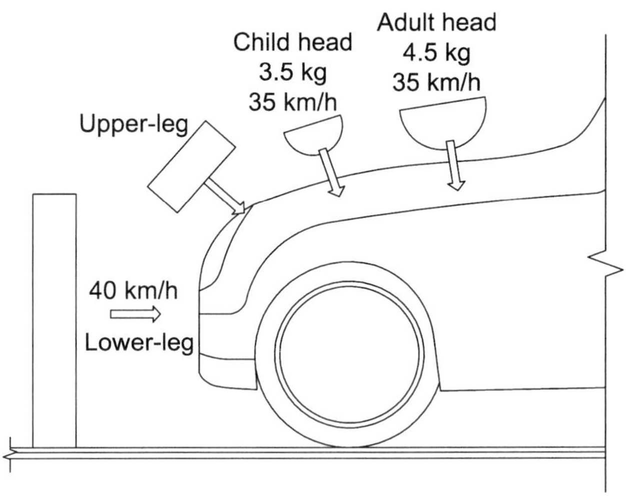

Fig. 2.1 Leg and head impact test configurations.

These tests were incorporated into the consumer tests conducted by Euro NCAP [2-11]. In Japan, the initial set of regulations (International Harmonized Research Activities [IHRA], formulated by the Ministry of

Land, Infrastructure, Transport and Tourism [MLIT]) became effective in September 2005; additional categories of vehicles were required to meet the regulations by 2007, 2010, and 2012 [2-12]. Legislators in China and Korea are also considering enactment of similar pedestrian protection regulations by 2014. Vehicle manufacturers in Europe, Japan, Korea, Canada, and the United States are involved in discussions on "harmonized" global technical regulations for pedestrian protection. The test procedure calls for a free-flight bumper impact at 40 km/h with a leg form developed by the Transport Research Laboratory (TRL), in Berkshire, UK [2-13]. This leg form is a simplified device that approximates human anthropometry using frangible steel knee ligament surrogates, which are designed to deform plastically during impact. The instrumentation of the leg form allows it to measure tibia acceleration, shear displacement, and bending angle at the knee.

2.3.2 Low-Speed Vehicle Damageability or Bumper Structural Tests

About 70% of total costs of damage are related to low-speed crashes [2-14]. Reducing vehicle damages in low-speed crashes could have a massive global financial benefit. In the case of a collision to the front or rear, with low-speed impact, the bumper components should absorb the energy to prevent or reduce damage to components such as the hood, lights, and cooling system in the front, and the tailgate and dec-lid in the rear of the car. In the United States, Federal Motor Vehicle Safety Standards (FMVSS) Part 581 [2-15] specifies requirements to minimize damage to the front and rear ends of passenger cars in low-speed collisions. In Europe, the Economic Commission for Europe (ECE) R-42 bumper test protocol [2-16] ensures that the front and rear ends of vehicles are designed in such a way as to allow contacts and small shocks to occur without causing any serious damage. These tests include impact by a pendulum or a barrier at speeds up to 4.0 km/h over the vehicle front and rear bumpers.

The Insurance Institute for Highway Safety (IIHS) has consumer information on a low-speed testing program [2-17] designed to evaluate bumper performance based on the cost of repairing the vehicle damage. The "Bumper Test and Rating Protocol" prescribes a new IIHS barrier, which has a rigid bumper-shaped barrier fitted with an energy-absorbing material and a cover, as shown in Fig. 2.2, which closely replicates the damage patterns observed in real-world low-speed crashes.

The new test protocols for the Research Council for Automobile Repairs (RCAR) in Europe [2-18] and IIHS with the new deformable barrier address three key aspects of bumper performance: geometry, bumper stability, and energy absorption.

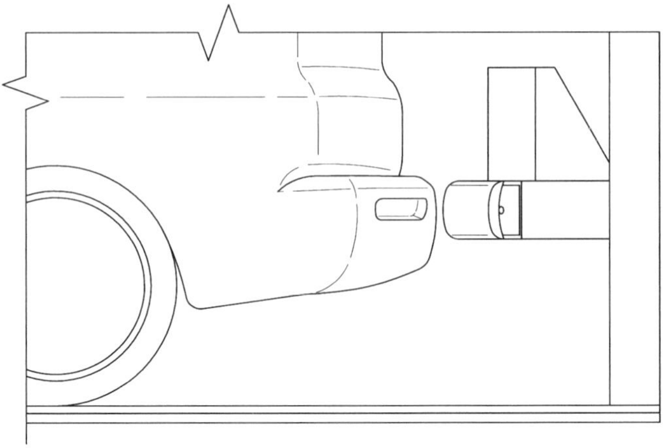

Fig. 2.2 IIHS full frontal test.

2.3.3 High-Speed Crashes for Occupant Protection

The common injury-producing crash types are frontal crashes, followed by side impacts, rear impacts, and rollovers. Figure 2.3 is a schematic representation of these tests.

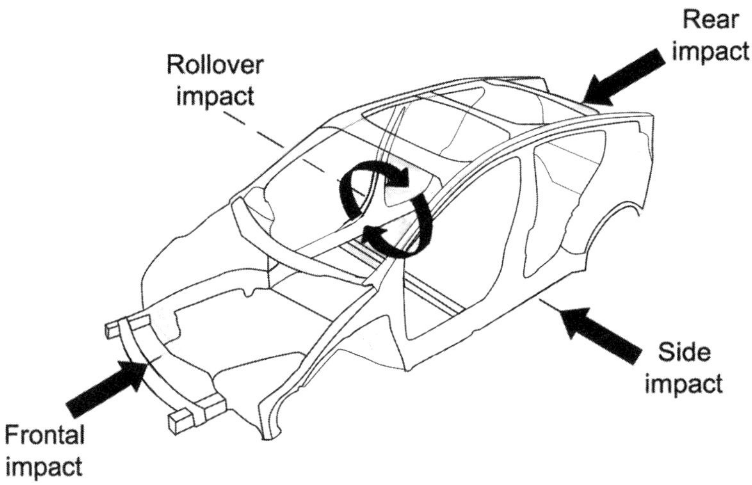

Fig. 2.3 A schematic representation of high-speed crash tests.

In Europe, the legislative tests cover the crash performance of new cars in front and side impacts. Euro NCAP consumer tests provide a star rating for crash performance in front and side impact tests based on legislative tests, a pole test, subsystem pedestrian tests, and inspection of aspects of the vehicle interior and restraint systems. In United States, the legislative tests cover the crash performance of new cars in front and side impact, 40% offset front impact test, a pole test for side impact, and a roof crush test for rollover crashes.

The vehicle body-in-white (BIW) components define the structural rigidity of the vehicle for crashworthiness, and the cushioning element in the vehicle helps to reduce injury to the occupants when they come in contact with the vehicle BIW. The best-designed vehicle on the road today provides crash protection for occupants wearing seat belts up to 70 km/h in frontal impacts and 50 km/h in side impacts [2-19].

2.4 Impact and Energy-Absorption Efficiency

An impact is generally defined as a high force or excitation happening over a short period of time when two or more bodies collide each other. It is estimated that the magnitude of force or acceleration generated during this time is significantly higher than that of an event where the same load is applied over relatively longer duration of time. The latter represents an inelastic collision, where the deformation of the bodies under collision absorbs a significant portion of the force of collision [2-20]. In other words, the kinetic energy of the colliding objects is dissipated as heat and sound energy as a result of the deformation and vibration induced. However, in the case of an impact resulting from a high-speed collision, the time period involved is so small that the objects do not have sufficient time to absorb most of the force of collision by their deformation, which in turn results in significantly high force and acceleration levels to be experienced by the colliding objects. As a result, the kinetic energy is either retained by the colliding objects or absorbed by the objects under collision by virtue of its material yielding. It is observed that most of the impact cases in the real world are combination of these two. Thus, impact energy, for all practical purposes, is defined as the sum of elastic and plastic energy absorbed by the deforming bodies.

An impact scenario usually consists of three phases: pre-impact, impact, and post-impact (schematically represented in Fig. 2.4).

During the pre-impact phase, two independent inertial systems with definite mass and velocity move closer at a relative velocity. As the impact is yet to take place, there is only one type of energy, that is, kinetic energy, associated with the systems. During the impact phase, some portion of the kinetic energy is stored as internal energy in either or both the systems. It is this portion of the impact energy that is used to break the bonds

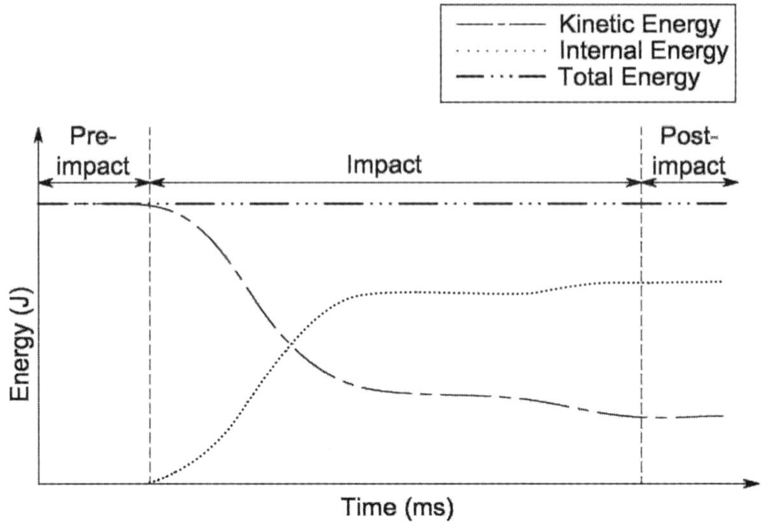

Fig. 2.4 Typical phases and types of energy involved in impact.

within in the material and causes the material failure. In other words, materials with higher yield stress and ductility are better energy-absorbing materials. The remaining portion of the kinetic energy is retained by the systems during the post-impact phase, which is also known as the rebound phase.

Several factors, such as the properties of the materials, impact conditions, ambient conditions, geometry, and mode of energy absorption, play an important role in determining the energy-absorption efficiency of a system or component [2-21]. For example, the material becomes more brittle and less efficient for energy absorption at low temperatures and higher strain rate, which typically is the scenario during impacts at high speeds. On the other hand, a highly brittle material can also be used for energy-absorbing elements such as a helical spring, where the elastic nature of the component is leveraged for energy absorption. Similarly, a foam block, which is highly weak in tension, can be used for energy absorption when it is operated in compression mode. Various energy-absorbing elements and energy-absorbing mechanisms are explained in detail in the next section.

The energy-absorption capabilities of a system and their effectiveness can be understood from the basic principles of physics applied to an impact scenario, as is schematically represented in Fig. 2.5.

Figure 2.5 shows an energy-absorbing element, which is mounted over fixed platform, being impacted with a rigid impactor at certain velocity

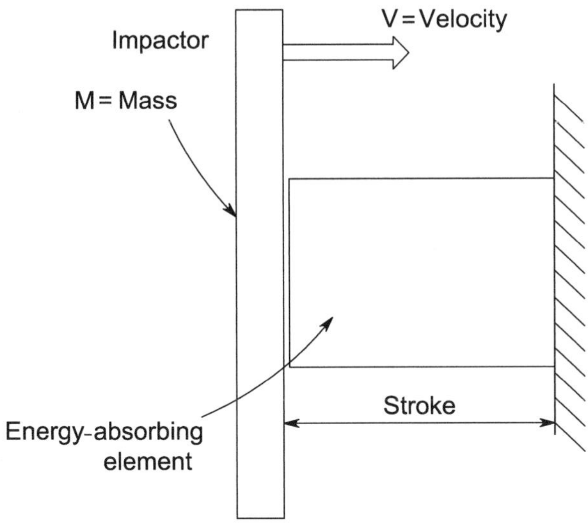

Fig. 2.5 A typical impact scenario.

and mass. The impactor can travel a maximum distance of "stroke," during which it also starts experiencing impact forces of varying magnitudes. This variation of force during the impact can be schematically represented using a force versus stroke curve, as shown in Fig. 2.6. The area under this curve gives a measure of energy absorption. Therefore, to maximize the energy absorption, the magnitude of the force experienced should be as high as possible. But this is not possible, as every energy-absorbing element is designed to control the force or acceleration levels experienced by the components undergoing the impact. So, the goal is to maintain the force level at its upper limit as much as possible throughout the impact phase. Therefore, an ideal energy-absorbing system is capable of achieving and maintaining the desired force levels throughout its stroke [2-22]. This is extremely difficult because the proper engagement of the two systems (under collision) does not happen at "zero" stroke. Hence, a more pragmatic and theoretically achievable curve generates peak value at around 10–15% of the maximum possible stroke. Having achieved this state, the challenge now is to maintain the force level at this value throughout the remaining portion of the stroke. A much more practical curve usually has a small dip or variation for the force value during the remaining part of the stroke. Typical low-density foams, because of their inherent microstructural limitations, do not generate the peak in the early stage, and the force levels gradually increase during impact. Representative curves for all four cases are shown in Fig. 2.6.

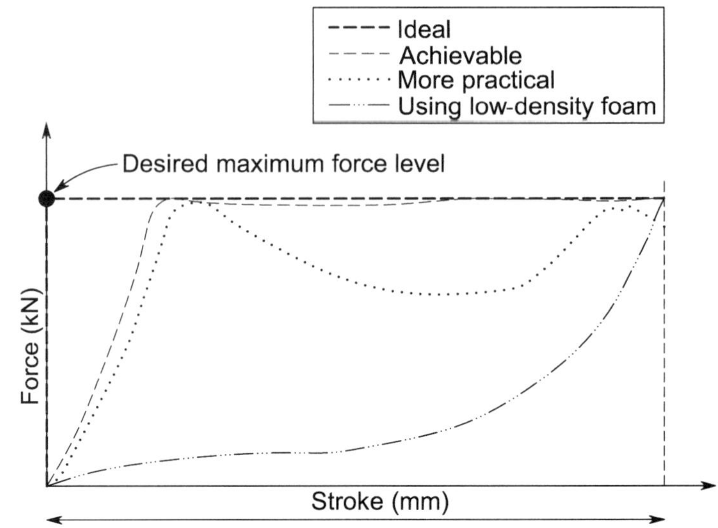

Fig. 2.6 Energy-absorption efficiency.

2.5 Design of Energy-Absorbing Elements

Material, design, and the respective modes of energy absorption need to be well understood to accurately model and design crashworthy energy-absorbing elements. While some of these elements are designed to absorb energy with the help of elastic deformation, some others undergo permanent deformation or failure during the impact. The former typically happens in the case of low-speed impacts, and the latter is generally associated with high-speed collisions. Helical springs that can absorb energy in both axial compression and torsional modes, foam blocks that get compressed and absorb the energy, and leaf springs that undergo significant amount of bending before failure are few examples of energy-absorbing members, which operate in the elastic regime. On the other hand, crash cans that undergo axial crushing, soft C-section geometries (mounted to relatively stiff structures/beams) that buckle during an impact at its front face, and so forth, represent those energy-absorbing members that undergo permanent deformation upon impact. Thermoplastics can be used in the design of most of the aforementioned energy-absorbing members that are required for an automobile.

During a crash or impact scenario, most of the thermoplastic energy-absorbing components absorb energy by making use of the impact energy and fail through bending, buckling, or crushing mode. For effective bumper energy management, during both high-speed and low-speed crashes, either all plastic structures or plastics-steel hybrid structures are used. For example, in rollover countermeasures, controlled energy absorption and structural safety can be achieved with steel structures that have plastic inserts as internal reinforcements. In the case of energy-absorbing members that are positioned in front of bumper beams for pedestrian safety and protection against low-speed vehicle damageability, controlled and efficient absorption is achieved by crushing of these members. Open-section, injection-molded, thermoformed, or blow-molded thin-walled structures are among the most common configurations of pedestrian-safe energy absorbers (EAs) [2-22, 2-23].

The axial folding of structures, which typically happens when plastic crush boxes are used, has been used for several decades as an excellent energy-absorbing mechanism. Today, these crush boxes are deployed during crash situations where high energy needs to be absorbed in a controlled way. Shell structures are used in many engineering applications because of their efficient load-carrying capability relative to material volume. Since thin-walled sections are one of the most efficient energy-absorbing components of an automobile, axial crushing of thin-walled structures has long been the subject of extensive research [2-24, 2-25]. More details of various thermoplastic energy-absorbing structures and their application are given in subsequent sections.

2.6 Pedestrian Protection

This section describes the details of diagnosing vehicle front stiffness for pedestrian protection and development of pedestrian impact countermeasures.

2.6.1 Vehicle Bumper Stiffness Profile

Conventionally, the term *bumper* is used to represent the relatively stiff beam in front and at the rear end of a vehicle. The primary function of these beams, which are typically mounted to the vehicle chassis front and rear side, is to safeguard the expensive and critical parts of a vehicle during impacts at low speed. Some of these important components are the engine and radiator in the front side of the vehicle and tailgate, as well as the back panel at the rear side of the vehicle. While these stiff beams provide necessary protection for the expensive vehicle parts, these will in no way guarantee pedestrian safety during a vehicle-to-pedestrian impact. Such is the case with a front-end module also, whose top half becomes hard points for a pedestrian's head during a head impact on the hood. With more functionalities and components integrated into vehicles, the term "bumper" is also being redefined. Now, the bumper system at the front-end of a vehicle consists of several key elements such as fascia, upper and lower grille, hood, spoiler or lower-leg protector, bumper beam and others [2-26, 2-27].

As briefly explained in section 2.3, four impact scenarios are considered to be the major cause of pedestrian injuries and fatalities during a vehicle-to-pedestrian impact. The first two include the impact of the bumper system with the pedestrian's leg. This impact can be either at the lower portion of the leg or with the thighs, the upper portion of the leg, depending on the height of the vehicle from the ground. This impact is generally managed by controlling the frontal stiffness profile of the vehicle.

The other two impact cases consist of a child head impact on the hood and the adult head impact either on the hood or the windshield depending on the geometry of the vehicle. While the former two impact cases involve direct impact of pedestrian with the bumper system, the scenario with head impact cases, is quite different. Therefore, to address these head impact cases, automobile OEMs generally use a variety of techniques such as active hood systems, arrangement of cushioning foam between the hood and the hard points such as front-end module, and so forth.

The most popular lower-leg model used now is a TRL leg form as explained in Section 2.3. This leg form consists of two steel rods, which emulates the bones present in the human leg, connected with appropriate joining mechanism that emulates the knee joint present in the human leg. Although few other leg models are used across the automobile industry, the fundamental mechanism based on which these dummy models are developed is basically the same. While these leg models are extensively used for measuring the injury criteria, there are some concerns as well with regard to their ability to quantify the bone flexibility [2-28]. Therefore, regulatory bodies and few other organizations in Japan and Europe are now involved in the development of a biofidelic flexible leg impactor (Flex-PLI) [2-29]. Researchers claim that these models can not only simulate the human body flexibility and human knee joint stiffness properly but also can measure many more injury parameters such as long bone strains at multiple locations, knee ligament elongations, compression forces between the femoral condyles and tibia plateau. These dummy models, however, are yet to be adopted in practice, across the industry. In any case, vehicle frontal stiffness profile plays a crucial role in determining the injury measures in these leg models.

Vehicle aggressiveness for pedestrian safety can be related to the stiffness profiles offered by vehicle front, which defines the leg kinematics, during pedestrian impact. During a vehicle-to-pedestrian collision, pedestrian's leg typically bends about the knee location, which also experiences high acceleration and sheer. This lower leg to vehicle bumper impact kinematics can be simplified by assuming three different load paths that is, upper, middle and lower load path as shown in Fig. 2.7.

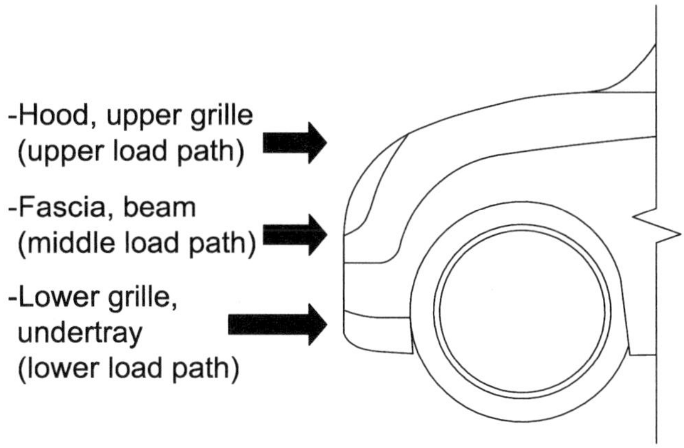

Fig. 2.7 Various load paths of a vehicle front.

Vehicle bumper stiffness is mapped as upper, middle and lower stiffness as shown in Fig. 2.8.

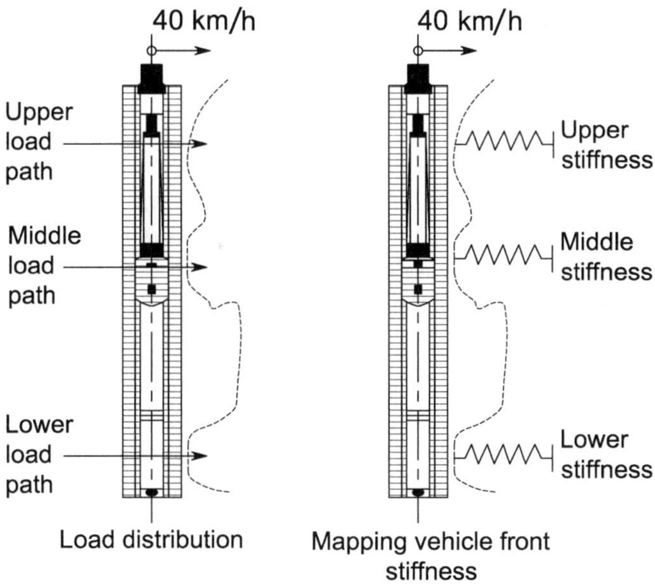

Fig. 2.8 Stiffness mapping to various load paths.

The upper stiffness is the combined stiffness provided by the grille, hood, front-end module, and the like; middle stiffness is provided by the fascia, bumper beam, EA, and so forth; and lower stiffness is provided by the lower-leg protector/undertray, and the like, to the lower-leg impactor. In reality, these stiffness values are obtained with the help of finite element analysis of the vehicle setup. A vehicle bumper stiffness profile should offer controlled support to the lower leg, during impact, for lesser injury to the knee and tibia. The stiffness profile of the vehicle front bumper can be used to diagnose vehicle aggressiveness for pedestrian safety. The advantage of such stiffness profile mapping is that one can get a rough idea of the vehicle stiffness and how it can be related to pedestrian leg impact performance. A study conducted [2-26] on the recent trends in bumper design for pedestrian impact also reveals that most of the research in this area by academic institutes and the automobile industry aims at understanding the aforementioned three load paths and designing appropriate components at these load paths to meet the required stiffness distributions.

In addition to the vehicle stiffness, the geometric parameters of the vehicle front also play an important role in quantifying pedestrian impact performance. Therefore, a simplified yet realistic model of a generic vehicle front profile can be obtained using three springs and a few geometric parameters representing the vertical distances of these springs from the ground and horizontal distances from the first impact point. The schematic representation of such a model is shown in Fig. 2.9.

H_u, H_m, and H_l represent the height or distance of the upper, middle, and lower load paths from the ground, respectively. The parameters B_u and B_l represent the horizontal distances of the upper load path and the lower load path from the first impact point, and P_S represents the packaging space, the distance between the bumper fascia and the bumper beam. It is worth noting that the first impact point in this model is assumed at the middle load path. This simplified model can be validated and used for the design of the appropriate energy-absorbing elements at various load paths, as explained in the next sections.

2.6.2 Design of Pedestrian-Safe Bumper Systems

As explained in the earlier section, design of pedestrian-safe bumper systems is all about achieving controlled kinematics for a pedestrian leg form during impact. This impact, depending upon the vehicle type, could involve the lower leg or the upper leg or a combination of both. Also, the pedestrian lower-leg kinematics is completely different in the case of sports utility vehicle (SUV) compared to that of a sedan, shown in Fig. 2.10.

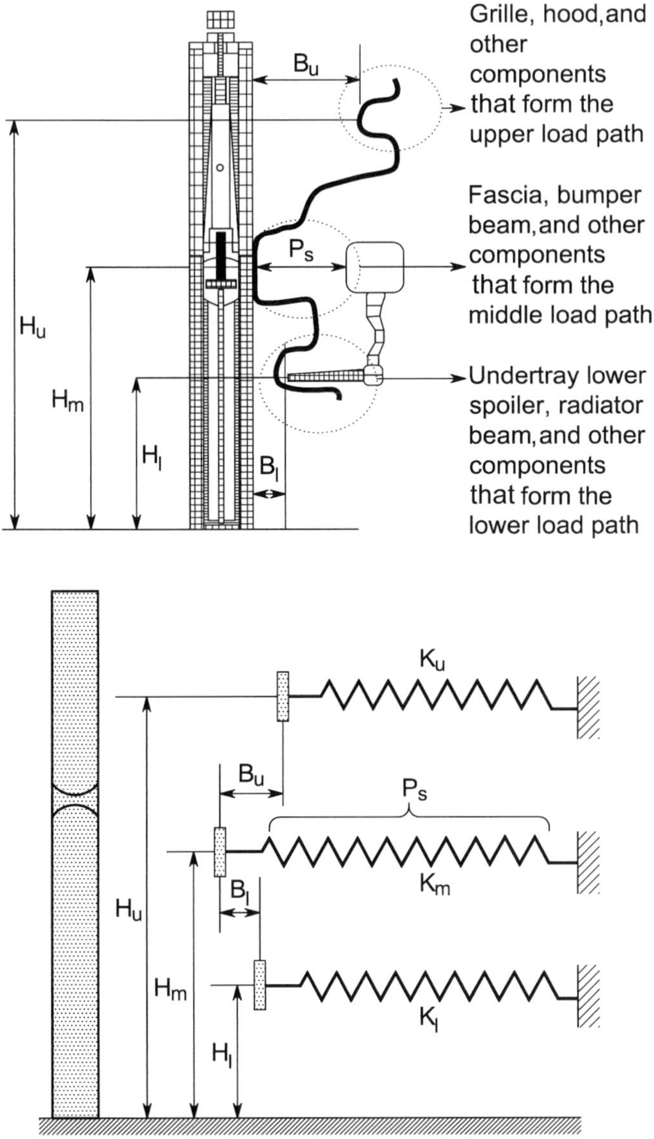

Fig. 2.9 Stiffness model of vehicle frontal stiffness profile.

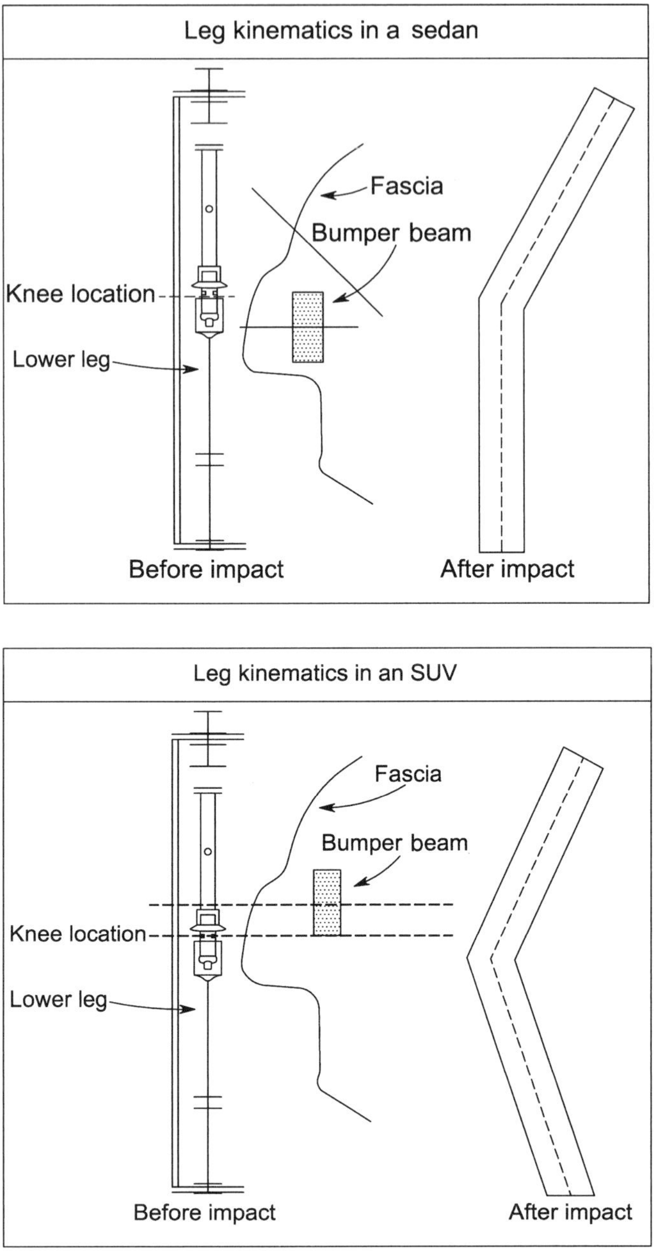

Fig. 2.10 Pedestrian leg-impact kinematics for various vehicles.

For instance, in the case of a compact car, the lower portion of the pedestrian's leg comes into contact with the bumper system, and the upper portion comes into contact with the vehicle hood during the impact. The leg kinematics is completely different in the case of an SUV, where the upper portion of the leg comes in direct contact with the vehicle bumper system. A detailed representation of pedestrian impact requirements is schematically represented in Fig. 2.1. More details of the regulatory requirements for the lower-leg impact and upper-leg impact are shown in Fig. 2.11 and Fig. 2.12, respectively.

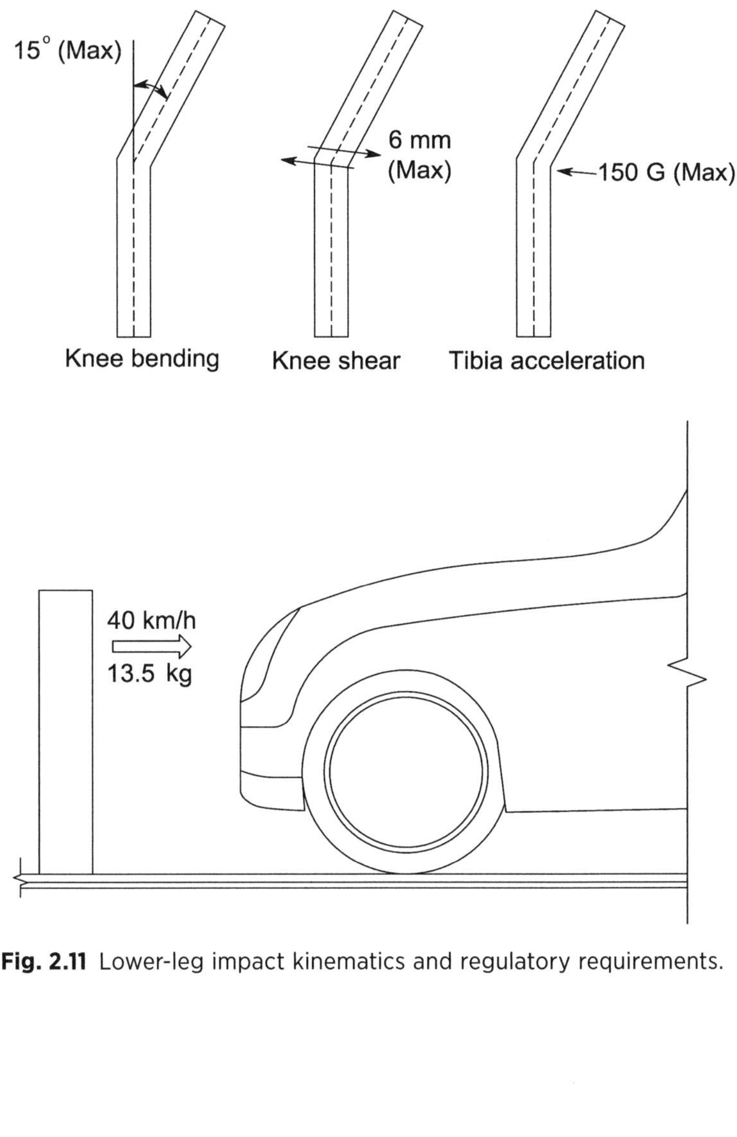

Fig. 2.11 Lower-leg impact kinematics and regulatory requirements.

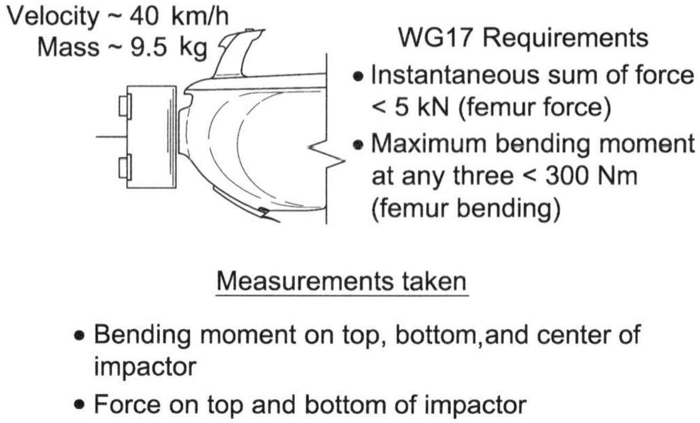

Fig. 2.12 Upper-leg impact to bumper and regulatory requirements.

As shown in Figs. 2.11 and 2.12, in general, a pedestrian-safe bumper system must satisfy the lower-leg impact requirements in case of a compact car, and it must meet the upper-leg impact requirements, in addition to the lower-leg impact requirements, in the case of an SUV. Upper-leg impact, however, is a relatively localized impact and can be met by arranging localized energy-absorbing elements across the vehicle width. This is generally not the case with lower-leg impact, which requires an optimized stiffness profile for the front bumper system, as explained in the previous section.

A typical representation of vehicle bumper profiles with the help of a simplified model that contains three springs at three different load paths and few geometrical parameters of the vehicle was explained in the previous section. Such a simplified model can be used to get a rough idea of the lower-leg impact performance of a given vehicle, as shown in Fig. 2.13.

Such a simplified numerical model for the lower-leg impact performance can be derived from a rigorous design of experiments (DoE) studies [2-26]. The performance parameters such as acceleration, rotation, and shear can be represented as nonlinear functions of the geometric parameters and the various stiffness values as shown in Eqs. 2.1–2.4. It is worth noting that G_1 and G_2 represent the first peak and second peak of the acceleration curves, and R and S represent the maximum value of rotation/bending and shear experienced by the leg form at the knee joint.

$$G_1 = f_1(K_l, K_m, K_u, H_l, H_m, H_u, P_s, B_u, B_l) \tag{2.1}$$

$$G_2 = f_2(K_l, K_m, K_u, H_l, H_m, H_u, P_s, B_u, B_l) \tag{2.2}$$

$$R = f_3(K_l, K_m, K_u, H_l, H_m, H_u, P_s, B_u, B_l) \tag{2.3}$$

$$S = f_4(K_l, K_m, K_u, H_l, H_m, H_u, P_s, B_u, B_l) \tag{2.4}$$

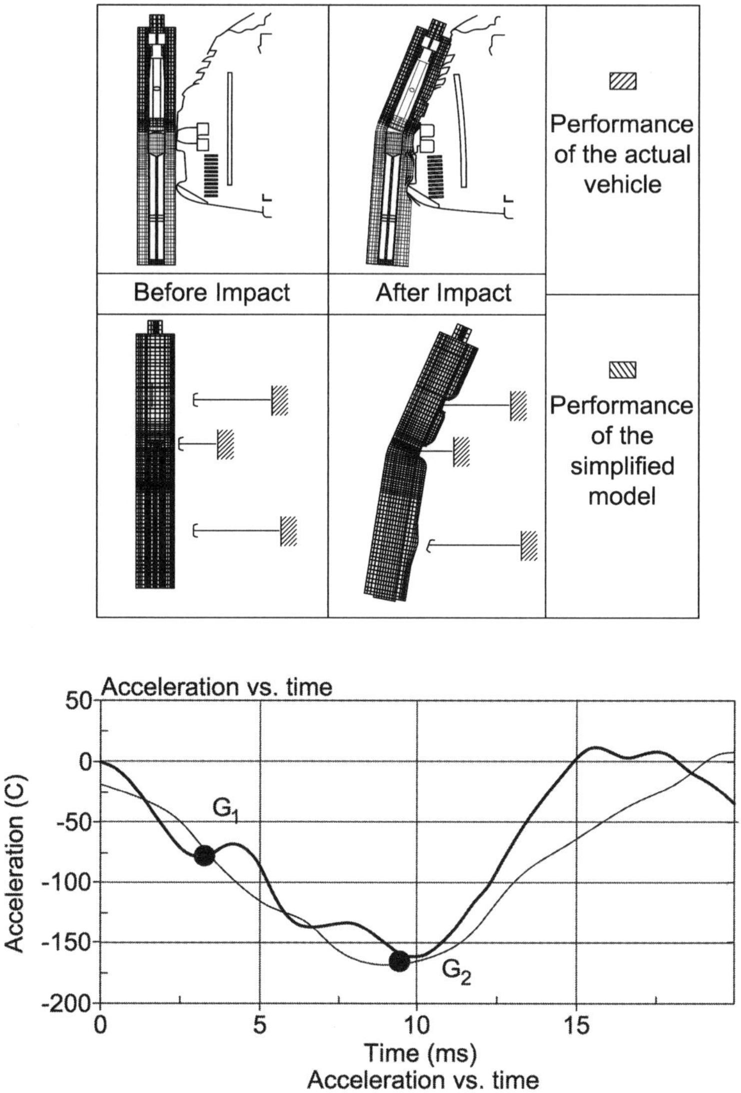

Fig. 2.13 Performance comparison of actual vehicle and simplified model.

(Continues)

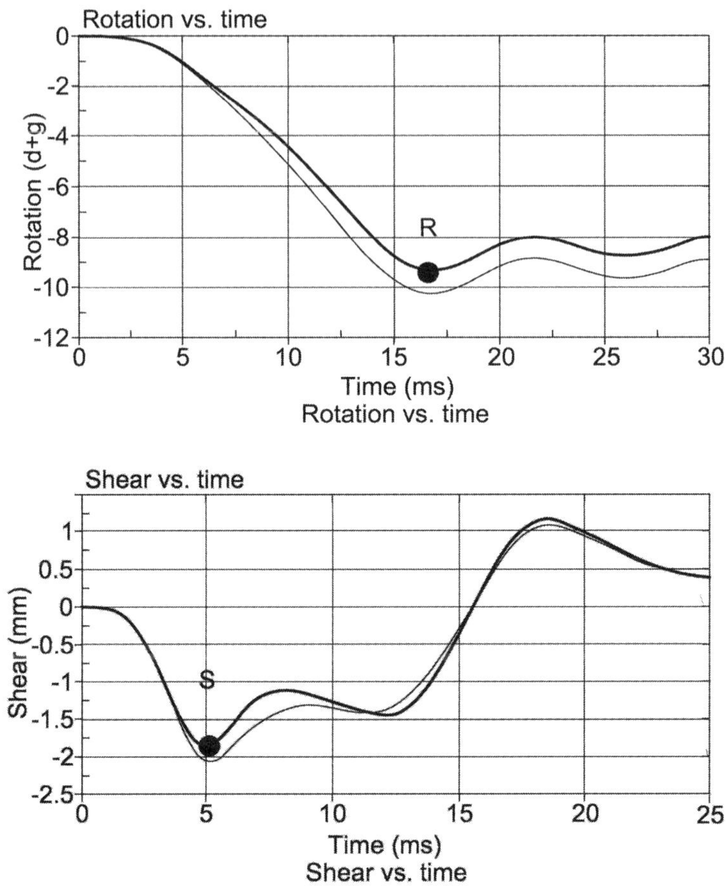

Fig. 2.13 Performance comparison of actual vehicle and simplified model. (*Continued*)

The design of pedestrian-safe bumper system for a given vehicle is limited to achieving the optimum values of these nine vehicle parameters. In most of the cases, six geometric parameters are frozen during the early phase of the vehicle design by the vehicle styling team. Therefore, the task now boils down to finding the optimum stiffness parameters at the upper, middle, and lower load paths. A systematic optimization procedure can now be used to find these stiffness values. One of the ways of formulating this optimization problem, out of many possible options, is represented by the following equations:

$$\min_{\{K_l, K_m, K_u\}} G_2 \tag{2.5}$$

subjected to

$$G_1 \leq 120\text{g}, R \leq 13\text{deg}, S \leq 4\text{mm} \tag{2.6}$$

It is also worthwhile to note that an experienced designer need not require the aforementioned numerical expressions and optimization procedures for the design of energy-absorbing elements: A combination of logical reasoning and trial-and-error methods would suffice to achieve the desired performance by designing separate energy-absorbing elements at these load paths.

2.6.3 Pedestrian Energy Absorbers

2.6.3.1 Pedestrian Energy Absorbers—Middle Load Path

Traditionally, pedestrian EAs were defined as a relatively soft component kept in between the bumper fascia and the bumper beam. These components absorb the maximum portion of the energy during a lower-leg pedestrian impact and thereby protect the pedestrian's knee from experiencing higher acceleration levels. Rotation of the knee is mainly controlled by the undertray or lower-leg protector and the component at the top portion of the vehicle, which is called SUV EA in this chapter. More details on the undertray and SUV EA and their roles are explained in detail in the following sections. In the absence of EA, the pedestrian's knee comes into direct contact with the stiff bumper beams, which are kept in front of the vehicle. Therefore, it is essential that these EAs absorb energy in a very efficient manner. Figure 2.14 shows the most common arrangement of few internal frontal bumper components at the front portion of a vehicle.

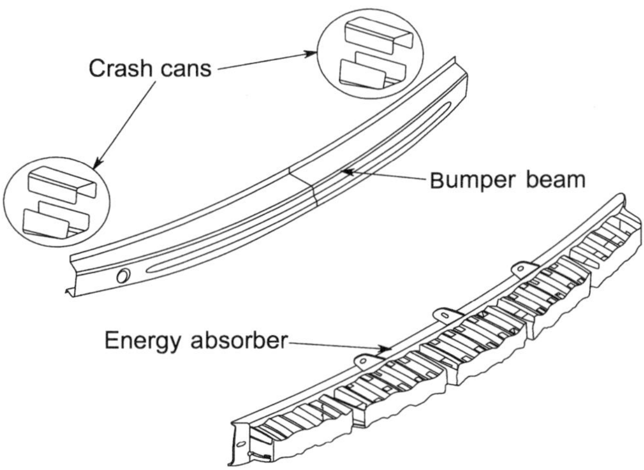

Fig. 2.14 Typical arrangement of an EA over a bumper beam.

As shown, the EA is mounted over the bumper beam, which is in turn mounted over the crash cans or rails. It is worth noting that the EA presented in the figure is a C-section injection-molded thermoplastic member. Foam and metal are also generally used to design these EAs (see Figs. 2.15 and 2.16).

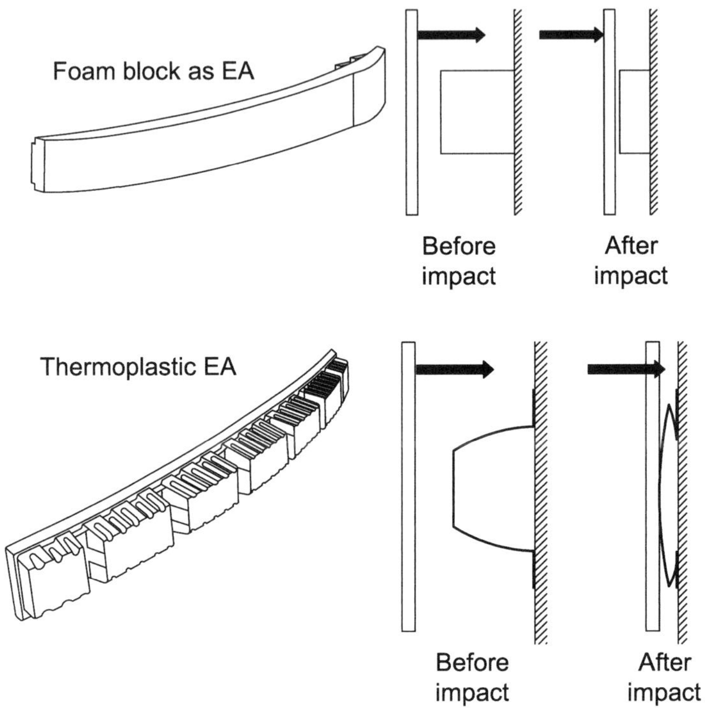

Fig. 2.15 Thermoplastic and foam EA.

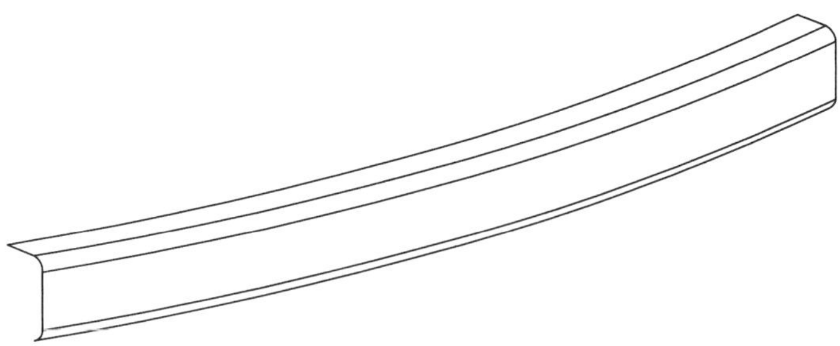

Fig. 2.16 A typical sheet metal EA.

Each of the solutions has its own pros and cons. Foam EAs are relatively cheap and elastic at room temperature, but are not efficient because of their inherent limitations. Usually foams are soft in the beginning of the impact, and they generate stack-up toward the end of impact. Therefore, they fail to generate the desired force levels and cannot make use of the complete space available for energy absorption. Metals EAs have uniform thickness and are manufactured using sheet metal forming. The metal EA is then separately welded to the bumper beam, and therefore the assembly cost increases. It is also difficult to tune the stiffness of the metal EAs at selective locations, as they have uniform thickness. An injection-molded thermoplastic EA, on the other hand, can offer excellent efficiency by generating the required force levels at the beginning of the impact phase itself. They also do not generate any stack-up, as the profile of the energy-absorbing walls of these EAs can be tuned in such a way that during the impact these walls buckle in outward direction. This type of EA also offers the freedom of varying the stiffness across its width, as the thickness and corrugation of the energy-absorbing walls can be easily varied, as shown in Fig. 2.17.

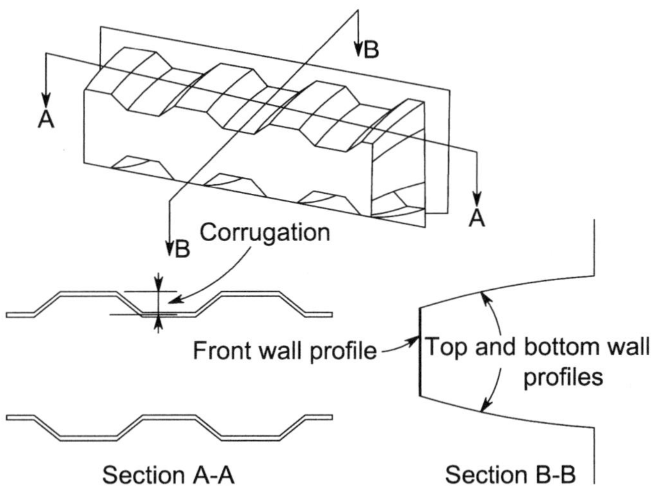

Fig. 2.17 Various geometric parameters of an injection-molded EA.

The assembly cost of these EAs is lower because they are mounted to the bumper beams with the help of few snap fits that match with the holes present in the bumper beams. A case study, given in the following paragraph, comparing the performance of a collapsible thermoplastic EA and foam block for a packaging space of 90 mm clearly demonstrates the efficiency of thermoplastic EA. The geometric details of the EA are shown in Fig. 2.18.

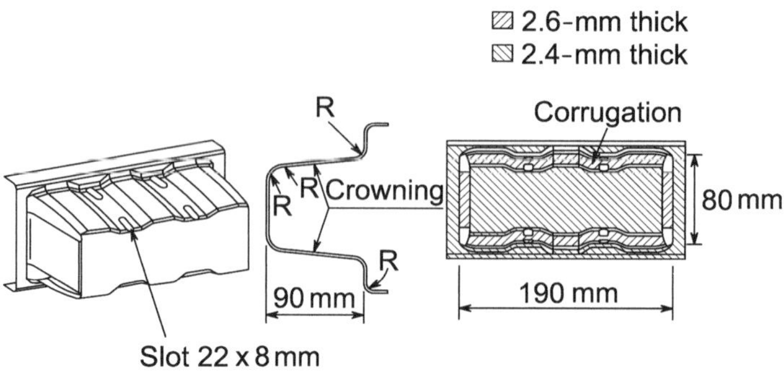

Fig. 2.18 Thermoplastic EA used for comparison.

The desired initial force level can be achieved by fine-tuning the structural stiffness through geometry configuration. Crowning (tuned curvature) is designed on upper and lower walls to initiate crushing. Corrugations are designed to give required bending stiffness during crushing. High corrugations are designed on the front of an EA to give initial force spike. Corrugations are reduced on the backside of the EA. Slots are designed to allow controlled collapsing of the walls, thus maintaining a constant force level during subsequent intrusion. The upper and lower walls used in this study are 2.6 mm thick, and the side straps are 2.4 mm thick. These thickness combinations are calculated after a series of optimization runs. Radii fillets of 5–10 mm are provided at sharp corners to avoid stress concentration and cracking of the part during crushing. Thickness of the EA and corrugations, and curvature of upper and lower walls and slots, can be fine-tuned to control the crushing behavior of the EA. Injection-molded thermoplastic EAs give greater design freedom to the engineer to achieve the required performance target. A solid block of foam (30 g/l) is placed between the beam and polypropylene (PP) plate. Depth of the foam EA used in this study is 90 mm. Force versus deformation comparison of for thermoplastic and 30 g/l foam EA are shown in Fig. 2.19.

The thermoplastic EA shows close to ideal performance. EA design features built up the required force level during the initial 10-mm intrusion. Subsequently, the design can maintain a constant force value from 10- to 80-mm stroke. Thermoplastic design showed stable and uniform crushing of the EA throughout the impact process. Geometric design parameters and the high ductility of thermoplastic can ensure that no cracking happens in the component during impact.

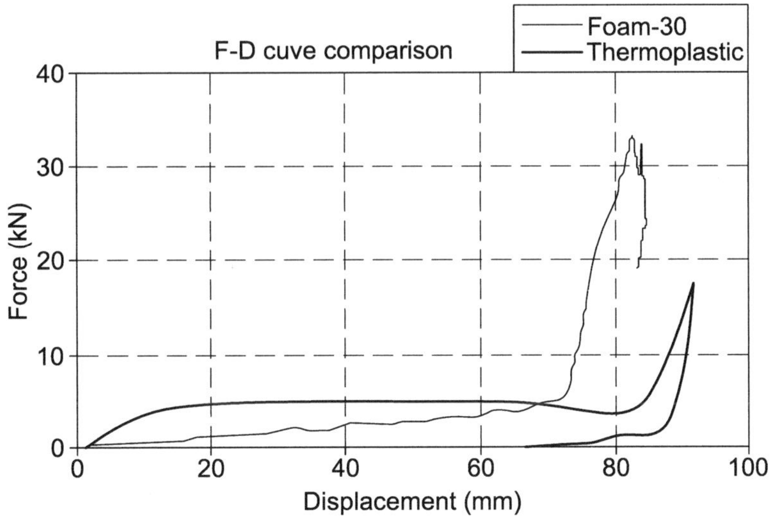

Fig. 2.19 Performance comparisons between the thermoplastic EA and foam EA.

2.6.3.2 SUV Energy Absorbers—Upper Load Path

In recent years, the number of SUVs with high bumper height had been increasing significantly. Since the bumper heights are different for a passenger vehicle and an SUV, the bumper beam location in an SUV is marginally above the knee location. Traditionally, the EAs are mounted on the vehicle bumper beam. Because of the high bumper height of SUVs, the large offset between knee location and EA front face results in a non-uniform support to the lower leg, and it increases the challenges to meet the lower-leg rotation requirements. A typical vehicle setup for an SUV is shown in Fig. 2.20.

The bumper beam location is marginally above the lower-leg knee location [2-30]. During the lower-leg impact case, the upper and lower portion of leg rotates about the knee without any control and results in higher bending of lower leg. A novel energy management approach is proposed for pedestrian-safe SUVs, as shown in Fig. 2.21, where the lower leg is supported from the top load path of the vehicle to minimize the bending.

An injection-molded component made of thermoplastic is designed, which follows the fascia geometry and is mounted on the grille opening reinforcement (GoR), and it carries the weight of the vehicle bumper fascia and supports the upper portion of leg during lower-leg impact case. A C-section injection-molded component made of thermoplastic, mounted on bumper beam, is integrated to the designed upper fascia support. This component absorbs the kinetic energy of lower leg by crushing and avoids the failure of knee ligament due to higher force values.

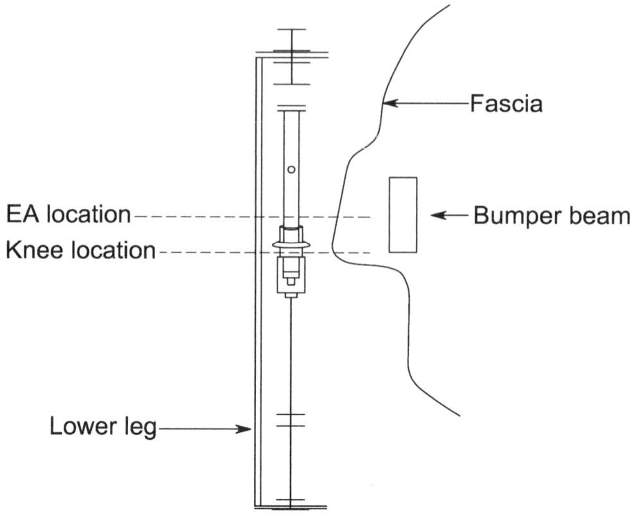

Fig. 2.20 Typical SUV setup.

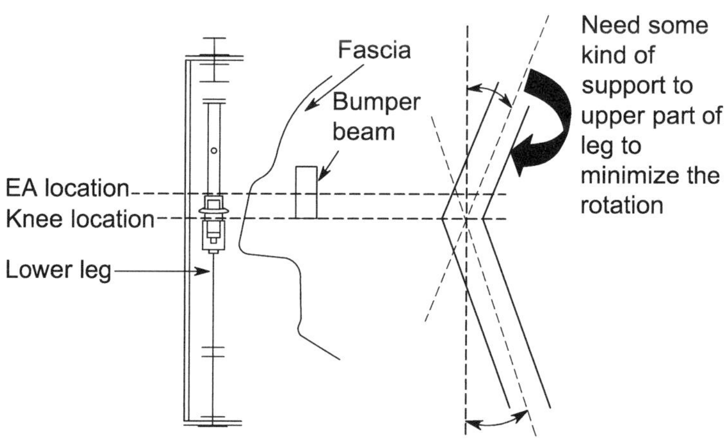

Fig. 2.21 Energy management approach for SUVs.

A C-section EA integrated with the upper-leg support is designed as a single-piece injection-molded component made of thermoplastic. Figure 2.22 shows the designed SUV EA system.

The upper-leg support is snap-fitted on the top portion of the front-end module at seven locations, and the C-section EA is snap-fitted on bumper beam at 10 locations. Figure 2.23 shows the SUV setup with a proposed EA system.

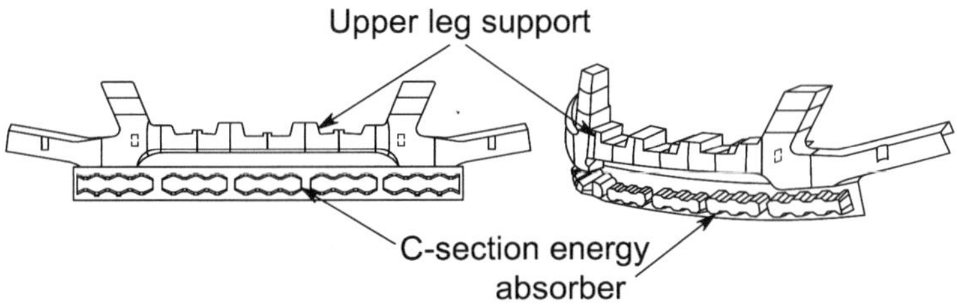

Fig. 2.22 Pedestrian-safe EA solution for SUVs.

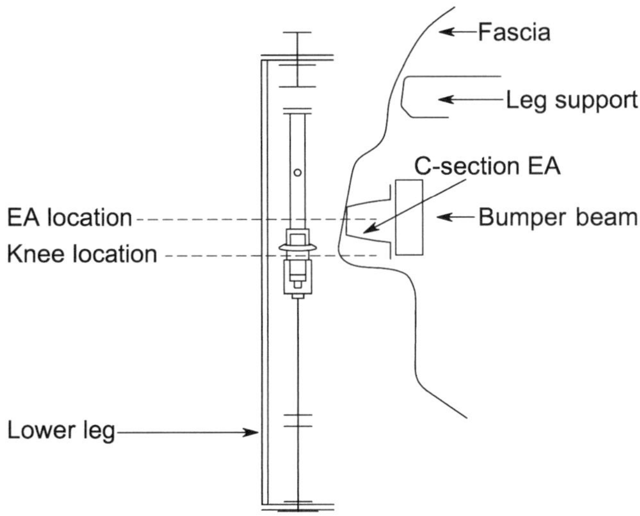

Fig. 2.23 Pedestrian-safe SUV setup.

The C-section EA is designed to crush during impact and absorb the kinetic energy of the leg, which reduces the chances of serious injury in the knee ligament due to higher forces. The upper-leg support helps to support the leg from the top during impact and helps to avoid severe bending of the knee joint. Thus, as a complete solution, the C-section EA with integrated leg support gives more controlled energy management for the SUV bumper system, for the lower-leg impact case.

2.6.3.3 Undertray—Lower Load Path

Generally, the height of the bumper beam is specified by OEMs and is specific for the type of the vehicle. It has been observed that, when the bumper is in front of or in line with the knee joint of leg, further optimization of upper EA lobes will not help in minimizing rotation values, and there is a need for additional reinforcement at the lower bumper to control or minimize the rotation of lower-leg members. So, in the lower load path, as discussed in an earlier section, there is a need for additional support at the lower part of the leg to minimize rotation. One commonly available solution is the use of metal spoilers/tubes, which extend to the front fascia of the vehicle; another solution comprises a simple plastic sheet or tray [2-30]. Such a thermoplastic undertray is shown in Fig. 2.24.

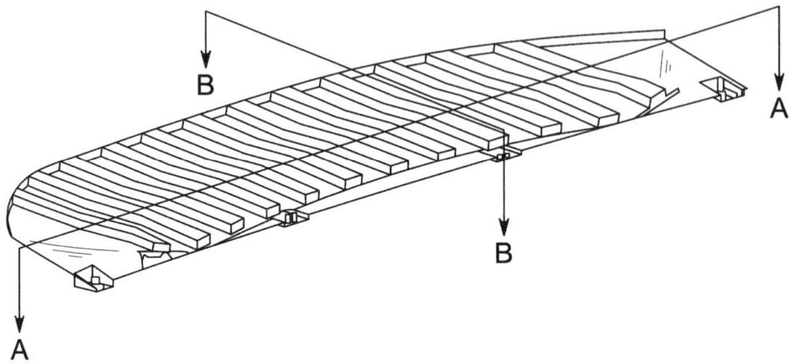

Fig. 2.24 A typical thermoplastic undertray.

It is worth noting that these undertrays are typically mounted to the lower portion of the front end module of the vehicle. Thermoplastic undertrays offer some extra benefits, as described in the following list, in addition to meeting the functional requirement of supporting the lower portion of the leg to control the rotation.

- The design of corrugated geometric structures allows the free flow of air, guiding air intake for the cooling radiator and the engine components under the hood.
- A closed structure that wraps the bumper prevents stones and other unwanted particles from entering inside and thereby damaging the parts under the hood.

Unique curvatures and angles provided for the component along the impact direction help to quickly achieve the optimum stiffness required for the lower-leg protector for a given vehicle. It provides the necessary spring-back effect without any major permanent damage to the part, as shown in Figs. 2.25 and 2.26, which display the cross-sectional views of the undertray from the side and the front, respectively.

Different values of the angles, heights, and lengths of corrugations can be tested to achieve the desired performance values. The optimized undertray component, when assembled over the vehicle, offers excellent pedestrian

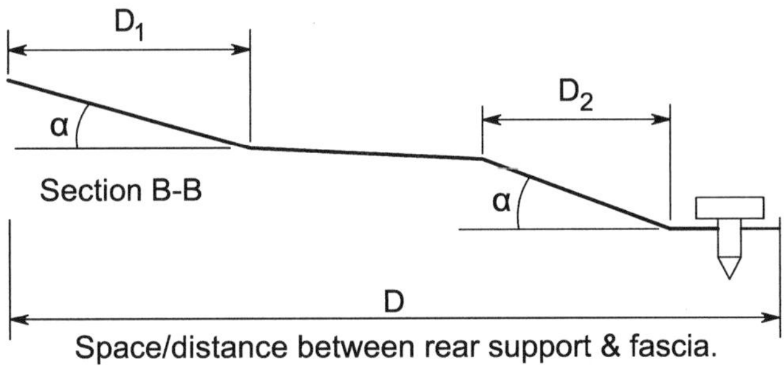

Fig. 2.25 Sectional view of undertray—side view.

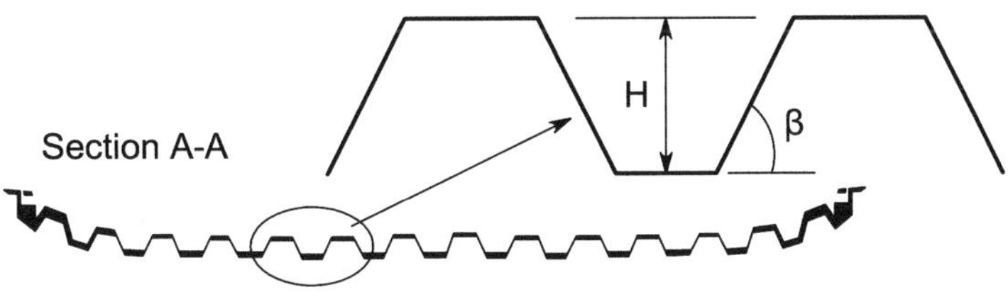

Fig. 2.26 Sectional view of undertray—front view.

lower-leg impact performance by absorbing a significant amount of energy and pushing the leg after the impact, which results in reduced rotation at the pedestrian leg knee location. The lower-leg protector also absorbs significant energy during a low-speed vehicle-to-barrier impact, thereby reducing the damage to the vehicle.

2.7 Countermeasures for Low-Speed Vehicle Damageability Tests

2.7.1 Bumper Design Challenges

Although low-speed automobile impact does not cause much injury to the occupant and pedestrian, it causes significant damage to the vehicle. Recent publications summarize that even low-speed impacts can cause considerable damage, and the repair cost could go up to US $4000 [2-31]. It is estimated that the cumulative damage cost to the motorists and insurance agencies for all the bodywork damage in United Kingdom amounts to US $2.4 billion. Similar numbers were reported out in other countries of Europe, the United States, and the Asia-Pacific region. To reduce these repair costs, regulatory bodies and insurance agencies together proposed various low-speed damageability requirements for the protection of vehicles during front and the rear impacts. FMVSS Part 581 regulations on bumper standards in the United States [2-32] and ECE R42 regulations [2-33] on vehicle front and rear protective devices in Europe and Korea are the outcomes of these regulatory measures. Figure 2.27 is a diagram shows various low-speed damageability regulatory requirements.

Metallic solutions are traditionally used in automobiles to make the vehicle compliant with these low-speed damageability regulations. Figure 2.27 shows two kinds of impacts, a 4.02 km/h (2.5 m/h) pendulum impact and a 15 km/h RCAR barrier impact. Any one of them or combinations of these impacts are used to evaluate the damageability criteria across various parts of the globe. Although metal beams are traditionally used to reduce the damage during pendulum impacts, crash cans take on the role of beams during RCAR impact. For any of these passive impact cases, the major challenge is to design simpler, more cost-effective, and lighter components that will meet the requirements for automobiles at varying heights from ground. This is important, as the colliding vehicles could have different bumper heights from the ground, as shown in Fig. 2.28.

Hence, these components are designed in such a way that during a low-speed impact, the vehicle will have the minimal damage and the repair cost will remain low [2-14, 2-34, 2-35]. As a result, the bumper system will absorb most of the energy without causing much damage to the other expensive components of the car such as hood, lighting, and cooling systems [2-36, 2-37]. The problem becomes much more complicated in the front side of the vehicle, where it also has to meet the pedestrian-safety requirements, which demands a relatively soft bumper system. As explained in the previous section, automobile manufacturers, however, manage to meet these requirements by providing a relatively soft member in front of the bumper beam that absorbs the energy in a controlled manner.

Many materials such as metals, composites, and plastics are being used in the industry for the design of bumper beams and crash cans that meet low-speed damageability requirements. While metal is a relatively cheaper solution, it has its own limitations such as high system mass and the need for additional welding operations. Composites, on the other hand, have excellent stiffness-to-weight ratio, but are relatively expensive both from the raw material and from the manufacturing point of view. Thermoplastic materials can offer excellent performance with lighter weight and at moderate cost. An additional advantage is that thermoplastics can be easily integrated with crash cans at either end of the vehicle and with the bumper fascia. The next section of this chapter gives more details on the thermoplastic bumper beams, which are also called solitary beams,

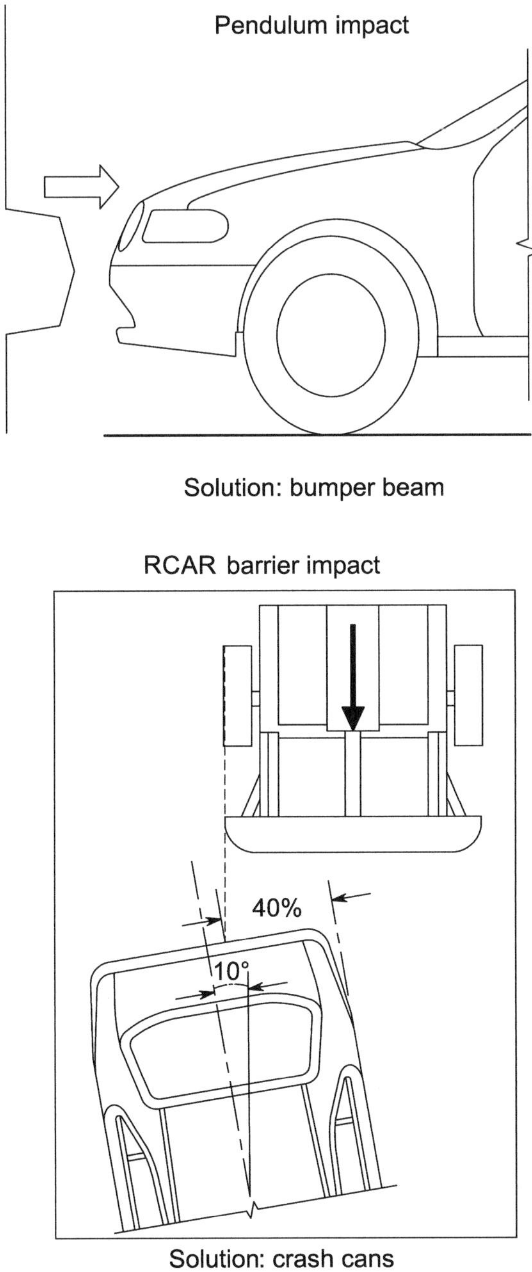

Fig. 2.27 Low-speed vehicle damageability requirements.

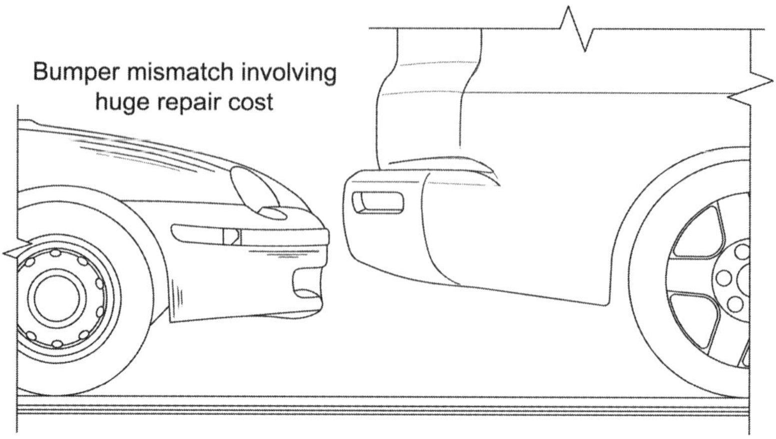

Fig. 2.28 Side views of vehicle during low-speed impact.

and the various aspects that need to be considered in designing these bumper beams.

2.7.2 Thermoplastic Solitary Beam Solutions

As with any design, several factors, such as the targeted performance, system mass, system cost, and possible material options, need to be kept in mind while designing a thermoplastic beam for low-speed damageability. A three-step approach is generally followed to arrive at an optimum solution that meets these requirements [2-38, 2-39]. The first step typically involves the identification and quantification of the requirements for vehicle damageability and reparability, and getting some understanding of the targeted system mass, system cost, and the design space available. Various possible solutions, including the approximate geometry, materials, and the manufacturing process corresponding to each material and geometry, are also explored in this stage. In the next step, each of the proposed solutions is evaluated for its performance, and appropriate modifications are performed to optimize it further for reduced mass, improved performance, and reduced cost. Several tools, such as computer-aided engineering (CAE) simulations, analytical methods (for simpler geometries), and other tools such as numerical models derived from detailed DoE [2-40], can be employed to quantify these performance measures and to optimize each solution. The performance and cost associated with each of these solutions are quantified at this stage. The optimum solution among the various solutions evaluated in the second step is then selected for the third step. Some compromising on either the system cost or the system performance is usually necessary at this stage. It is estimated that a thermoplastic beam solution, in general, results in approximately a 33% reduction in the assembly time and assembly cost [2-39]. This is mainly because thermoplastic beams can be easily integrated with crash cans and the few brackets that are typically used to hold the bumper fascia. Three common thermoplastic bumper beam solutions that can be used to meet the low-speed damageability requirements are shown in Fig. 2.29.

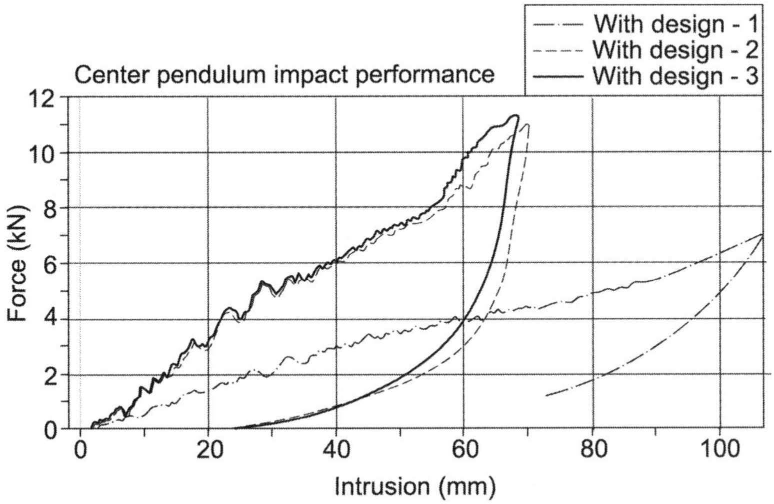

Fig. 2.29 A few thermoplastic bumper beam solutions for low-speed damageability requirements.

It is worth noting that the last two bumper beams in Fig. 2.29 have relatively stiff ribs/crush boxes integrated at either ends. All these beams can be manufactured using injection molding. Blow-molded beams can also be used, but it has limitations in terms of its limited geometric configurations and high processing time. The objective is to make the beams as stiff as possible so that the intrusion level during an impact at low speed is as low as possible. In the case of a rear beam of a car, such low intrusions would offer adequate protection to the back panel of the tailgate, which is typically placed behind such beams. Performance of each of these beams for the center pendulum impact, the most critical impact requirement according to ECE R-42/FMVSS Part 518 regulatory requirements, is shown in Fig. 2.30.

The force-deformation curves for design 2 and design 3 show higher efficiency of energy absorption than design 1. Figure 2.31 is a diagram shows the top view of the one of these beams before and after collision.

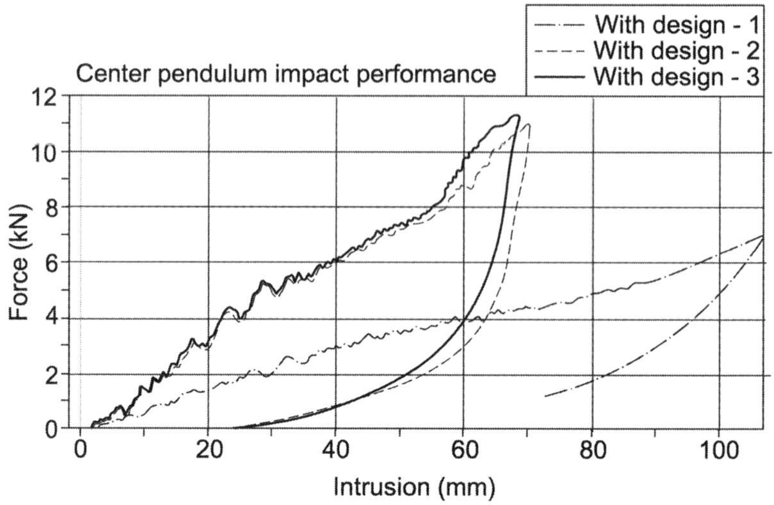

Fig. 2.30 Center pendulum impact performances of the three beams.

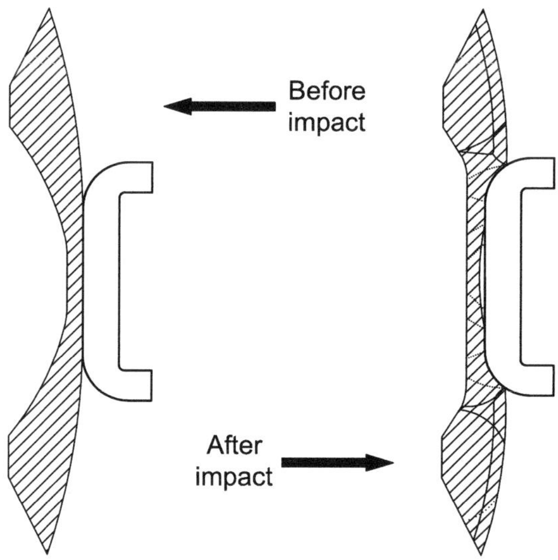

Fig. 2.31 Top view of beam before and after impact.

A DoE-based methodology [2-40], as explained earlier, can be effectively used to quantify the performance and to arrive out at the optimum mass and performance for a given beam geometry, shown in Fig. 2.32 and Table 2.1.

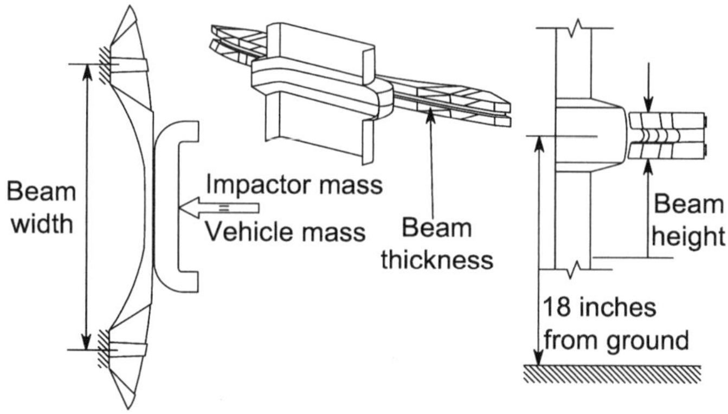

Fig. 2.32 Schematic representation of variable considered for DoE.

To achieve this, first a robust transfer function is developed that relates various independent parameters such as the vehicle mass, beam thickness, beam height, and beam width to performance parameters such as the reaction force generated during the impact and the intrusion level. Although the numerical models (from this study) are limited to this beam geometry and specifications, observations from the study (Table 2.1) can be effectively used in the design of thermoplastic beams for a generic vehicle. The beam stiffness reduces linearly as the beam width increases, and it results in the increase in intrusion levels.

Table 2.1 Variables Considered for Design of Experiment Study and Their Range

Independent Parameters (*X*)

Actual parameter	Parameter mentioned in DoE	Range		Levels
		Low	High	
Vehicle mass	Mass (kg)	900	1200	2 levels
Beam thickness	Thickness (mm)	2.2	3.5	2 levels
Beam height	Height (mm)	100	120	3 levels
Beam width	Width (mm)	900	1100	3 levels

Dependent Parameters (*Y*)

Force (kN)	Intrusion (mm)

- Beam stiffness increases as the thickness is increased and the intrusion levels are reduced.
- Vehicle mass and beam height have some effect (though not as high as the beam thickness and beam width), and the intrusion level increases as these parameters are increased.
- Reaction force value (the force offered by the vehicle bumper to the impactor during impact) decreases in a quadratic manner, as the beam width is increased.
- Reaction force increases linearly as the vehicle mass is increased.

An optimization to minimize the intrusion values as shown in Eq. 2.7 subjected to constraints shown in Eq. 2.8 can be performed using these numerical models to calculate with the minimum achievable intrusion value for a given beam geometry, beam material, vehicle mass, and vehicle width. Results of such an optimization study [2-41] are shown in Table 2.2.

Table 2.2 Matrix Generated for Computing the Minimum Achievable Intrusion Values

	Vehicle width (mm)		
Vehicle mass	900	1000	1100
900	40	59	62
1200	46	68	94

$$\min_{\{\text{Mass},\text{Thickness},\text{Width},\text{Height}\}} \text{Intrusion} \tag{2.7}$$

Subjected to

$$\text{Mass} = \text{Mass}_{\text{spec}},\ \text{Width} = \text{Width}_{\text{spec}},\ \text{Thickness} \leq \text{Thickness}_{\text{spec}} \tag{2.8}$$

2.7.3 Hybrid Plastic-Metal Bumper Beam Solutions

Car manufacturers design their vehicles to perform well in IIHS and RCAR structural tests with countermeasures that need to exhibit good crash behavior in real-world accidents. Inclusion of new deformable or bumper barrier impact with IIHS and RCAR bumper test protocols redefines the development of countermeasures for low-speed damage-ability. Most common energy management solutions include crash cans and bumper beams, both made of thermoplastics or metals, but they may not exhibit good crush behavior for real-life accidents and for the new IIHS or RCAR barrier impact [2-42]. A hybrid thermoplastic-steel solitary bumper beam solution (Fig. 2.33) has been proposed to meet new IIHS and RCAR impact requirements.

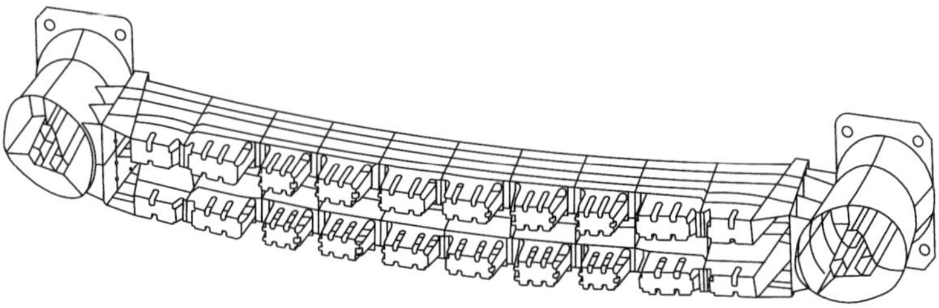

Fig. 2.33 Hybrid bumper beam—front view.

This bumper beam includes two components: one W-section thermoplastic bumper beam and a small rectangular steel beam, shown in Fig. 2.34.

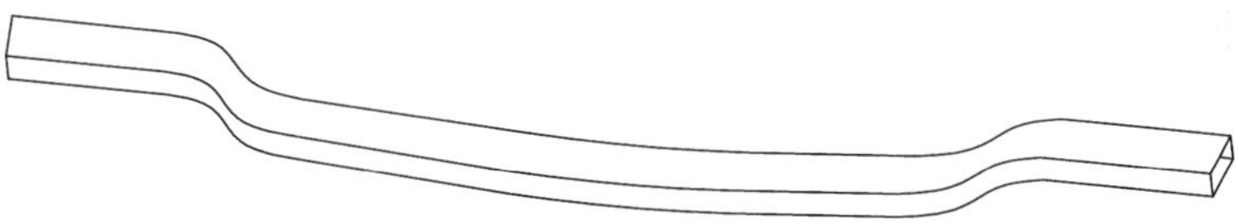

Fig. 2.34 Metal piece used in hybrid bumper beam.

The thermoplastic beam has a provision (a narrow channel) at the rear to accommodate a steel beam. It is worth noting that the hybrid beam has conical crush boxes (Fig. 2.35) at the either end of the beam, and these crush boxes reduce the damage to the vehicle during 40% overlap 15 km/h RCAR impact.

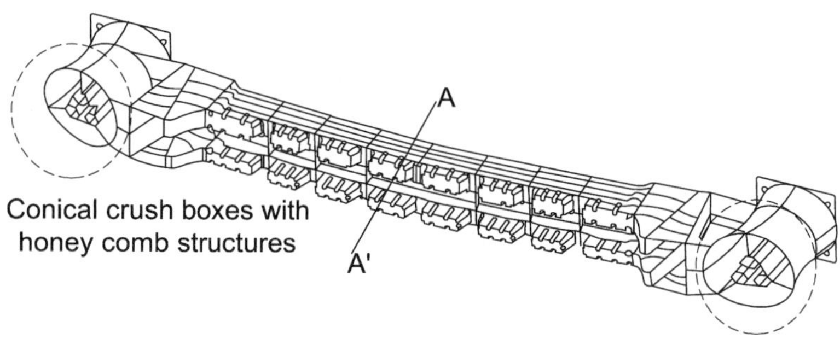

Fig. 2.35 Representation of conical crush boxes in hybrid beam.

The complete assembly is mounted over vehicle rails and absorbs energy by structural bending and crushing of the thermoplastic walls during impact. In general, the bumper beam is positioned at a vertical height sufficient to offer proper engagement during vehicle override and underride conditions. The overhang thermoplastic walls take reactions from the top and bottom faces of the steel beam for crushing during override and underride conditions. Figure 2.36 shows the geometry and cross-sectional details of a hybrid solitary beam designed using resin polycarbonate (PC) plus polybutylene terephthalate (PBT) resin and high-strength steel.

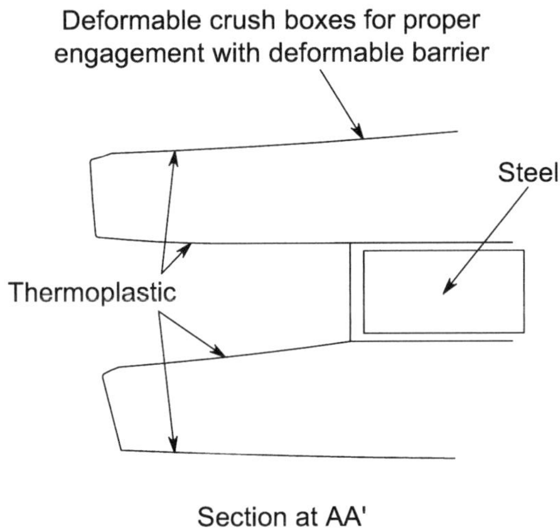

Fig. 2.36 Cross-sectional view of bumper beam.

The front walls of the plastic beam (a blend of PC and PBT) include deformable crush boxes, which offer proper engagement with new IIHS/RCAR deformable/bumper barriers during a 10 km/h speed impact. This significantly avoids the risk of slippage between two bumpers during an impact.

The proposed hybrid solitary beam must be evaluated with CAE tools to validate the performance using generic vehicle platform and observed to be meeting all performance requirements of IIHS/RCAR/ECE42 structural test protocols for bumpers as reported in Reference [2-42].

2.8 Low-Speed Damageability and Lower-Leg Impact-Compliant Bumper System

In the United States, bumper systems are designed for the Part 581 bumper standard and IIHS bumper structural test protocols. There has also been discussion in the North American automotive industry about the merits of incorporating some measure of pedestrian protection into their systems as well. Compliance with the potential pedestrian leg requirements creates a design conflict with current bumper damageability standards and possibly Corporate Average Fuel Economy (CAFE) laws [2-43]. The difficulties of designing a bumper system that is rigid enough to protect the vehicle in low-speed crashes and, at the same time, compliant enough to protect a pedestrian raise questions as to whether these ideas are compatible. This section explores reasonable means by which one can manage lower-leg impact, Part 581 bumper standards, and IIHS bumper damageability assessment criteria from a styling and cost structure standpoint, which is the design trend of today's automotive industry.

2.8.1 Conflicting Energy-Absorbing Requirements for Bumpers

A lower-leg test procedure calls for a free-flight bumper impact at 40 km/h with a leg form weighing ~13.8 kg and involves impact energy level of ~825 joules (J). The Part 581 test procedures include a series of pendulum impacts at 4 km/h on the vehicle bumper. Depending upon the vehicle mass, the impact energy levels may vary from 750 to 1100 J. The IIHS bumper-like barrier impact at a speed of 10 km/h on the vehicle bumper may involve energy levels of ~7000 J. Figure 2.37 represents a schematic view of different energy levels to be considered for designing vehicle bumper system.

An EA designed for lower-leg impact requirements may be significantly softer than the EA designed for IIHS bumper like barrier impact case.

Figure 2.38 shows a generic response of a bumper system equipped with different EAs for the different impact cases of lower-leg, Part 581, and IIHS bumper-like barriers.

A bumper equipped with an EA for pedestrian protection is intended to offer softer stiffness to lower-leg impact. The maximum force value is generally <15 kN, and higher deformations cushion the knee during impact. For Part 581 impact cases, lower deformation of the bumper is intended to minimize damage to expensive components like the hood and headlamps. A typical bumper equipped with an EA designed for Part 581

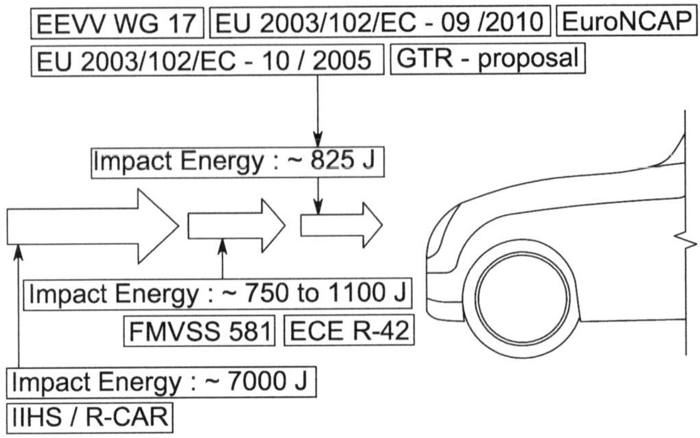

Fig. 2.37 Conflicting energy requirements for automotive bumpers.

IIHS EA	Bumper to absorb ~6000J	Force <120kN	Package ~100 mm Stiff
Part 581 EA	Bumper to absorb ~900J	Force <45kN	Package ~85 mm Stiff
Ped. EA	Bumper to absorb ~450J	Force <15kN	Package ~65 mm Soft

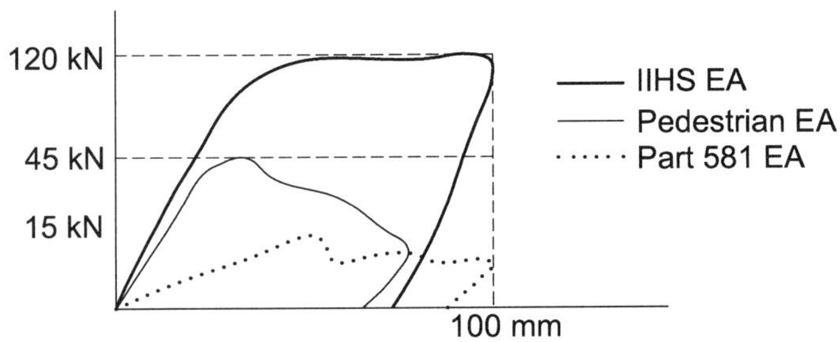

Fig. 2.38 Force versus displacement curves of bumper for conflicting impact requirements.

impact cases may offer 30–45 kN force value and bumper deformation <60 mm. An IIHS bumper-like barrier impact at a speed of 10 km/h requires the bumper to be stiffer in order to manage higher impact energy levels.

A single-bumper system equipped with an EA designed for lower-leg impact cases may not serve the purpose for the Part 581 and/or IIHS bumper-like barrier impact case. Accordingly, the need exists for an energy-absorbing system that addresses the above mentioned conflicting energy-absorption-level requirements while not adversely affecting the cost and performance of the vehicle.

2.8.2 Dual-Stage Energy-Absorber Approach

A multistage EA system has been explored that includes at least two energy-absorbing stages that provide different energy-absorption characteristics to the system. The first energy-absorbing stage can provide an energy absorption characteristic that is generally useful for low-impact and/or low-speed collisions. Subsequent energy-absorbing stages provide an energy-absorption characteristic that is generally useful for impact and/or collisions at a higher energy level or speed, according to bumper structural test procedures. Figure 2.39 is a schematic representation of the cross section of a multistage EA design.

A dual-stage energy-absorption approach, as shown in the Fig. 2.40, has been proposed to meet lower-leg impact and IIHS structural test requirements.

The first stage of energy absorption is tuned softer to cushion the knee during lower-leg impact, and the second stage is tuned stiffer to absorb a higher energy level and minimize bumper damage during the IIHS bumper-like barrier impact procedure.

A novel geometric configuration has been designed, as shown in Fig. 2.41, in which the front stage elements of the EA are tuned to be softer, and second stage is tuned to be stiffer.

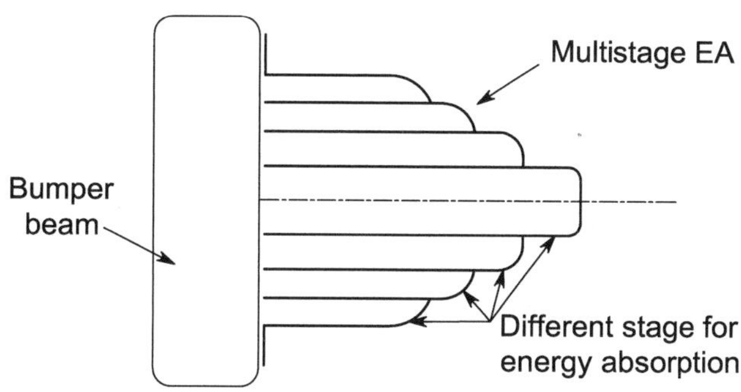

Fig. 2.39 Multistage EA design.

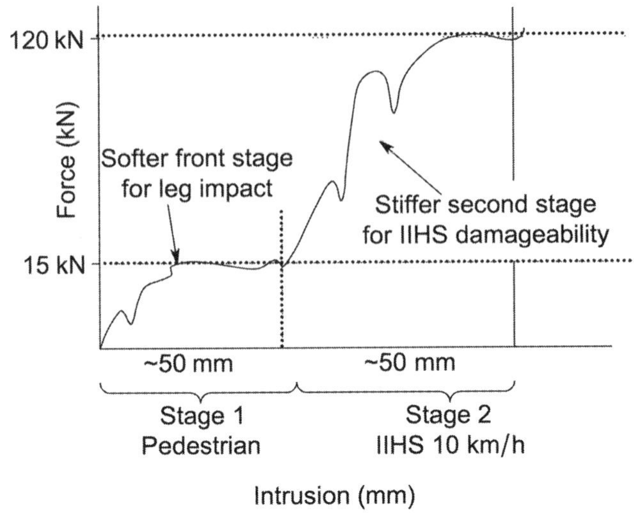

Fig. 2.40 Dual-stage energy-absorption approach.

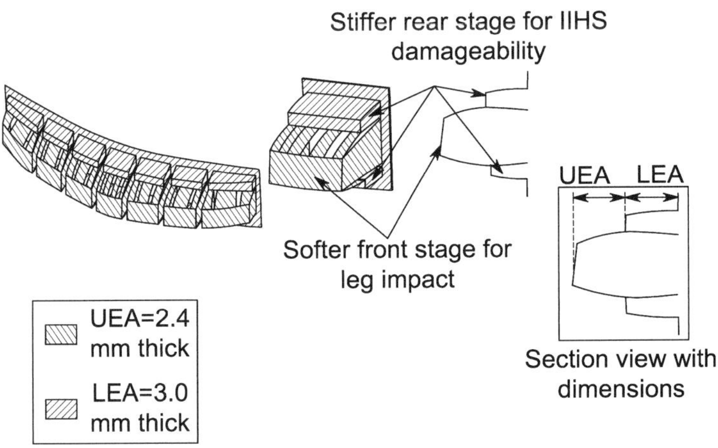

Fig. 2.41 Novel dual-stage EA design.

A C-section geometry with engineered curvature on slots at the top and bottom load walls has been designed as the first stage of the EA. The second stage of the EA includes a stiffer closed-section structure, which helps in absorbing higher impact energy levels during IIHS bumper test protocols.

2.8.3 Performance Evaluation

A generic vehicle front bumper system, in conjunction with the proposed thermoplastic dual-stage EA is studied for lower-leg impact, Part 581, and IIHS bumper-like barrier impact procedures. A free-flight bumper impact at 40 km/h with a leg form is studied for two impact location, first at the center of vehicle and second at the most outboard impact location, that is, 470 mm outboard from center. The performance numbers for the lower-leg impact were observed to meet the lower-leg impact Phase II targets (acceleration < 150 G, knee rotation < 15°, and tibia-shear < 6 mm) with a good safety margin. Figure 2.42 shows the crushing of only first-stage elements of the EA, which are softer than second-stage elements.

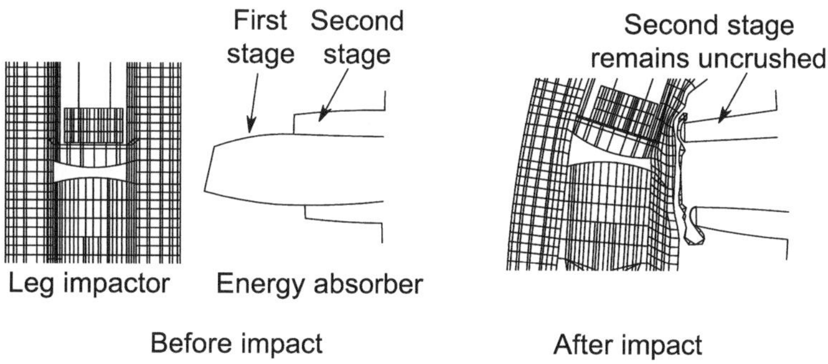

Fig. 2.42 Crushing of an EA during a leg impact case.

Figure 2.42 clearly demonstrates the dual-stage EA functions for pedestrian safety.

Sequential impact cases according to the Part 581 impact procedure is simulated for the complete bumper system. In addition to pendulum impacts per the Part 581 procedure, an additional flat barrier impact is also considered. It has been observed that there is no damage to hood and headlamp, which is the typical performance requirement from the OEMs and insurance agencies.

With different vehicle heights, bumper mismatch during impact may result in more damage to vehicle bumpers. The IIHS has introduced a bumper-like barrier impact in its structural test procedure, to estimate the damage closer to real-life impact cases. In this test the complete vehicle bumper system is impacted over the IIHS bumper-like barrier at a speed of 10 km/h. No damage to it has been observed on the vehicle bumper components like the headlamp and hood. The maximum back of beam deflection has been observed less than the space available between the beam and the cooling tower, and no damage to radiator assembly has been predicted.

2.9 Vehicle Structural Integrity for High-Speed Crashes

Vehicle structural integrity or crashworthiness connotes a measure of the vehicle's structural ability to plastically deform and yet maintain a sufficient survival space for its occupants in crashes involving reasonable deceleration loads. Restraint systems and occupant packaging provide additional protection to reduce severe injuries and fatalities. Vehicle BIW, as shown in Fig. 2.43; crush boxes; front rails; A, B, C, D pillars; roof rails; floor rockers; floor cross bars; and the ladder of the chassis play major roles in energy absorption during high-speed front, side, rear, and rollover crashes.

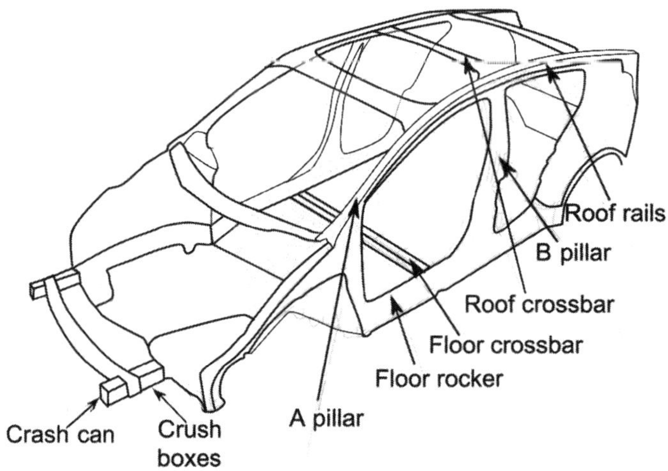

Fig. 2.43 Vehicle BIW components.

As described in Section 2.3.3, vehicle crashworthiness is now evaluated in four distinct modes: frontal, side, rear, and rollover crashes. A crashworthy BIW structure of a vehicle may have following characteristics:

- Deformable, yet stiff, front structure with crumple zones to absorb kinetic energy
- Deformable rear structure to maintain integrity

- Properly designed side structures (B pillar, floor rocker, and floor cross bar) and doors to minimize intrusion
- Strong roof structure (roof panel and roof cross bars) and stiff A and B pillars for rollover protection

The BIW components contribute to the major portion of vehicle mass. Upcoming CAFE requirements target making vehicles lighter to reduce CO_2 emissions. Many attempts have been made to make vehicles lighter without compromising crashworthiness. Use of high-strength steel and ultra-high-strength steel in the BIW components has increased rapidly. Lighter materials like aluminum and magnesium have also been used extensively in recent years.

This section focuses on the plastic-metal hybrid (PMH) solutions for lighter BIW components and maintaining good crashworthiness of vehicle. PMH rail extensions have been discussed for frontal impacts and the plastic-reinforced BIW components for the vehicle rollover and side impact requirements.

2.9.1 Hybrid Rail Extensions for Frontal Crashes

Rails extensions are mounted at back of the beam to support from both sides, as shown in Fig. 2.44.

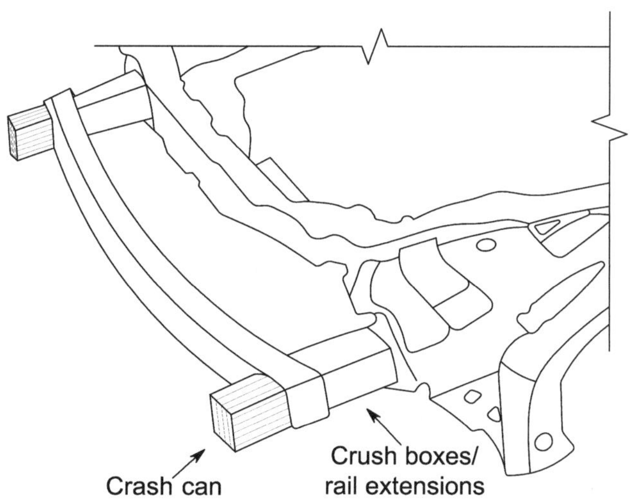

Fig. 2.44 Vehicle rail extensions.

These rail extensions are designed to absorb energy in high-speed frontal crashes (~50 km/h) and carry the weight of the bumper beam. During a high-speed frontal crash, each of the rail extensions is intended to absorb ~10,000–14,000 J of energy. These extensions are designed to be softer than the vehicle rails to achieve progressive crushing and absorb significant energy with plastic deformation, thereby reducing rail intrusion. A typical crash performance requirement for rails extension is shown in Fig. 2.45.

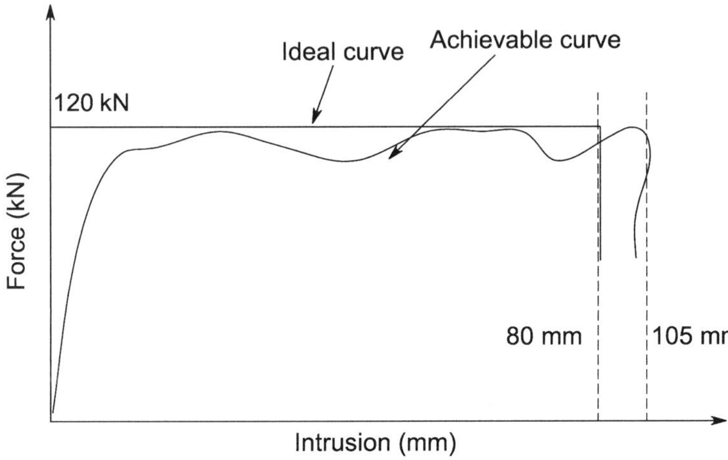

Fig. 2.45 Vehicle rail extensions' crash performance requirement.

Many attempts have been made to provide bumpers for automotive vehicles, which could be able to absorb a major portion of impact energy during a crash. Metal crush boxes are the most common solutions, because of good impact and creep performance. But metal solutions may be heavier and involve secondary operations for manufacturing. Various impact energy-absorption devices exist; one in particular is made of alveolar structures and is commonly called "honeycomb." These structures are known for their good impact energy-absorption properties. The first type of honeycomb is made of metals. These honeycombs can be manufactured by gluing flat sheets along preset lines and widened to form cavities, or by gluing preshaped metal sheets. These metal honeycombs also have good energy-absorption characteristics as well as better creep performance. The second type of honeycomb is made of plastics. The extruded plastic tubes are glued together to form a honeycomb shape, or in other words, the injection-molded plastic honeycombs. The extruded plastic honeycombs involve high cost and limited fields of application because of their relatively thin walls. The injection-molded plastic honeycombs are relatively cheaper, and their impact energy-absorption performance is relatively better, than those of extruded plastic honeycombs. Both extruded and injection-molded plastic honeycombs have inferior creep performance and undergo significant deformation in the traverse direction when employed to support relatively heavier components such as bumper beams. The third type of honeycomb solution is crush boxes made of laminated composite material. This solution may offer good impact and creep performance. It is relatively lighter in weight but involves high manufacturing and system costs. Figure 2.46 shows a comparison of different crush boxes and rail extension solutions.

	Impact performance	Creep performance	System weight	System cost	Manufac-turability
Metal crush box	●	●	⊜	○	○
Metal honeycomb	●	●	⊜	○	⊜
Plastics extruded	⊜	⊜	●	●	○
Plastics molded	●	⊜	●	○	○
Laminated composites	●	●	●	⊜	⊜

● Good ⊜ Bad ○ Challenging

Fig. 2.46 Different rail extension solutions and their limitations.

A hybrid rail extension is designed as a deformable unit consisting thermoplastic and metal components manufactured as a single-piece assembly by the co-molding process. The thermoplastic unit of the hybrid rail extension is an array of columns and channels, and the metallic unit is a box-shaped structure (open or closed) with localized crash initiators, which is wrapped around the thermoplastic unit. The hybrid crush box unit is crushable in the axial direction and intended to absorb a significant amount of impact energy while subjected to axial loading. It is also intended to perform a better than plastic solution for creep, when used as a supporting member to other components and subjected to load in transverse direction. Figure 2.47 is a diagram shows a typical PMH rail extension design with an open-section metal component.

Lancing features [2-44] are designed on the edges of metal structure for flow of plastic, and they achieve bonding between these bimaterial systems. A cross-ribbed plastic structure offers a high stiffness-to-weight

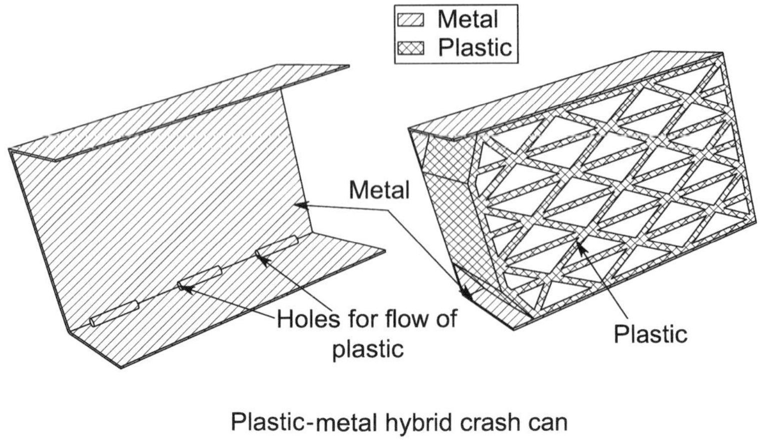

Fig. 2.47 Hybrid rail extension with open-channel metal structure.

ratio and helps in achieving highly efficient energy absorption. Figure 2.48 is a diagram shows a plastic-metal rail extension with a closed box metal structure.

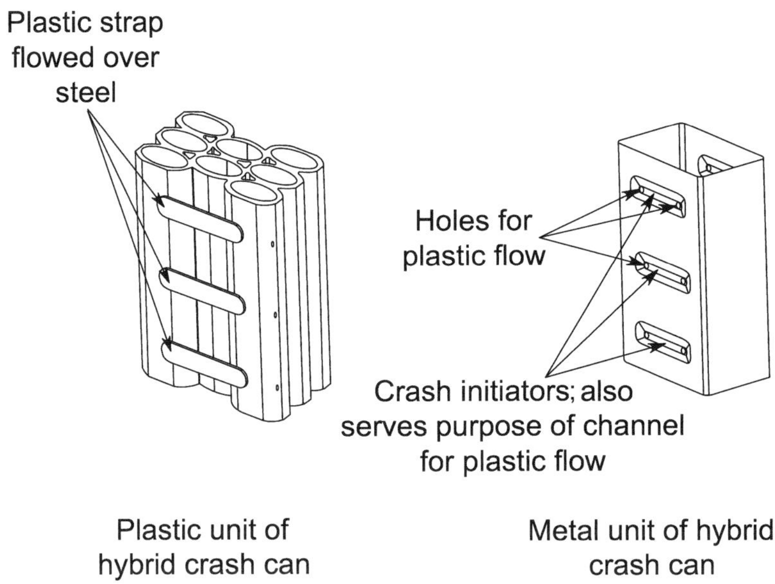

Fig. 2.48 Hybrid rail extension with closed box metal structure.

Crash initiators are designed on the flat faces of metal components, which also serve the purpose of filling plastic through the holes designed therein. A honeycomb structure with plastic is designed to offer highly efficient energy absorption during axial crushing. Figure 2.49 is a diagram shows the second design of plastic-metal rail extension with a closed box metal structure.

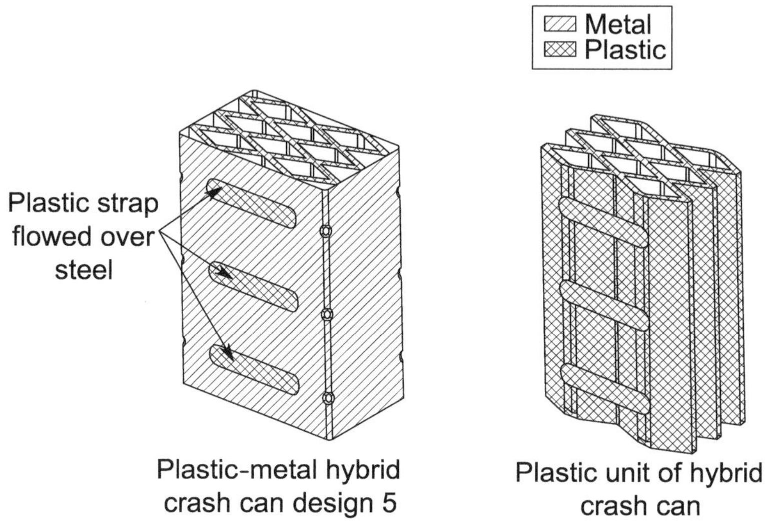

Fig. 2.49 Hybrid rail extension Design 2, with closed box metal structure.

Crash initiators are designed on the flat faces of metal components, which also serve the purpose of filling plastic through the holes designed therein. Additional crash initiators are designed on the edges of metal structure to further enhance progressing crushing. The thermoplastic unit of the hybrid rail extension is an array of diamond-shaped columns or channels.

Different honeycomb or ribbed structures with plastic can be designed to achieve required performance levels. Figure 2.50 is a representative picture, showing the insert or overmolding process for manufacturing a hybrid rail extension as a single-piece unit. Plastic is allowed to flow through the holes in the metal box cover and lock the metal and plastic pieces together.

The different geometric configurations listed above are evaluated through rigorous CAE studies for impact performance. The PMH rail extensions show 80–90% efficiency of energy absorption during impact cases. These rails extensions are intended to be used between the bumper beam and vehicle chassis. Another functionality of this device is to hold the bumper beam, which applies continuous loading on rail extensions resulting

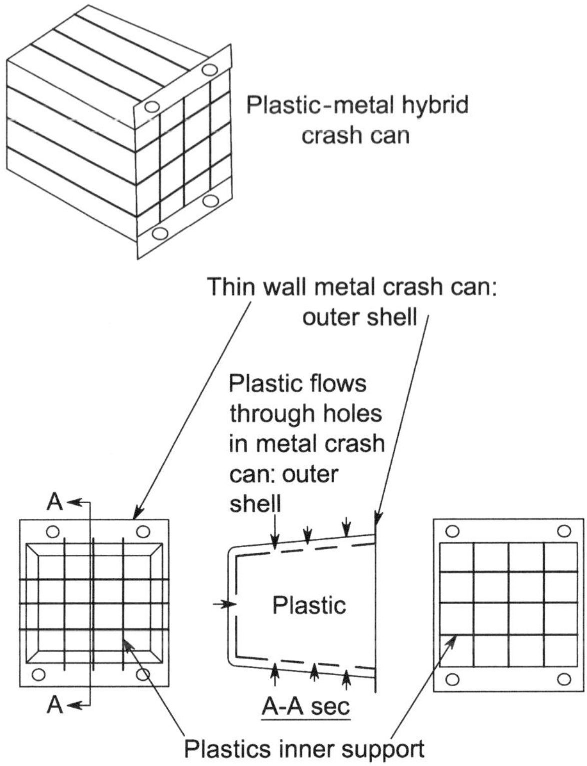

Fig. 2.50 Method of manufacturing hybrid rail extensions.

from the gravitational load of bumper beam. Plastics, traditionally, have inferior creep performance behavior compared to metals and undergo high deformations when subjected to transverse loading. The designed PMH rail extensions are expected to perform better than extruded or injection-molded plastic devices, because of the metal cover that surrounds the plastic crash-can structure, given that metals have much better creep performance.

2.9.2 Plastic Reinforced Body-in-White Structures

Use of reinforcement in motor vehicle structures, specifically in BIW, is a known technology. These structures are designed to fit in the hollow space of BIW metal channels. One of the solutions is a load-bearing article made of steel in conjunction with expandable foam materials in hollow channel. The load-bearing article primarily retains the foam material in the hollow part. The foam material expands during the e-coat cycle and provides the connection to the load-bearing walls. The reinforcement characteristics are dependent on the mechanical properties of foam. However, these reinforcement systems require a chemical reaction that must be adapted to the production process of the vehicles, particularly in terms of their ability to withstand high incident temperatures. The other limitation of this approach is that the structural parts can no longer be disconnected (disassembled) from one another easily, making recycling more difficult. The second type of reinforcement system includes steel stampings, which are fixed in the hollow section of BIW sheet channels using thermoset adhesives. The adhesive activates and expands as the BIW goes through the curing oven that bakes the paint. Typically, these stampings are heavy, and excessive adhesive is applied to ensure a solid bond from the reinforcement to the BIW.

PMH reinforcement to BIW components is designed to be placed at strategic locations based on the impact requirements. The plastic unit of the hybrid solution is an array of columns and channels, and the metal unit is the open channel of steel. This hybrid reinforcement provides good stiffness and chemical and heat resistance up to 204 °C (399.2 °F) for 30 minutes, which enables the plastic unit to maintain integrity when the part goes with the BIW for the bake cycle of painting.

Figure 2.51 is a representation of hybrid reinforcement at identified strategic locations of A and B pillars, floor rocker, and the roof rail.

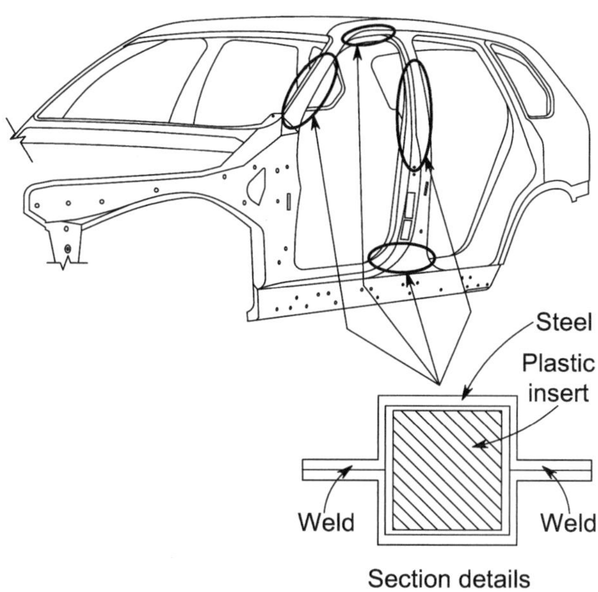

Fig. 2.51 Hybrid reinforcement to the BIW components.

The cross-sectional view shows two steel channels that are welded together to form the hollow space of a BIW component, and a hybrid reinforcement is placed in that empty space. There could be different strategic locations for these reinforcements in the BIW assembly, which is dependent on the BIW architecture and the kind of impact loading. Figure 2.52 is a diagram shows another design of PMH reinforcement.

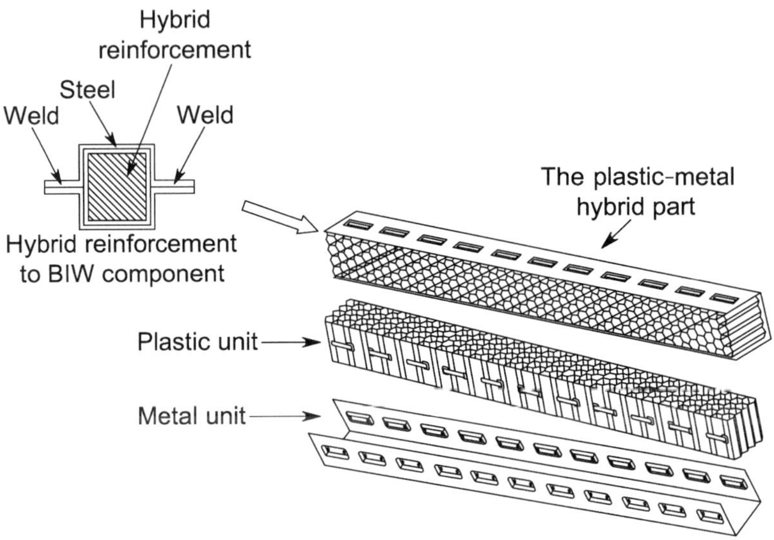

Fig. 2.52 Plastic-metal hybrid reinforcement.

Figure 2.52 includes an open-section steel channel and hexagonal honeycombs of plastic. Holes are designed in the vertical walls of steel channel to allow the flow of plastic.

2.9.3 A Case Study on Roof Crush Countermeasures

Rollover crash is an important issue in vehicle safety. Federal Motor Vehicle Safety Standard (FMVSS) No. 216 has been in effect since 1971 and describes the roof crush resistance requirement for vehicles in various weight classes. In 2009, the National Highway Traffic Safety Administration (NHTSA) upgraded the current FMVSS 216 standard [2-45]. The new FMVSS 216 standard requires a significantly stronger roof structure and now covers larger SUVs as well, which were otherwise exempted from the earlier rollover requirements.

A full vehicle finite element model of a generic SUV is used in this study. Figure 2.53 shows the vehicle setup along with BIW and other structural

Fig. 2.53 Rollover crush test setup over a generic vehicle.

components, which contribute to the energy absorption during the rollover incident.

Nonlinear dynamic analysis software [2-46] is used for computational analysis. Shell elements are used for modeling most of the components. The rigid plate is defined and is positioned as per the FMVSS 216 test protocol. The longitudinal rails of the vehicle are fixed in all degrees of freedom during the quasi-static simulation. The contact force between the rigid plate and the vehicle provides the crush resistance of the roof. Force-deformation performance of different countermeasure is used to compare the crush resistance. A typical steel reinforcement fitted inside the B pillar to improve the crush resistance of the roof during the rollover is shown in Fig. 2.54.

These steel countermeasures in Fig. 2.54 are made from two steel stampings that are spot-welded to form a closed section insert. The thickness of the inner and outer reinforcement of these countermeasures is 1.3 mm. The weight of each countermeasure is ~1090 g. High-strength steel (370 MPa yield stress) is used for the steel countermeasure. These countermeasures are fixed to the B pillar using structural adhesive.

Two alternative lightweight solutions have been proposed to match the roof crush performance of the traditional steel countermeasure design: (1) a complete plastic reinforcement and (2) a hybrid plastic-metal reinforcement. An unfilled thermoplastic alloy of polyphenylene oxide (PPO) and a polyamide (PA6) is selected as the material to meet the requirements of e-coat bath cycle of BIW. These countermeasures are designed within the package space defined by the current B-pillar reinforcements. Figure 2.55 shows the full thermoplastic countermeasure design.

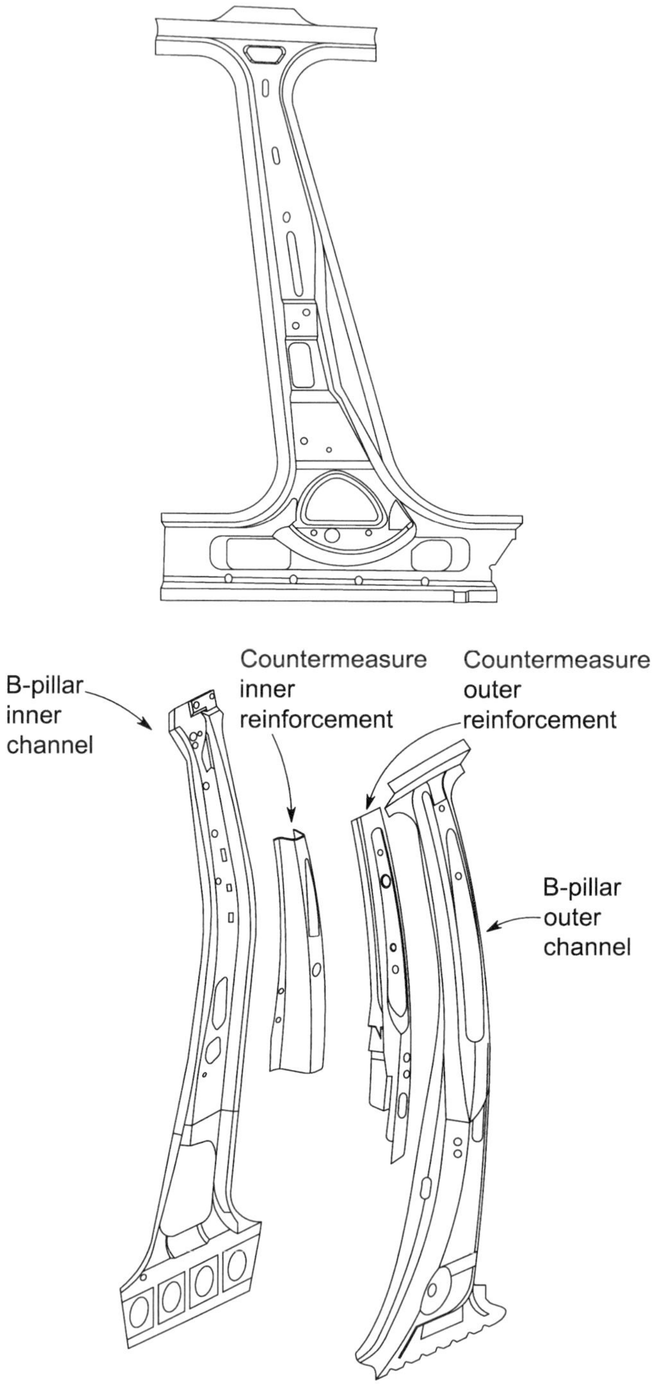

Fig. 2.54 Traditional steel reinforcement to B pillar.

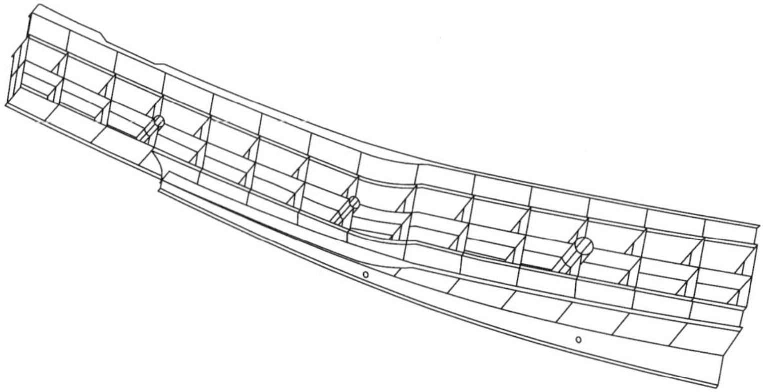

Fig. 2.55 A complete plastic reinforcement to B pillar.

This design consists of a main longitudinal rib along the bending direction stiffened by cross ribs to strengthen local buckle points. Rib design and thickness of the ribs are optimized to get the required bending stiffness from the system. Thickness of the section varies from 2.8 to 3.7 mm. The countermeasure is adhesively bonded to the inner reinforcement of the B pillar. The plastic-metal hybrid reinforcement design is shown in Fig. 2.56.

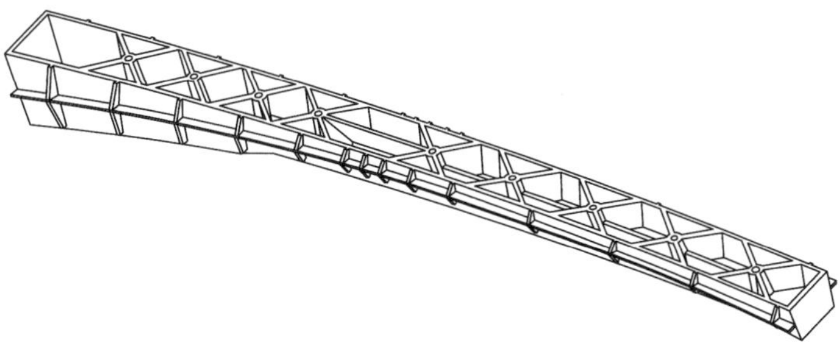

Fig. 2.56 A PMH reinforcement to B pillar.

The hybrid consists of steel reinforcement overmolded with a thermoplastic rib structure. The stamped steel section has local openings to allow plastic to flow in and create a mechanical bond between the steel and plastic. High-strength steel (350 MPa yield stress) is used for the plastic-metal hybrid countermeasure. Thickness of the steel is 1 mm and the plastic is 2.7 mm (average) in the hybrid design. The hybrid countermeasure is spot-welded in the B pillar, eliminating the need for adhesive. Figure 2.57 shows the performance comparison of all plastic reinforcement and hybrid plastic-metal reinforcements with steel countermeasures.

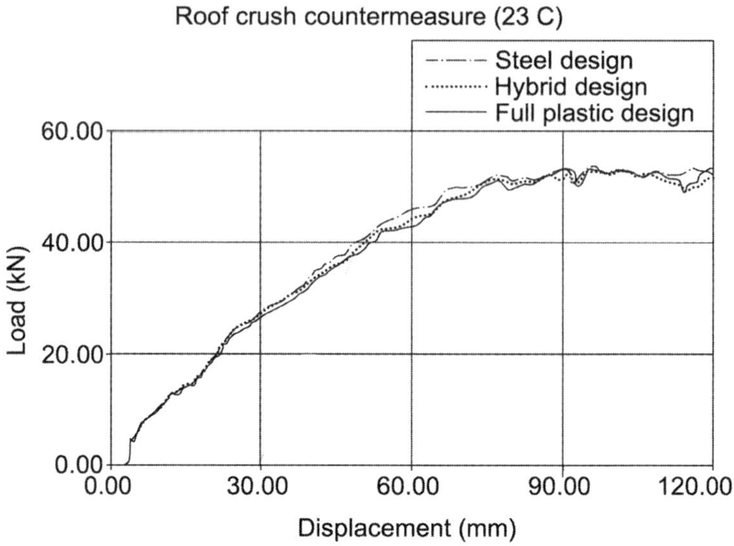

Fig. 2.57 Performance comparison of different reinforcements to the B pillar.

Both of the designed reinforcements (plastic and hybrid) provide crush resistance equivalent to the steel design and are observed to be effective in building a load-bearing structure in these applications. The all-plastic and hybrid reinforcement options offer 40–50% weight savings compared to the traditional steel countermeasures without sacrificing the strength and stiffness of the roof structure.

2.10 Summary

Different energy management systems have been discussed in this chapter to prevent or reduce the severity of injury to occupants and vulnerable road users during vehicle crashes. Plastic-based crash countermeasures help in meeting the design challenges of deformable structural systems stiff enough to absorb kinetic energy of impact, enhance pedestrian safety during crashes, and reduce the costs associated with damage at low-speed vehicle crashes and occupant protection at high-speed impacts. Tuned energy-absorbing elements, strategically positioned in the vehicle's front bumper, helps to make the vehicle pedestrian-safe, through controlling the kinematics of the pedestrian leg during impact. Lightweight plastic beam and crash-can elements at the front and rear bumper protect the expensive components during low-speed impact cases and reduce system mass. Plastic-metal hybrid reinforcements in vehicle BIW components improve the structural integrity of the vehicle frame, enable the components to deform plastically, and yet maintain a sufficient survival space for occupants in crashes involving reasonable deceleration load.

2.11 Trends

By 2030 road crashes are predicted to become the fifth leading cause of death, rising to 2.4 million fatalities per year [2-47]. The regulatory "push" and market "pull" may lead to sustained innovation, research, and development of safety technologies. For example, Euro NCAP released new safety rating plan, in which the weight-to-pedestrian-protection score will increase to 65% in calculations of overall star rating of vehicles by 2015 [2-48]. To make cars safer for vulnerable road users, vehicle manufacturers have already started looking for pelvis protection and active hood systems to minimize pedestrian head injuries [2-49, 2-50, 2-51]. To improve the occupant protection, IIHS has introduced a new, small, overlap crash test protocol [2-52]. A few car manufacturers have started implementing countermeasures to address these new structural test protocols [2-53]. NHTSA is developing a dynamic test on rollovers by motor vehicles in order to carry out a program of conducting such tests and, as these tests are being developed, to conduct a rulemaking process to determine how best to disseminate test results to the public [2-54].

CAFE is a requirement for average fuel economy performance of the entire fleet of passenger cars and light trucks for each vehicle manufacturer for the model year. The CAFE standards will increase fuel economy to the equivalent of 23.17 km/l (54.5 mpg) for cars and light-duty trucks by model year 2025 [2-55]. This new standard may lead to nearly double the fuel efficiency of those vehicles compared to new vehicles now on our roads. Achieving the new fuel efficiency standards will encourage innovation and investment in advanced technologies for vehicle lightweighting, increasing engine and powertrain performance, and improving vehicle electrification. Multimaterial design may be one of the key technologies to develop future lightweight CAFE-compliant vehicles. Use of composite materials for automotive and alternate energy markets are on the verge of implementation. Car manufacturers are already implementing and launching carbon-fiber-composite development programs [2-56] and working with both domestic and foreign producers of carbon fibers and composites. With the significant weight savings associated with composites, it is essential for automotive engineers to become knowledgeable about this technology, which may revolutionize the way carbon fiber is used in automobiles and ultimately be instrumental in meeting government mandates for fuel economy [2-57].

There is a need for a systems approach when developing federal regulations, shifting from a fragmented, single-issue focus (e.g., safety, energy, or emissions) to a broader, more integrated approach to achieve reductions in fossil fuel use, greenhouse gases and other emissions, and road injuries and fatalities, while simultaneously increasing the efficiency of our national transportation system [2-58]. Future motor vehicle regulatory policies and initiatives may be shifted from a single-issue, independent focus on safety, energy, and emissions to a more efficient systems approach that considers all these issues before independent regulatory changes on any one of these subjects are made.

Novel designs, processes, and material solutions lead to meeting the challenging requirements of styling, reducing weight, and enhancing vehicle safety. Continuous research on the development of crash test protocols by legislative bodies to simulate real-life crashes, development of innovative materials, evolution of modern design solutions by researchers, and the confirmed trend of new technologies being employed and accepted by car manufactures will help to produce vehicles of the highest safety standards.

2.12 References

2-1. "IRCOBI [International Research Council on Biomechanics of Injury] Conference Proceedings," IRCOBI, http://www.ircobi.org/proceedings.php. Accessed January 21, 2012.

2-2. "Stapp Car Crash Journal and Technical Papers," Stapp Car Crash Conference, http://www.stapp.org/pubs.shtml. Accessed January 21, 2012.

2-3. *NHTSA Vehicle Safety and Fuel Economy Rulemaking and Research Priority Plan 2011–2013*, http://www.nhtsa.gov/staticfiles/rulemaking/pdf/2011-2013_Vehicle_Safety-Fuel_Economy_Rulemaking-Research_Priority_Plan.pdf. Accessed January 19, 2012.

2-4. "Status Report, Federal Republic of Germany, 22nd ESV Conference Washington, DC, 2011," http://www-nrd.nhtsa.dot.gov/pdf/esv/esv22/22ESV-000455.pdf. Accessed April 29, 2013.

2-5. "Government Status Report of Japan, 22nd ESV Conference Washington, DC, 2011," http://www-nrd.nhtsa.dot.gov/pdf/esv/esv22/22ESV-000459.pdf. Accessed April 29, 2013.

2-6. "Status Report of Republic of Korea, 22nd ESV Conference Washington, DC, 2011," http://www-nrd.nhtsa.dot.gov/pdf/esv/esv22/22ESV-000461.pdf. Accessed on January 19, 2012.

2-7. "Developing a Policy Framework for Vulnerable Road Users," AVV Transport Research Centre, www.ictct.org/dlObject.php?document_nr=159&/Methorst.pdf. Accessed April 29, 2013.

2-8. "World Forum for Harmonization of Vehicle Regulations, 156th Session," United Nations Economic and Social Council, http://www.unece.org/fileadmin/DAM/trans/doc/2011/wp29/ECE-TRANS-WP29-1094e.pdf. Accessed on January 21, 2012.

2-9. "Driving Speeds and Pedestrian Safety," Walk21, http://www.walk21.com/papers/San%20Sebastian%2002%20Pasanen%20Driving%20Speeds%20and%20Pedestrian%20Safet(1).pdf. Accessed on 9 November, 2012.

2-10. "Directive 2003/102/EC of the European Parliament and of the Council of 17 November 2003 Relating to the Protection of Pedestrians and Other Vulnerable Road Users Before and in the Event of a Collision with a Motor Vehicle and Amending Council Directive 70/156/EEC," Official Journal of the European Union, http://eurlex.europa.eu/LexUriServ/LexUriServ.do?uri=OJ:L:2003:321:0015:0015:EN:PDF. Accessed January 19, 2012.

2-11. "European New Car Assessment Programme (Euro NCAP), Pedestrian Testing Protocol," http://www.euroncap.com/files/Euro-NCAP-Pedestrian-Protocol-Version-5.3.1---0-0560739b-595f-4a38-96b2-eeda5050dee8.pdf. Accessed on January 21, 2012.

2-12. "Establishment of Regulations for Pedestrian Safety in Japan," http://www.unece.org/fileadmin/DAM/trans/doc/2002/wp29grsp/TRANS-WP29-GRSP-32-inf05e.pdf. Accessed January 22, 2012.

2-13. "Development and Drawings of the TRL Pedestrian Legform and Upper Legform Impactors," TRL Reports, http://www.trl.co.uk/online_store/reports_publications/trl_reports/cat_road_user_safety/report_development_and_drawings_of_the_trl_pedestrian_legform_and_upper_legform_impactors.htm. Accessed April 29, 2012.

2-14. Nolan, J., Brumbelow, M., Zuby, D., and Avery, M. 2004. "Important Considerations in the Development of a Test to Promote Stable Bumper Engagement in Low-Speed Crashes." SAE Paper No. 2004-01-1319. SAE International, Warrendale, PA.

2-15. "Laboratory Test Procedure for Regulation Part 581 Bumper Standard," U.S. Department of Transportation NHTSA, http://www.nhtsa.gov/DOT/NHTSA/Vehicle%20Safety/Test%20Procedures/Associated%20Files/TP-581-01.pdf. Accessed on April 29, 2012.

2-16. "ECE Uniform Provisions Concerning the Approval of Vehicles with Regard to their Front and Rear Protective Devices (Bumpers, etc.)," http://www.unece.org/fileadmin/DAM/trans/main/wp29/wp29regs/r042e.pdf. Accessed on January 21, 2012.

2-17. "IIHS Low-Speed Crash Test," May 2006," Insurance Institute for Highway Safety, http://www.iihs.org/ratings/protocols/pdf/archived_versions/test_protocol_low.pdf. Accessed on January 19, 2012.

2-18. "RCAR Low-Speed Structural Crash Test Protocol," RCAR, http://www.rcar.org/Papers/Procedures/rcar_LowSpeedCrashTest2_2.pdf. Accessed on January 21, 2012.

2-19. Tingvall, C. and Haworth, N. 1999. "Vision Zero—An Ethical Approach to Safety and Mobility." 6th ITE International Conference Road Safety & Traffic Enforcement: Beyond 2000, Melbourne, Australia.

2-20. Goldsmith, W. 2001. *Impact: The Theory and Physical Behavior of Colliding Solids*. Toronto: Dover, p. 180.

2-21. "Determination of Impact Energy Using Charpy Test," Azom.com, http://www.azom.com/article.aspx?ArticleID=2763. Accessed on January 19, 2012.

2-22. Sofi, F., Kulkarni, S., Haarda, M., and Takaaki, N. 2008. "A Novel Energy Absorber Design Technique for an Idealized Force-Deformation performance." SAE Paper No. 2008-01-0184. SAE International, Warrendale, PA.

2-23. Schuster, P.J. 2006. "Current Trends in Bumper Design for Pedestrian Impact." SAE Paper No. 2006-01-0464. SAE International, Warrendale, PA.

2-24. Magee, C.L. and Thornton, P.H. 1978. "Design Considerations in Energy Absorption by Structural Collapse." SAE Paper No. 1978-02-780434. SAE International, Warrendale, PA.

2-25. Yucheng, L. and Michael, L.D. 2007. "Simplified Modeling of Thin-Walled Tubes with Octagonal Cross Section—Axial Crushing." *Proceedings of the World Congress on Engineering and Computer Science*, San Francisco, CA.

2-26. Nagwanshi, D., Mana, D., Allen, K., and Bobba, S. 2010. "Diagnosing Vehicle Aggressiveness for Pedestrian Leg Impact and Development of Efficient Front End Energy Management Systems." SAE Paper No. 2010-01-1168. SAE International, Warrendale, PA.

2-27. Mooijman, F., Schuler, S. and Nanda, A. 2005. "Bumper Assembly Including an Energy Absorber." Patent Publication No. US 6866313 B2.

2-28. Atsuhiro, K. and Masaaki, T. 2004. "Development of a Biofidelic Flexible Pedestrian Legform Impactor." *JARI Research Journal* 26(3): 139–142.

2-29. "Information on the Flexible Pedestrian Legform Impactor (Flex-PLI) from J-MLIT Research," J-MILT, http://www.unece.org/fileadmin/DAM/trans/doc/2004/wp29grsp/ps-106.pdf. Accessed January 12, 2012.

2-30. Bhagat, M. 2011. "Influence of Vehicle Front End Design on Pedestrian Lower Leg Performance for SUV Class Vehicle." *SAE International Journal of Passenger Cars—Mechanical Systems* 4: 12–21.

2-31. "Don't Bump," Thatham home page, http://www.thatcham.org/research/pdfs/bumpers.pdf. Accessed on April 29, 2013.

2-32. "Federal Motor Vehicle Safety Standards and Regulations," U.S. Department of Transportation NHTSA, http://www.nhtsa.gov/cars/rules/import/fmvss/index.html. Accessed on January 19, 2012.

2-33. "European Regulations Related to Crash Testing," Crash Network, http://www.crash-network.com/Regulations/ECE_Regulations/ece_regulations.html. Accessed on January 19, 2012.

2-34. "Appendix I: Mini Guide to the Federal Motor Vehicle Safety Standards and Related Regulations," *Handbook—Standards and Regulations, Montana Office of Public Instructions,* http://opi.mt.gov/pdf/PupilTransport/PTHandbook.pdf. Accessed on April 29, 2012.

2-35. "Welcome to Tokyo," *RCAR Newsletter* (2006 Annual RCAR Conference issue) http://www.rcar.org/Newsletters/2006_oct_newsletter.pdf. Accessed on 9 November, 2012.

2-36. Ramon, L.V. and Enderich, T. 2007. *Advance Simulation Techniques for Low Speed Vehicle Impacts.* Frankenthal, Germany: LS-DYNA Anwenderforum.

2-37. Isaksson, E. 2006. *Simulation Methods for Bumper System Development.* Licentiate Thesis: Luleå University of Technology, Sweden.

2-38. Mana, D., Nagwanshi, D., Marks, M. and Arunachala, P. 2011. "Thermoplastic Rear Bumper Beams for Automobile Low-Speed Rear Impact." SAE Paper No. 2011-01-0544. SAE International, Warrendale, PA.

2-39. Nagwanshi, D.K. and Kulkarni, S. 2009. "Light weight Solitary Beam Design to Meet Low Speed Vehicle Damageability and RCAR Impact Requirements." SAE Paper No. 2009-26-0007. SAE International, Warrendale, PA. Paper presented at Symposium on International Automotive Technology (SIAT), Automotive Research Association of India (ARAI): Pune, India.

2-40. Montgomery, D.C. 2004. *Design and Analysis of Experiments.* Singapore: Wiley, p. 168.

2-41. Design-Expert Software, version 6.0, 2000. Stat-Ease, Minneapolis, MN.

2-42. Nagwanshi, D., Chaturvedi, M., Marur, S. and Allen, K. 2010. "Hybrid 'Thermoplastics and Steel' Bumper Beam Solution to Protect the Vehicle in Low Speed Crashes." SAE Paper No. 2010-01-1009. SAE International, Warrendale, PA.

2-43. "CAFE—Fuel Economy," NHTSA, http://www.nhtsa.gov/fuel-economy. Accessed on April 29, 2012.

2-44. Mooijman, F.R., Prasad, P.S. and Poovanna, T.K. 2008. "Hybrid structure and method." Patent Publication No. US 2008/0138586 A1.

2-45. "Department of Transportation, National Highway Traffic Safety Administration, 49 CFR Part 571, Docket No. NHTSA-2005-22143, RIN 2127-AG51," Federal Motor Vehicle Safety Standards; Roof Crush Resistance, http://www.nhtsa.gov/cars/rules/rulings/roofcrushnotice/216nprm-to-fr.html. Accessed on April 29, 2012.

2-46. *LS-DYNA Keyword User's Manual, 2007,* Version 971. Livermore, CA: Livermore Software Technology Corporation.

2-47. "Global Trends 2020 and Beyond: Automobile Use, Environment & Safety Challenges," FIA Foundation, http://www.acea.be/images/uploads/files/Presentation_Ward.pdf. Accessed on November 14, 2012.

2-48. "European New Car Assessment Programme: Assessment Protocol—Overall Rating, Version 6.0," Euro NCAP, http://www.euroncap.com/files/Euro-NCAP-Assessment-Protocol---Overall-Rating---v6.0---0-4e7fab4e-8606-4d35-be22-07e5b41529dc.pdf. Accessed on November 15, 2012.

2-49. "Active Bonnet—Protection for Pedestrians in an Accident," Daimler, http://www.daimler.com/dccom/0-5-1210222-1-1210363-1-0-0-1210342-0-0-135-0-0-0-0-0-0-0-0.html. Accessed on November 15, 2012.

2-50. "Panamera GTS Safety," Porsche, http://www.porschecentre brisbane.com.au/models_panamera_gts.html#safety. Accessed on April 29, 2012.

2-51. "Technical Bulletin (M19) Citroen C6 Active Bonnet System," IAG Insurance Australia Group, http://iagresearch.com.au/index .php?option=com_docman&task=doc_view&gid=19. Accessed April 29, 2013.

2-52. "An Update on the IIHS Small Overlap Research Program," Insurance Institute for Highway Safety, http://www.sae.org/events/gim/ presentations/2012/sherwood.pdf. Accessed on November 15, 2012.

2-53. "XC60 and S60—IIHS Top Safety Pick + Award," Gengras Volvo, http://www.gengrasvolvohartford.com/iihs-new-small-overlap-frontal-crash-test.htm. Accessed on November 15, 2012.

2-54. "National Highway Traffic Safety Administration, 49 CFR Part 575 [Docket No. NHTSA-2001-9663; Notice 3], RIN 2127-AI81," Consumer Information, New Car Assessment Program, Rollover Resistance, http://www.nhtsa.gov/cars/rules/rulings/rollfinal/index.html. Accessed on November 15, 2012.

2-55. "Obama Administration Finalizes Historic 54.5 mpg Fuel Efficiency Standards," NHTSA, http://www.nhtsa.gov/About+NHTSA/ Press+Releases/2012/Obama+Administration+Finalizes+Historic+54 .5+mpg+Fuel+Efficiency+Standards. Accessed on November 15, 2012.

2-56. "Ford Develops Prototype Carbon Composite Bonnet," ReinforcedPlastics.com, http://www.reinforcedplastics.com/ view/28778/ford-develops-prototype-carbon-composite-bonnet/. Accessed on November 12, 2012.

2-57. "Frost & Sullivan: Automotive Carbon Fiber Composites to Grow at a Staggering Growth Rate until 2017," Frost & Sullivan, http://www .frost.com/prod/servlet/press-release.pag?Src=RSS&docid=263051000. Accessed on November 12, 2012.

2-58. "Analysis of Future Issues and Changing Demands on the System: Commission Briefing Paper 4J-06," National Surface Transportation Policy and Revenue Study Commission, http://transportationfor tomorrow.com/final_report/volume_3_html/technical_issues _papers/paperae99.htm?name=4j_06. Accessed on November 14, 2012.

Chapter 3
Interiors

N. Krishnamoorthy, P. Arunachala, N. Venkatesha, K.R.K. Ravikanth, S.M. Kumar, and Sudhakar R. Marur

3.1 Introduction

Engineering thermoplastic materials are widely used in automotive interior applications. Typically, the interior of a car is expected to be comfortable for the occupants, with low noise levels from the rattling of parts and with long durability. Several of the interior parts act as the key interface to vehicle operation and hence are important. The aesthetics of the interior parts play a significant role in the customer's assessment of the vehicle.

Some of the possible interior applications with engineering thermoplastics (ETP) are the instrument panel (IP) carrier, bezels, steering wheel, seating system, center console, door handle, cup holders, door pull cup, door trims, pillar trims, and so forth, as shown in Fig. 3.1.

Generally, the structural rigidity, stiffness, and visually appealing surface finish are some of the key requirements in these applications. With thermoplastic materials, several types of surface finish can be achieved based on the theme of the interior design. The surface could have a low-gloss, high-gloss, or matte finish with different textures, which can be accomplished through appropriate surface treatment of the injection-molding tool. Wood finish, which is commonly seen in

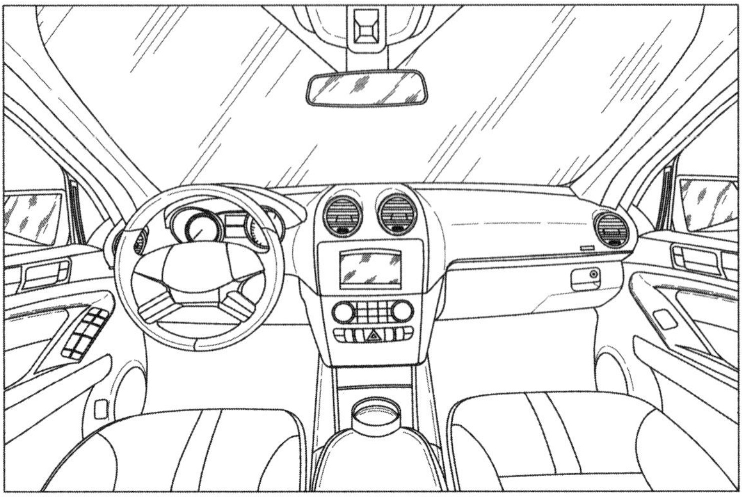

Fig. 3.1 Interior of an automobile.

automobiles, can be achieved through painting or in-mold decoration. Thermoplastics have made it possible to integrate several different parts into a single part. IP is a classic example of such integration. With the use of thermoplastics, several functional requirements such as head impact and seamless airbag opening could be integrated with the IP, through appropriate design.

New thermoplastic materials that can withstand structural loads at low temperatures of −40 °C (−40 °F) have enabled exploration of new applications in interiors such as the steering wheel and parts of the seating system. Highly scratch-resistant materials are finding applications in interior displays.

IP carriers are one of the largest parts made from thermoplastics. The center console, air duct, louvers, airbag assembly, and brackets are some of the components of an IP made with plastics. IP is mounted on the cross car beam and takes impact loads such as knee and head impact. These parts have been made from several thermoplastics such as PP (polypropylene), PC-ABS (polycarbonate–acrylonitrile butadiene styrene), and PC. Besides IP, steering wheel is a structural part that is made from thermoplastics. With the use of advanced material blends and design, plastic steering wheels can be designed to meet the structural and impact requirements.

3.2 Instrument Panel

3.2.1 Key Drivers in Instrument Panel Design

The IP (also known as the dashboard) used in an automobile is a control panel mounted in front of a driver. It houses instruments and control systems for the automobile. Typical features of an IP include a carrier, seamless tear seam for airbag, knee bolster, and center console.

3.2.2 Automotive Instrument Panel Carriers

The IP carriers are the parts that hold the automobile's dashboard in place, as shown in Fig. 3.2. They are semistructural parts and typically are not visible within a vehicle. Traditionally, they have been made of a frame within a polyurethane foam structure that is then covered with a skin of (polyvinyl chloride (PVC). The latest trend in the automotive industry is that the OEMs (original equipment manufacturers) are constantly looking for higher-stiffness and lower-density materials for this application. Additionally, the materials with excellent flow are preferred, so that this component can be designed with a wall thickness of less than 2 mm and thereby use less material without compromising on dimensional stability, stiffness, and strength requirements.

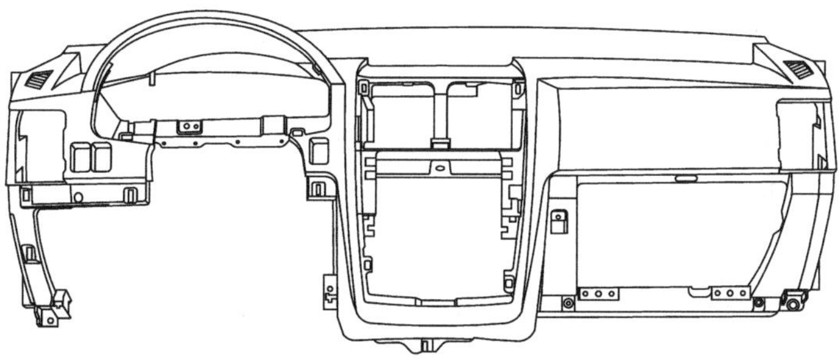

Fig. 3.2 IP carrier.

3.2.2.1 Occupant Safety: Head and Knee Impact

The key aspect of IP design is occupant safety in the event of any crash, governed by the Federal Motor Vehicle Safety Standards (FMVSS) Regulations 201 and 208 for head [3-1] and knee [3-2] impact, respectively.

This FMVSS 201 specifies the impact protection of an occupant of a passenger car or multipurpose passenger vehicle, truck, and bus with a gross vehicle weight rating (GVWR) of 4,536 kg or less. (GVWR is the maximum weight value of a vehicle as specified by the manufacturer.) For the IPs, this standard test procedure requires that a head form 165 mm in diameter and 6.8 kg in mass impacts the defined impact zones with an initial velocity of 19.2 km/h (11.9 mph) with an airbag and 24.1 km/h (15 mph) without an airbag. The standard stipulates that accelerations of the head form should not exceed 80 g during any 3-millisecond interval. The head-impact setup on IP is as shown in Fig. 3.3.

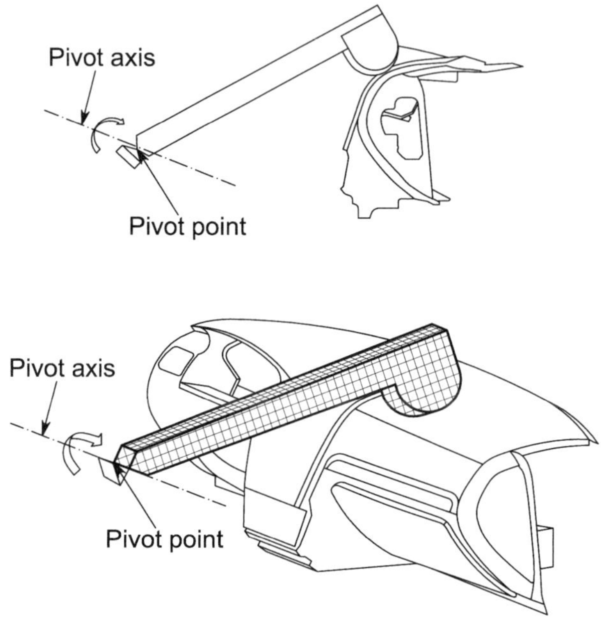

Fig. 3.3 Head-impact setup on IP according to FMVSS 201.

Using the rigid pendulum impact to replicate the head impact on a generic IP topper assembly with majority of components present, the system performance is evaluated through head injury criteria (HIC) and 3-millisecond clip acceleration and peak accelerations through acceleration versus time curves, at different temperatures (–30 °C [–22 °F]) to ambient), for various thickness and geometric parameters of the IP.

The knee impact analysis for an IP assembly, as prescribed by FMVSS 208, is done by measuring the energy in joules, load or force on the femur, and knee intrusion on the driver as well as passenger side. The knee impact on an IP setup is as shown in Fig. 3.4.

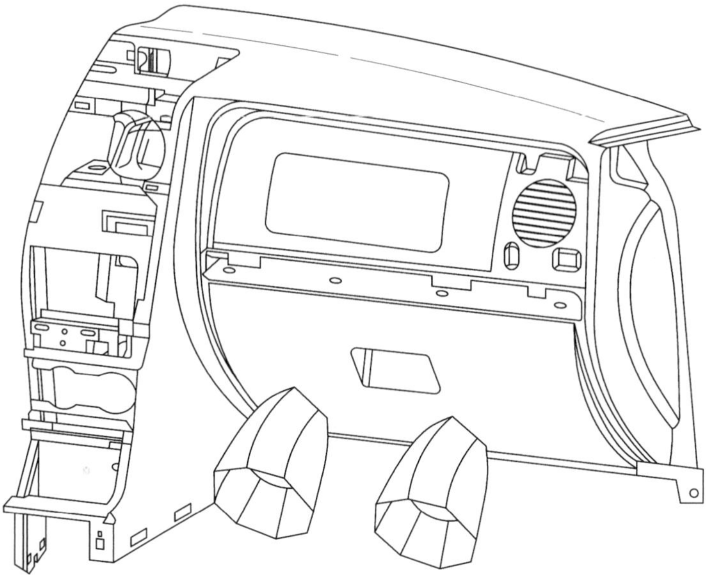

Fig. 3.4 Knee impact on IP as per FMVSS 208.

3.2.2.2 Processing Challenges of Instrument Panel Carriers

In the automotive industry, the latest IP carriers are generally designed for thin wall with a thickness of about 1.8 to 2.0 mm. The typical materials used are long glass-fiber-reinforced plastic resins, such as polypropylene. This structure is usually big, with many openings for housing the dashboard components. The dimensional stability is critical as multiple components are assembled onto it. The stiffness and strength are equally critical, as it is a semistructural part.

The following are processing challenges faced during injection molding of a typical IP carrier:

1. To design the optimum number of gates for balanced filling
2. Injection molding without any aesthetic defects such as flow marks, gate blushes, knit lines on critical areas, burn marks, warpage, and so forth
3. Thin-wall molding with minimum possible injection pressure and clamp tonnage to achieve lower residual stresses with in the molded part
4. To achieve higher knit line strength

3.2.2.3 Mold-Filling Simulations of Instrument Panel Carriers

The cost-effective and efficient method to solve the processing challenges of an IP carrier is to use the predictive engineering tools and perform detailed mold-filling simulations during the design phase, before tooling. The mold-filling simulations are very useful to optimize the part design, tool design, and molding process. They can also help in identifying the best choice of thermoplastic material for easy processing.

In mold-filling simulations, the IP carrier model will be set up for a virtual injection-molding process. It is a three-step iterative process consisting of preprocessing, analysis, and postprocessing to obtain optimum results. The typical steps are the following:

1. The finite element shell model of an IP carrier is built, and the required thickness is assigned.
2. A feed system with sequential valve gating is modeled, and a suitable thermoplastic material is chosen with properties such as rheology, processing, thermal, shrinkage and mechanical.
3. The mold-filling simulation is then run by setting all the processing parameters such as injection time, mold and melt temperature, cooling time, holding pressure, and holding time.
4. The simulation results are interpreted for mold filling, pressure, temperature, clamp tonnage, knit line, air-trap locations, and so forth, along with deflections in the part.
5. The calculations are repeated until the part, tool, and process parameters are optimized.

3.2.3 Seamless Airbag Design

The function of an airbag in an automobile is to protect the passengers against head and chest injuries during an impact. The typical passenger airbag consists of a large bag connected to a high-speed gas generator and its surrounding structure. In seamless airbag design [3-3], the inflatable airbag system is mounted directly behind the IP on the passenger side of the vehicle and provides better aesthetics without any style-interrupting lines (Fig. 3.5).

The traditional airbag system design included an injection-molded door with welded metal or plastic reaction plates as reinforcements to prevent any possible shatter or fragmentation of the door at the seams (which could happen if the material lacked low-temperature ductility). Also, secondary operations used to include woven cloths and scrims, and the

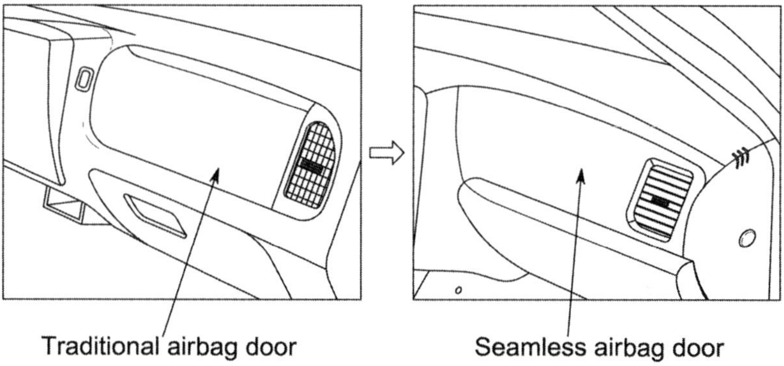

Fig. 3.5 Traditional versus seamless airbags.

process involved an expensive quality check for the multicomponent assembly process.

With the advent of the polycarbonate-siloxane (PC-Si) copolymer, IPs with integrated chute and seamless airbag systems became feasible at lower cost and mass, compared to traditional IPs [3-4]. In this approach, the chute, which contains the airbag assembly, is molded as part of the IP retainer, resulting in a one-piece product, thus avoiding the secondary operation of having to weld the chute to the retainer top.

As this material demonstrates significant low-temperature ductility at −40 °C (−40 °F), high flexural modulus and heat resistance, better resistance to hydrolytic aging (of 1000 hours at 900 °C [1652 °F] and 95% humidity), and improved chemical resistance, the single-piece airbag door with laser scored seams has become feasible [3-3]. By engineering the laser score geometry suitably, safe airbag deployment has been achieved without any fragmentation over a range of temperature and environmental conditions.

In the automotive industry, two types of IP are used: hard IPs, which are hard to the touch and generally used in low-end cars and trucks, and soft IPs, which are covered with a layer of foam and skin, giving them a soft feel, and generally used in high-end cars. One of the trends is to use seamless airbags in which the airbag door is invisible. In soft IPs, the airbag door can easily be hidden behind the layers of foam and skin. In hard IPs, seamless technology requires a tear seam in the material.

The geometry and parameters of the tear seam determine the opening behavior of the airbag door. The tear seam could either be directly molded-in during the injection-molding process of the IPs, or it could be laser scored, as a secondary process, after the molding. This extensive process consists of molding, laser scoring, deployments, design, and parameter modification, which is repeated until the desired result is obtained. Therefore, to determine the optimal design of the airbag door and the best geometry of the tear seam, plaque studies need to be performed. In this study different tear seam designs can be tested and observed.

3.2.3.1 Tear Seam Plaque Study

The opening behavior of an airbag door strongly depends on the design of the tear seam geometry and the way it is scored. Three different tear seam geometries (Fig. 3.6) are tested with a main goal of the door opening uniformly from the center of the door and without any rupture or fracture of the part.

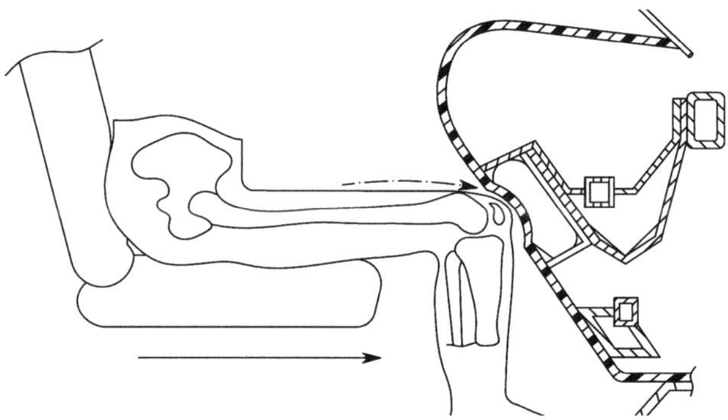

Fig. 3.6 Different tear seam geometry designs.

The second design in Fig. 3.6 has the most stable opening behavior. In the first design, the corners of the hinges were weak, and in third one, the opening behavior was not uniform; that is, the door did not always open from the centerline.

3.2.4 Knee Bolster

Knee bolsters are metal plates covered with foam that are installed under the IP assembly and primarily aimed at reducing knee fractures during a collision. The knee bolster and IP assembly are shown in Fig. 3.7.

Though they have reduced the knee fractures, bolsters have led to an increase in hip fractures and dislocations. Therefore, airbags have been placed on the bolsters to redistribute the forces against the occupants and help them maintain their upright position [3-5].

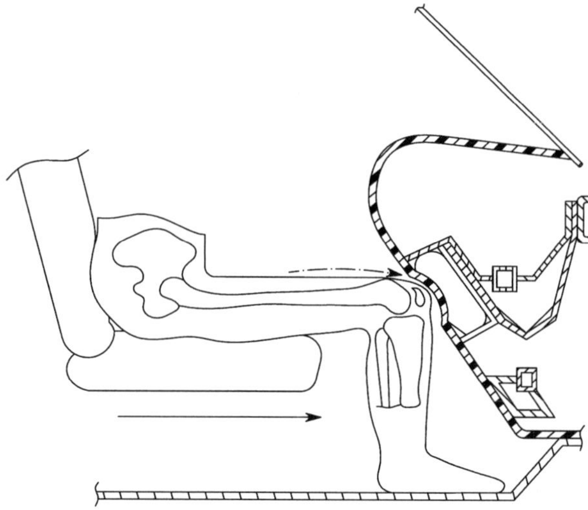

Fig. 3.7 Knee bolster and IP assembly.

Instead of the traditional two-piece sheet metal knee bolster, a single piece with plastic material reinforced with rib patterns and energy-absorbing brackets could be designed [3-6] to achieve the required intrusion under a specific impact force (Fig. 3.8).

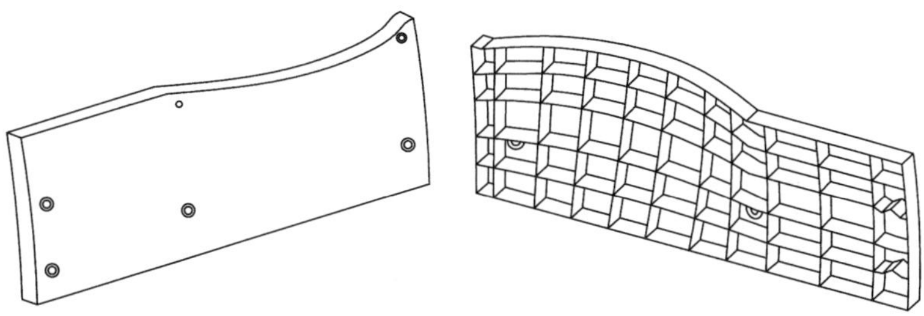

Fig. 3.8 Single-piece plastic knee bolster with rib structure.

3.2.5 Center Console

The center console is one of the most important parts of the car interior [3-7]. This console consists of the surfaces in the center of the front of the vehicle interior. It contains the entire dashboard including the part beneath the dashboard. Generally, in many cars this part often merges with the transmission tunnel located in between the driver and passenger seats.

The center console includes a wide variety of storage compartments and cup holders. In high-end cars, beneath the center console, the air conditioning duct is mounted. This duct cools the rear portion of the car. In addition to conventional use, the center console surface is used for electronic controls such as temperature display, car audio controls, and the power-charging socket. In geared vehicles, the gear stick is mounted where the console and tunnel merge. Typically, the center console is covered with a lid, and this lid acts as an armrest, as shown in Fig. 3.9.

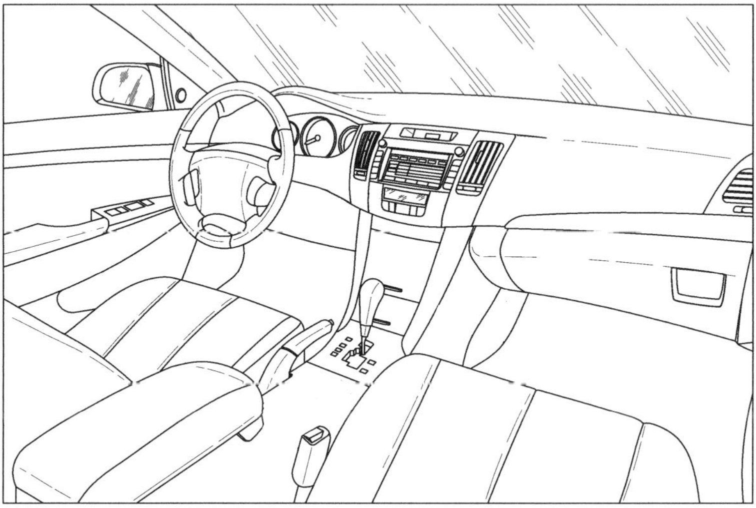

Fig. 3.9 Center console of an automobile.

The center console is basically a functional part. Those who spend a great deal of time in their car while traveling appreciate the convenience of an appropriate console in their vehicle. The design of this part is very important, as several components such as the DVD player and navigation screen are mounted on it. A lot of cutouts are designed in center console, which may pose challenges in terms of processing and strength at weld lines. Though this part is not a structural component, it is important from the point of NVH (noise, vibration, and harshness) performance.

Both functional and aesthetic aspects need to be taken in to account before designing the center console. Functionally, it should be ergonomically designed. Structurally, it should withstand the loads expected to be borne by this part. The center console portion, which is in dashboard area, should be designed for withstanding head-impact loads as well. So, to get the part to perform against impact testing, the weld lines should be placed where the stresses are least. The center console retainer, which retains the armrest, cup holder, and so forth, should withstand different types of loads. When the armrest is tilted to the open position, there will be a load acting on the retainer. When the driver or passenger places a hand on armrest and pushes or pulls it, the retainer should withstand such torsional loads. The armrest should withstand the weight of the occupant's arm. Also, the part should meet NVH requirements.

Aesthetically, this part should be molded in such a way that the weld lines are not located in the highly visible surface. It would be a challenge to push all the weld lines to hidden areas or less visible areas, as there would be so many cutouts in the part. To achieve this, multiple gates are required, which may also have to open sequentially to eliminate weld lines. Given that the center console has to be assembled in IP retainer, the warpage in the part should be as low as possible; otherwise, there would be a large gap between IP retainer and the center console.

The center console retainer is not a visible part; hence the location of weld line is not very important. This part can be molded with two direct drops located in such a way that the filling is balanced. Generally, warpage will be in the car's *Y* coordinate direction, and it should be optimized to be less than 1 mm.

In conclusion, before proceeding to the feed design, one should know the following information for molding the part successfully:

1. Different kinds of loads on armrest
2. Locations where weld lines are permitted
3. Assembly tolerances

3.3 Steering Wheel

3.3.1 Introduction

The steering wheel in an automobile is an important functional part that serves as the human–machine interface to drive the vehicle. It has also evolved as a critical safety part, protecting occupants from impact with the collapsible steering wheel and airbags. Over the years, it has evolved from only performing the steering function to integrating several features such as the steering mount audio controls and cruise control switches. Earlier, steering wheels were made mostly with steel. Aluminum and magnesium replaced steel over time. However, the current trend is to use steel for low-end cars without airbags in developing nations and to use aluminum or magnesium with airbags, where regulations are mandatory [3-8].

3.3.2 Metal versus Plastic

A traditional steel steering wheel armature assembly comprises a steel hub/bushing, hub plate, metal spoke, and rim. The entire assembly is overmolded with polypropylene or polyurethane foam to give the required surface touch and feel, as shown in Fig. 3.10.

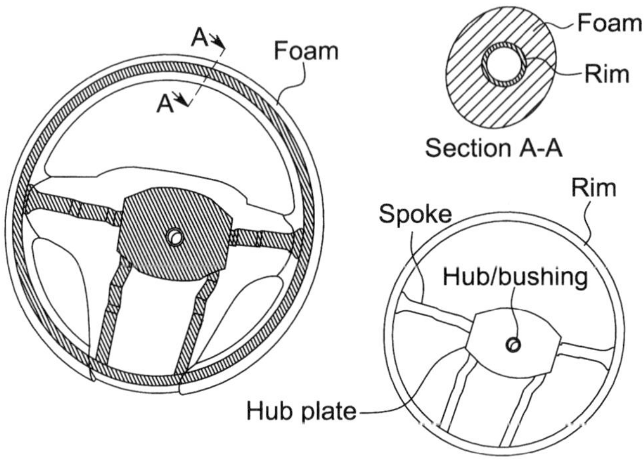

Fig 3.10 Steel steering wheel armature assembly.

In some cases, foam is overmolded up to the spoke area, and rest of the portion is covered by a plastic back cover. The main disadvantages of the traditional steel steering wheel are its weight and the need for many secondary operations such as bending and welding to get the final steel armature. The current trend is moving toward magnesium or aluminum alloy armature because of the abundance of magnesium at a competitive price in the global market and single-shot die casting process to get Mg/Al alloy armature. As shown in Fig. 3.11 this process integrates the rim, spoke, hub plate, and even the hub/ bushing component, which offers a substantial weight saving compared to steel armature [3-9].

Reducing the manufacturing costs by improving recycling capabilities is a key strategy that could make magnesium components less expensive.

Engineering a thermoplastic steering wheel is a recent innovation that offers significant mass reduction, part integration opportunity, and a new degree of freedom for styling. It also results in reduced production costs because of the well-established injection-molding process. Plastic steering wheels also give greater design flexibility for improving performance [3-10].

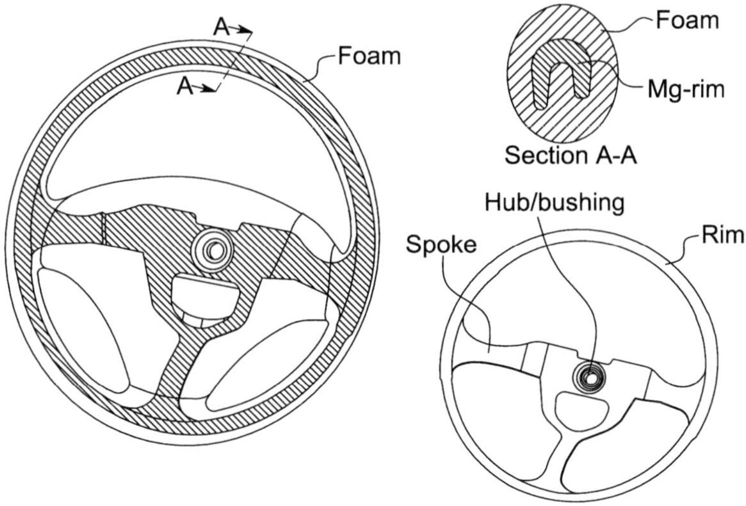

Fig 3.11 Magnesium steering wheel armature assembly.

3.3.3 Design Technology

Engineering a thermoplastic steering wheel design relies on two fundamental concepts: the single-piece and the two-piece design. The design is selected based on criteria such as styling, weight, cost, and structural performance specifications [3-11]. In the single-piece design, the entire plastic armature is manufactured as one piece, and the steel hub and bushing are insert-molded during the injection-molding process, and the entire plastic armature is overmolded with polyurethane (PU) foam, as shown in Fig. 3.12.

In a single-piece steering wheel, a U-shaped rim is commonly used because of easy tooling and manufacturing. Different rim section shapes can also be used, as long as they do not complicate the tooling and manufacturing processes [3-12]. The main advantage of single-piece technology is the one-shot process, which reduces the component cost. The main disadvantage is that it needs more PU foam and more plastic mass to meet the structural requirements.

In the two-piece design, the two injection-molded components, namely, the top half and bottom half, are manufactured, as shown in Fig. 3.13.

The bottom half has a hub plate and bushing, which are insert-molded with a structural rim, spokes, and a seating space for the airbag module. The top half is molded separately with structural ribs to further stiffen the steering wheel assembly, which covers the rim and spoke section of

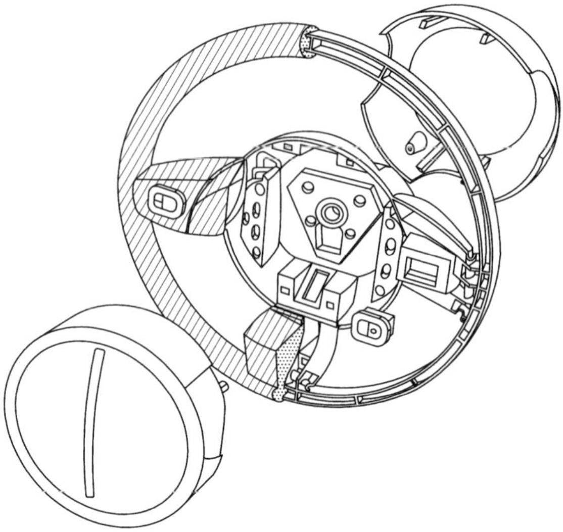

Fig. 3.12 Single-piece steering wheel.

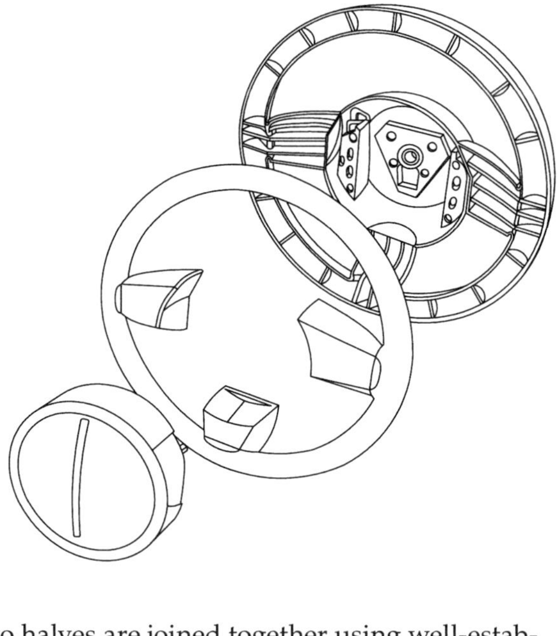

Fig. 3.13 Two-piece steering wheel.

the bottom half. The two halves are joined together using well-established vibration welding techniques. Vibration welding is done at the rim, spokes section, and other critical areas. The hollow section created through the integration of the two pieces gives high specific strength with a weight-efficient solution compared to the single-piece solution. The other advantage of the two-piece solution is that it requires less foam mass. On the other hand, the two-piece technology involves a secondary operation such as vibration welding and thus increases the cost, because of the initial capital investment in the vibration welding machine.

This disadvantage can be overcome by the use of alternate joining techniques such as adhesives to bond two halves, snap fits, or mechanical fasteners. Advanced injection-molding techniques such as gas- or water-assist molding can also be tried to get a hollow section in the rim portion. However, these techniques need further evaluation against the performance requirements of the steering wheel.

3.3.4 Materials

There are several thermoplastic materials that can be used to meet the requirements of a steering wheel. The selected material should have a good modulus to meet the stiffness requirements and good impact strength, while at the same time it should have cold- and hot-temperature ductility (60 °C [140 °F] to −30 °C [−22 °F]) to meet the heat-aging performance, according to OEM specifications. It should also have good chemical resistance against polyurethane foam, which is later overmolded. PA 6 / PA 66 (PA = polyamide) with glass filling is a potential candidate, given its good modulus for mechanical performance. But it does not have good ductility at negative temperatures, and hence during impact, there is a greater chance of failure of the part. A copolymer of polycarbonate with siloxane with 6% glass fiber material specifically developed to meet the contradicting requirements of stiffness and ductile failure would be an ideal candidate for steering wheels [3-13, 3-14].

3.3.5 Performance Requirements

Automotive steering wheel assembly is subjected to both OEM-specific requirements and regulatory requirements. For the North American market, steering wheels have to meet FMVSS 203 regulations, and in Europe, ECE R12 regulations [3-15, 3-16]. All these regulations are designed to minimize injuries to the occupants in frontal impacts. According to ECE R12, the body block test carried at 24.1 km/h is one of the important criteria to qualify steering wheels. Another critical test is the head-impact test, in which a head form strikes at various points of a steering wheel, at the speed of 24.1 km/h. There should not to be any failure in these dynamic impact test conditions.

In addition to regulatory requirements, there are several OEM-specific requirements for steering wheels. Many tests on steering wheels are structure related, and the rest are thermal and chemical requirements, such as heat aging, sun load, and flammability, which vary from one OEM to another. One such critical requirement is the bending stiffness test. In this test, a static load of around 400 N is applied at the 12 o'clock location on the steering wheel, and the deformation is measured. Generally, the deformation should not exceed 15 mm, with a permanent deformation of not more than 3 mm. Similarly, the test is repeated at 3 and 6 o'clock locations with different loading conditions and acceptance criteria. The pictorial representation of the test setup is shown in Fig. 3.14.

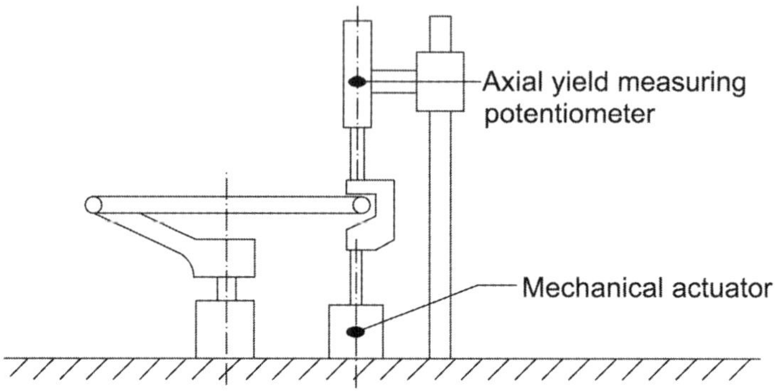

Fig. 3.14 Pictorial test setup.

Several other structural tests to ensure safe operation of the steering wheel under different conditions include torsional fatigue, bending fatigue, axial push, and 30° inclination test. Besides the mechanical performance testing, the steering wheel is also evaluated for natural frequency to make sure the frequency is more than the critical value specified by OEMs (generally, greater than 65 Hz). Airbag deployment is also tested to ensure that no failure of steering wheel occurs during the normal functioning of airbag. Another critical test is the heat-aging cycle test, in which the steering wheel has to withstand cyclic changes in temperature over time. After this treatment, the steering wheel undergoes a specific set of structural tests such as dynamic impact to meet the performance criteria. It is also important to keep the polar moment of inertia lower to avoid wobbling and provide comfortable feel to the user.

3.3.5.1 Role of Predictive Engineering

Predictive engineering plays a major role in simulating the wheel performance and optimizing it for all the requirements. These physical tests can be divided into three major categories: modal, static, and dynamic impact analysis [3-17].

Modal analysis looks at the natural frequency of vibration of the structure. In the case of the steering wheel, the objective is to avoid possible resonance with frequencies at idle speed as well as typical cruising speeds.

Static analysis is commonly of two types: The first one looks at the flexing of the wheel under repeated hand loading by the driver, and the second one studies the bending stiffness and deformed shape of the wheel at two different locations on the wheel.

Finally, dynamic impact analysis is done to simulate torso and head impacts at 24.1 km/h. The resulting reaction forces and decelerations are studied to avoid possible injury to the driver to meet the ECE R12 or FMVSS 203 regulations. The setup and simulation results are shown in Fig. 3.15.

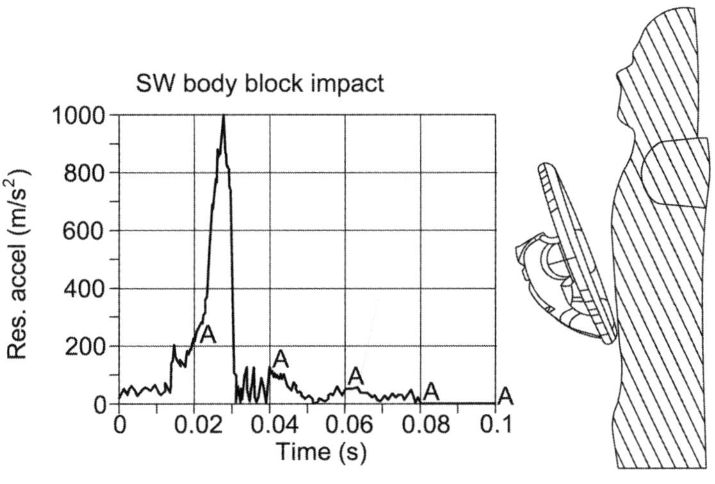

Fig. 3.15 Torso impact setup and results.

With the help of predictive engineering, the design is optimized to meet all the performance requirements. CAD (computer-aided design) data are generated to proceed with tool cutting.

3.3.6 Prototyping and Testing

Prototype tools can be built for a single- or two-piece design for standard injection molding. The bushing and flange are welded together to form the insert, which is kept in the injection-molding tool. In the single-piece design, the armature molded with a metal insert is ready for foam over-molding. In the two-piece design, the metal insert is molded to the bottom half and the top half is molded separately. These two parts can then be joined using any of the techniques explained earlier.

To optimize the process and the tool, a mold-filling simulation on plastic armature is carried out to evaluate the filling feasibility, predict the possible molding defects upfront, and suggest corrective action. Predictive analysis helps at different stages of the product development cycle, such as thickness optimization in part design, gate location, and feed system optimization for mold design and optimization of process parameters to be used during the molding process. A feed system design with four gates and a hot runner system and two-plate mold suitable for molding plastic armature is shown in Fig. 3.16. This type of design offers less flow length and greater flexibility during molding.

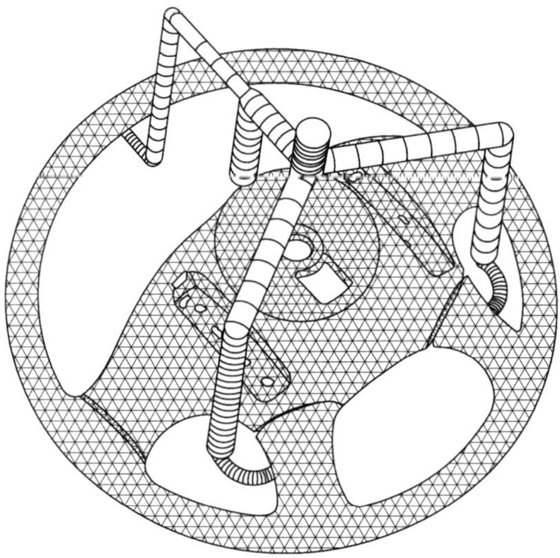

Fig 3.16 Mold flow feed system.

After the parts are molded, they are overmolded with foam and ready for testing. All the above performance criteria are tested with the help of appropriate jigs and fixture arrangements. Data are collected to validate each test criteria and recorded for qualification purposes. A typical bending performance test is as shown in Fig. 3.17.

The test results are compared with those of CAE (computer-aided engineering) to evaluate the accuracy of the predictions, improve the confidence level, and validate the CAE models. Figure 3.18 shows one such correlation between the actual testing and simulation results.

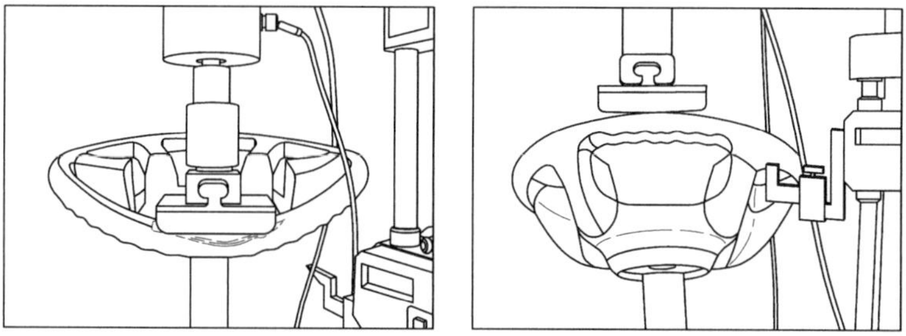

12 o'clock 6 o'clock

Fig. 3.17 Bending test 12 and 6 o'clock.

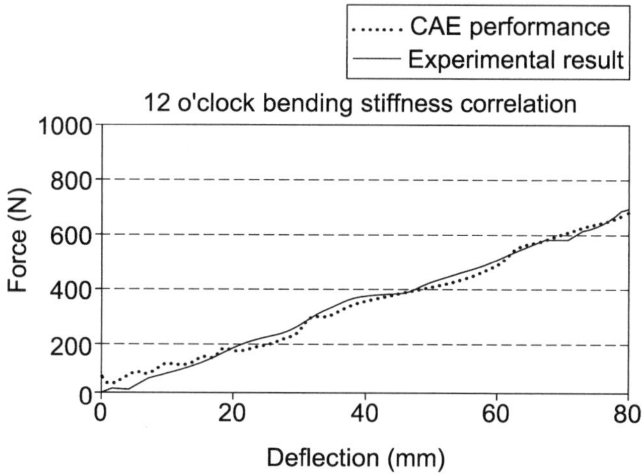

Fig. 3.18 Experiments versus CAE prediction comparison.

Similarly, all other tests are carried out and evaluated against the quality criteria, so that approval can be given for mass production [3-18].

3.4 Interior Components

Apart from the IP, center console, and steering wheel, there are several other interior components [3-19] that are molded with ETP materials. Door handles and door pull cups are typically molded with PC or PC-ABS. These two parts are common to all cars, with variations in design. In many cars, the window open/close buttons are mounted in the door pull cups. These parts are generally painted, but some cars use molded-in colors. Beneath the door handle cup and door handle lock, the inner garnish is located, which adds to the aesthetics of interiors. This part is also generally molded with

PC-ABS. The speaker grilles, which are mounted in the door module, are also aesthetic components. Each car has a different grille design, and there are challenges in processing this part, because of the tiny holes in them. However, this part can be molded successfully by using a design optimization process. In addition to the visible interiors, there are invisible parts that add to the structural rigidity of the vehicle and contribute to the safety of the passenger. The energy absorbers fixed in the roof of a car contribute to the passenger safety in case of a rollover. The crush box, located in between the passenger and driver seats, absorbs energy whenever there is an impact from the side. The energy-absorbing components are molded with PC/PBT (polycarbonate/polybutylene terephthalate) material and are designed to meet loads per regulations, which vary from country to country. Figure 3.19 shows different interior parts of an automobile.

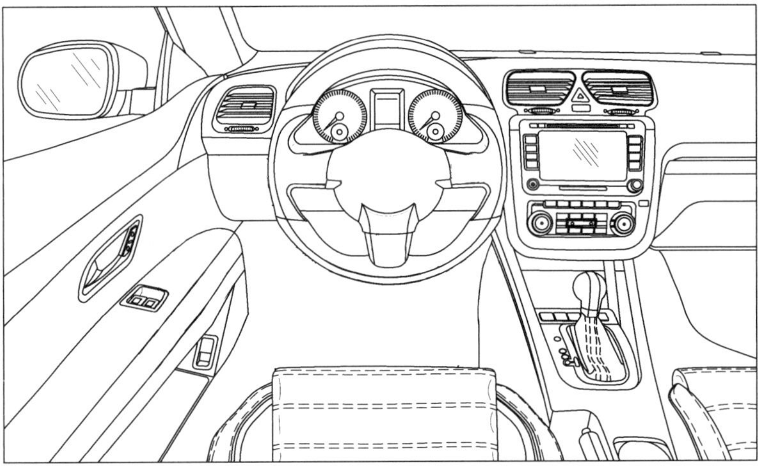

Fig. 3.19 Interior components in an automobile.

3.4.1 Roof Energy Absorber

During an automobile collision, the head of an occupant can hit the stiff automotive interior components. To reduce injury, regulations prescribe head injury criteria (HIC) [3-20], which necessitate an energy absorber (EA) mechanism that can be placed at the critical locations near the roof to absorb impact energy.

Transport Research Laboratory (TRL) has done a study to find the critical head-impact zones during a vehicle collision [3-21]. The collision may be frontal, rear, or side impact. The findings of the study are shown in Fig. 3.20, where critical impact locations for front seat occupants and rear seat occupants are shown.

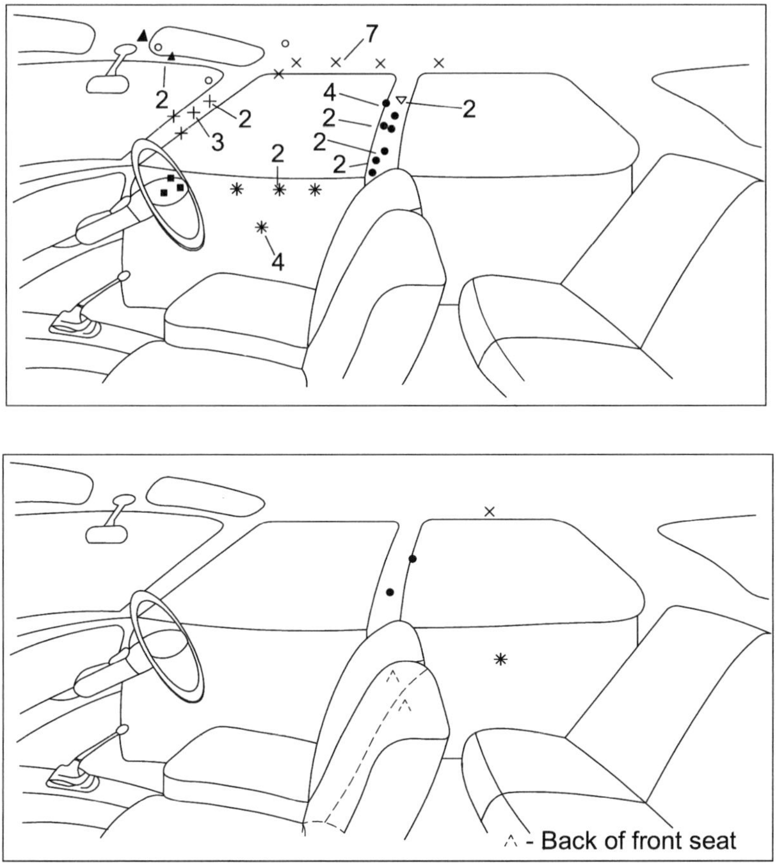

Fig. 3.20 Critical head-impact location (a) for all front seat occupants (b) for all rear seat occupants.

Small, modular EA strips (Fig. 3.21) can be placed at critical locations to provide necessary softness and energy-absorption capabilities [3-22]. The EA strips can be integrated with a low-cost headliner to reduce the overall roof EA cost. The EA strip will have crushable and support walls. Crushable walls are supposed to deform in an elastic and plastic manner and absorb the energy. The crushing of these walls can be optimized in such a way that the deformation of the roof is also reduced. The support walls have two functions: to connect crushable walls, ensuring the integrity of EA strips; and to provide the required reaction surface and load transfer to the crushable walls. Support walls are designed with a connection mechanism (snap fits or other alternatives), so that the EA strips can be assembled to their surrounding components.

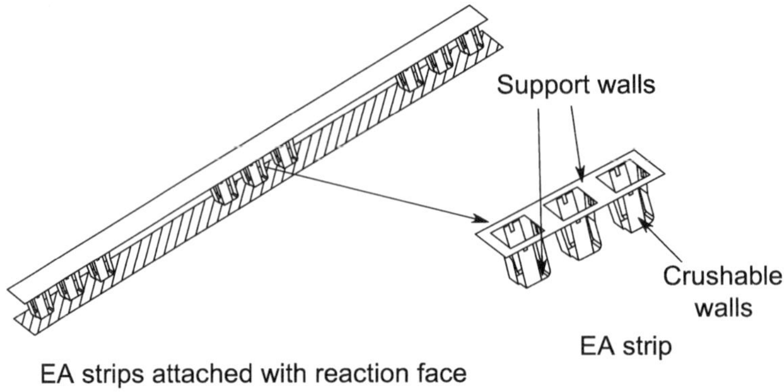

Fig. 3.21 Representative EA strips with reaction face.

Crushable walls can be designed with a variety of cross-sectional geometries. Some of the typical cross sections are shown in Fig. 3.22.

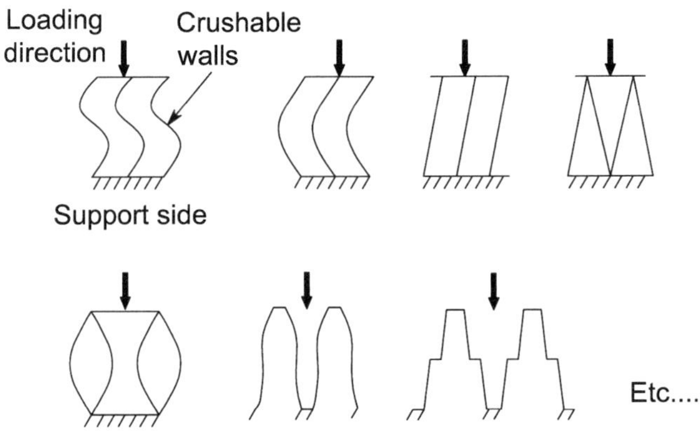

Fig. 3.22 Possible cross-sectional design for EA strips.

Crushable walls can have any one of the cross sections shown in Fig. 3.22 or an optimum combination of two or more cross sections to achieve the desired energy absorption. A similar energy-absorption strip can be designed for any other interior automotive components as well. Examples of other interior automotive components where this concept can be applied are the knee bolster, side door, door handle, and glazing panels. The roof EA strips can be produced by injection molding, thermoforming, blow molding, or any other alternative thermoplastic-molding techniques.

To understand the effectiveness of the aforementioned roof EA configurations, a study is performed on a roof model with a standard adult head form with a mass of 3.5 kg and nominal radius of 83 mm. The head form, along with one particular roof EA configuration whose front surface faces a rigid fixed plate, is shown in Fig. 3.23.

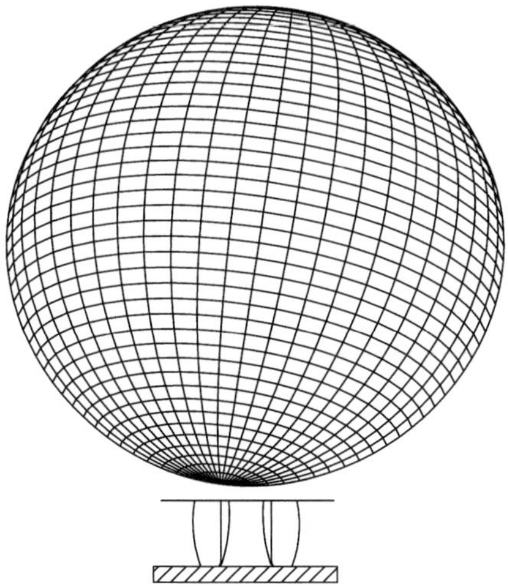

Fig. 3.23 Head form used for the impact study.

It is worth noting that the fixed rigid plate replicates the stiff metal structures of the BIW (body in white) that forms the roof frame, and the head form represents the occupant head. The head form is projected with a velocity of 24.1 km/h onto the roof EA, which is made of PC and PBT resin. Figure 3.24 shows the intrusion and the reaction force developed during the impact.

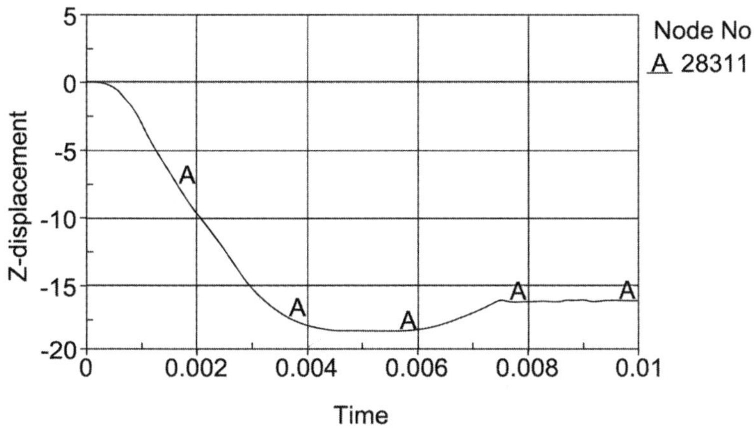

Fig. 3.24 Displacement and force curves versus time.

(Continues)

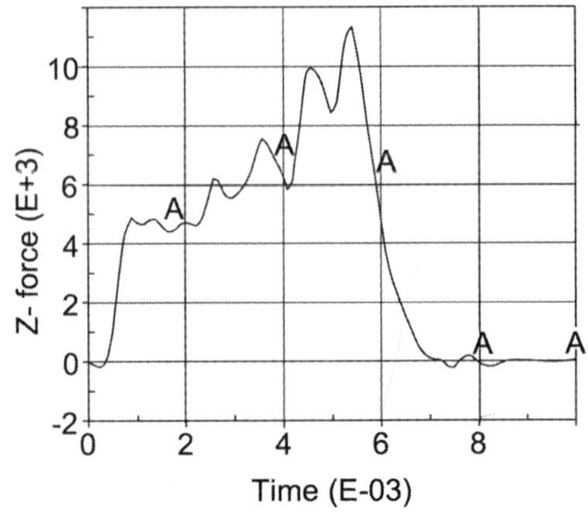

Fig. 3.24 Displacement and force curves versus time. (*Continued*)

It can be observed that the maximum force and maximum intrusion levels are generated at approximately 5.5 ms, after which the force curves start dipping down. It is also worth noting that the force curves are more or less flat during the impact, thus absorbing the required amount of energy with a high efficiency.

3.4.2 Door Handle and Door Pull Cup

The door handle and door pull cup are assembled in the door module. These two parts are typically molded in PBT because structurally these parts should be stiff. The door handle is an aesthetic part, and functionally the touch of the part should be comfortable. Every time the passenger gets out of the car, this door handle lever will be used to open the door. Thus, the door handle lever is designed with large fillets, which makes the part relatively chunky in cross section. A typical interior door handle and door pull cup are shown in Fig. 3.25.

Since these door handles are repeatedly operated, the parts are tested with cyclic loads. Endurance tests are carried out on a test rig. The part should not fail for 40,000 cycles in passenger cars and 100,000 cycles in sports utility vehicles or multi-utility vehicles (SUVs/MUVs).

When the door is manually locked, it cannot be opened by door handle lever, but there is chance that the passenger may abuse the part by pulling it very hard. To represent the abusive load, a pull test is also carried out by applying an approximately 500-N load on this part. Each OEM may have a different specification for these tests.

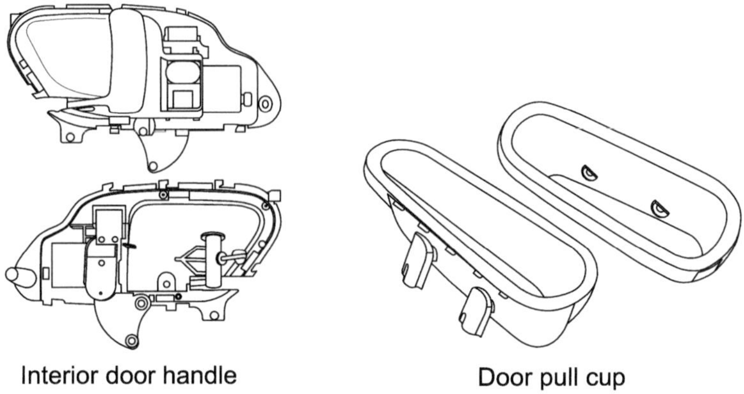

Fig. 3.25 Door handle and pull cup views.

The part geometry is generally thick, and filling predictions with midplane mesh are not possible. A three-dimensional (3-D) tetrahedral mesh should be used for carrying out filling predictions for this application. Because this part is aesthetic and given its 3-D nature, predicting the location of the weld line is very difficult. Thus, one has to locate the weld line very carefully by looking at the 3-D flow front. The gating location is very important for hiding the weld lines in a nonvisible area. A single gate is sufficient to fill this part, and the gate should be simple edge gate at a nonvisible area. In some cases, part geometry may need modification to eliminate or hide the weld lines, and the CAD file has to be changed iteratively. A 3-D mesh should be generated to carry out a new processing simulation to verify the elimination of weld lines from visible areas. Other than the weld line problem, filling of this part should not be difficult with proper gating.

The door pull cup is also a relatively simple part. A direct gate on the bottom surface will solve all problems. However, to reduce the complexity and cost of the mold, an edge gate is designed for a two-cavity RH-LH (right-hand–left-hand) family tool by some OEMs. Typically, gating is provided on the flange of the part, and given the geometry of door pull cup, there could be a weld line on the other side of the gate. The gate should be located such that the weld line will be positioned at a less visible area.

Quite often, there will be some ribs on the B-surface for strengthening the part, and these ribs often cause sink marks on the part. The thickness of the ribs should be minimized, and proper packing of the part will ensure that the sink mark depth is at acceptable levels.

3.4.3 Speaker Grille Cover

The primary function of the speaker grille cover in a car is to protect the speaker, while allowing the sound to clearly pass through it. Hence, during the design phase, a suitable compromise has to be made between protection of the speaker and quality of sound. Also, the grilles serve an aesthetic function by contributing to the overall appearance of the car. Grilles come in various sizes and are located in different locations in a car, such as on the IP, on the door module, or behind the rear seat. Generally, speaker grilles are made of either metallic mesh fixed to the ring or an integrated plastic mesh and rim design, as shown in Fig. 3.26.

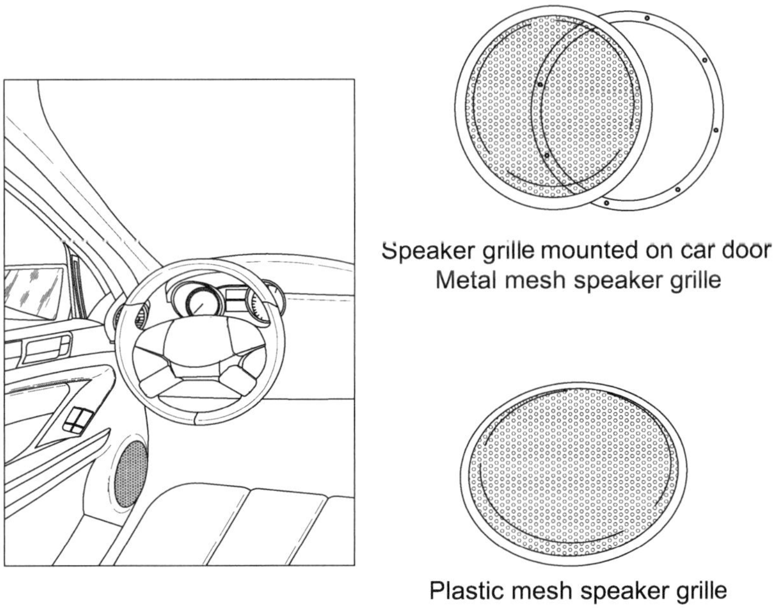

Fig. 3.26 Speaker grille.

Generally, speaker grilles are manufactured through an injection-molding process. In the plastic speaker grille design, the mesh pattern is integrated with the outer rim, which contributes to part consolidation and reduction of system cost. For good-quality sound output, the mesh pattern needs to be as fine as possible. This makes processing difficult, because finer the mesh, the greater the resistance to flow during the injection-molding process.

The typical processing challenges and guidelines for suitable solutions are described in the next paragraphs.

The grille design and gate mark restrictions at the center for aesthetic purposes dictate that the gates be on the outer rim. Therefore, the outer rim gets filled first, and the material starts flowing inward, causing potential air entrapment and burn marks at the center. To avoid these defects, the metal inserts are generally split into two or four segments, and suitable vents are provided between them. A high-flow material grade needs to be selected for this application because the material has to flow through the small grille sections. Inside the tool, the speaker grille region inserts need to be maintained at higher mold temperatures to help with better flow of material and to avoid any short shots caused by frozen-melt-flow front temperatures. It is advisable to have separated cooling lines designed for the grille areas in the tool for better temperature control during processing.

Performing detailed mold-filling simulations to optimize the flow pattern, or to design flow leaders or deflector and gate locations will greatly help in predicting potential molding defects, so that corrective actions can be planned before cutting the tool. It is also very challenging to model the grille pattern in the finite element model. The number of elements required will be huge, and the analysis time could be very high. The easiest solution for this problem is to use a shape factor concept [3-23] during the mold-filling simulation. This improves the accuracy of filling results and results in better correlation with the actual molding condition.

3.5 Summary

Thermoplastics will help enhance the feel, fit, and finish of the interiors of an automobile, while meeting the safety and regulatory performance requirements with lower mass compared to other materials. It opens up opportunities for better styling, improved ergonomics, and several integration possibilities. Many advancements have been made in material technology, such as low-temperature ductility, improved scratch resistance, and low-gloss finish, thus making plastics the ideal material for several interior components.

3.6 Trends

Automotive interior design is one of the major deciding factors for buying a particular vehicle, in addition to exterior styling and performance. Hence, OEMs and their tiers are working on differentiating their products from others to win a better market share. Recent trends in automotive interior design are the use of dual-tone or multicolor IPs and the shift toward brightly colored interiors. Other, future trends could be variable lighting, based on the mood of the occupant, and use of more human–machine interfaces such as touch screen controls for the radio and AC instead of knobs and switches. Also, fly-by-wire technology is likely to take over the gearshift and steering controls. In all these technology trends, thermoplastics will continue to play a dominant role in meeting regulatory, safety, and aesthetic requirements.

3.7 References

3-1. "Federal Motor Vehicle Safety Standards and Regulations," Standard No. 201—Occupant Protection in Interior Impact, http://www.nhtsa.gov/cars/rules/import/fmvss/index.html#SN201. Accessed January 1, 2012.

3-2. "Federal Motor Vehicle Safety Standards and Regulations," Standard No. 208—Occupant Crash Protection, http://www.nhtsa.gov/cars/rules/import/fmvss/index.html#SN208. Accessed January 1, 2012.

3-3. Wieczorek, T., Trappe, A., Meyers, A., Goyette, J., Lanning, V. and Chaturvedi, M. 2005. "An Efficient Solution for Hard or Soft Seamless Airbag Systems." SAE Paper No. 2005-01-1223. SAE International, Warrendale, PA.

3-4. Jaarda, E., Chaturvedi, M., Wieczorek, T., Meyers, A. and Chitteti, R. 2008. "Development of an Instrument Panel with an Integrated Chute and Seamless Airbag Door." SAE Paper No. 2008-01-1339. SAE International, Warrendale, PA.

3-5. Cooper, E.O. 2005. "Exploring Ways to Prevent Injuries During Vehicular Crashes." *Quest, Old Dominion University Magazine* 8(1): 10–13.

3-6. Chitetti, R. 2005. "Performance Evaluation of Knee Bolster." *Six Sigma Green Belt Report*. Bangalore, India: GE John F Welch Technology Center.

3-7. Bhise, V., Boufelliga, R., Roney, T. and Dowd, J. 2006. "Development of Innovative Design Concepts for Automotive Center Consoles." SAE Paper No. 2006-01-1474. SAE International, Warrendale, PA.

3-8. "Steering Wheel," Wikipedia, http://en.wikipedia.org/wiki/Steering_wheel. Accessed March 16, 2012.

3-9. Kawase, Y., Shinto, H. and Yoshida, T. 1991. "Development of Magnesium Steering Wheel." SAE Paper No. 910549. SAE International, Warrendale, PA.

3-10. "Plastic Steering Wheels—Technology in Waiting," ATZ online, http://www.atzonline.com/index.php;do=show/site=a4e/sid=3966caf52c34f9957d1aa7a705c0c5be/alloc=3/id=11682. Accessed March 16, 2012.

3-11. Jaarda, E., Chaturvedi, M., Surisetty, G. and Garg, A. 2009. "High Performance Thermoplastic Steering Wheel." SAE Paper No. 2009-26-0074. SAE International, Warrendale, PA.

3-12. Nayak, S., Garg, A., Kumar, S. and Krishnamoorthy, N. 2011. "Virtual Design Optimization of Thermoplastic Steering Wheel." SAE Paper No. 2011-01-0023. SAE International, Warrendale, PA.

3-13. "Lexan* EXL Resin: For Tough Conditions," http://www.sabic-ip.com/gep/Plastics/en/ProductsAndServices/ProductFamily/exl_lexan.html. Accessed March 16, 2012.

3-14. "Ultramid," BASF, http://www.basf.com/group/corporate/en/brand/ULTRAMID. Accessed March 16, 2012.

3-15. "NHTSA Regulations FMVSS 203," NHTSA, http://www.nhtsa.gov/DOT/NHTSA/Vehicle%20Safety/Test%20Procedures/Associated%20Files/TP-203-02.pdf. Accessed March 16, 2012.

3-16. "UNECE Vehicle Regulations—1958 Agreement," UNECE, http://www.unece.org/trans/main/wp29/wp29regs1-20.html. Accessed March 16, 2012.

3-17. Yang, Z., Raman, S. and Ma, D. 2005. "Virtual Tests for Facilitating Steering Wheel Development." SAE Paper No. 2005-01-1072. SAE International: Warrendale, PA.

3-18. Nayak, S., Garg, A., Chaturvedi, M. and Wieczorek, T. 2010. "Performance Evaluation of PU Over-molded Thermoplastic Steering Wheel." SAE Paper No. 2010-01-0916. SAE International: Warrendale, PA.

3-19. "Car Interior Parts," Auto Interior, http://www.plastics-car.com/Todays-Automobiles/Auto-Interior. Accessed April 30, 2013.

3-20. "European New Car Assessment Programme (Euro NCAP), Pedestrian Testing Protocol," http://www.euroncap.com/files/Euro-NCAP-Pedestrian-Protocol-Version-5.3.1---0-0560739b-595f-4a38-96b2-eeda5050dee8.pdf. Accessed on March 16, 2012.

3-21. "EEVC Working Group 13 Report. Head Contacts In Side Impact—An Accident Analysis," http://eevc.org/publicdocs/WG13_WD101_Hd_Contact_AccData_SC_Rev1.PDF. Accessed March 16, 2012.

3-22. Marur, S.R., Garg, A.K., Nayak, S., Tankala, T.C. and Wilson, J.R. 2011. "Energy Absorber Elements and Vehicle Systems." US Patent No. US 2011/0210579 A1

3-23. *Autodesk Moldflow User Manual AMI 2012*. San Rafael, CA: Autodesk.

Chapter 4
Glazing Applications

N. Krishnamoorthy, Triloka Tankala, and
Sudhakar R. Marur

4.1 Automotive Glazing Overview

Automotive glazing consists of parts that provide visibility to the driver and occupants and protect them from external environmental factors such as dust, heat, and rain. These parts include the front windshield, front quarter windows, door windows, rear quarter windows, rear door glazing otherwise known as *back-lite,* and the panoramic roofs. In early days, glazing in cars was limited to front windshields and door windows to some extent. The evolution of automotive engineering resulted in more styling and complex-shaped cars. Users always preferred to have more visible regions to have a better feel as well as safety. This led to the development of cars with more transparent regions such as large back-lite, front and rear quarter windows, movable sunroofs, or fixed panoramic roofs. Figure 4.1 illustrates glazing parts in a modern automobile.

Traditionally, glass used to be the material of choice for these parts because of its transparency, stiffness, and formability. Laminated glass is used in the front windshield to avoid splintering during a crash. However, tempered glass is used in other areas. In some high-end cars, laminated glass is used even for rear windows.

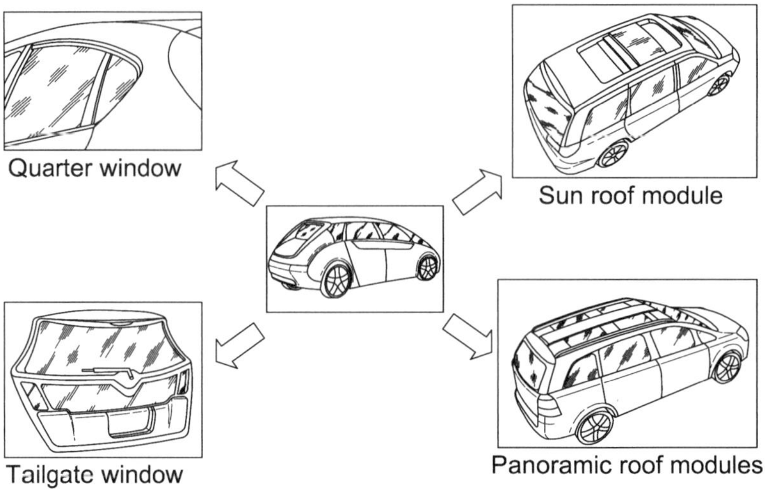

Fig. 4.1 Various glazing parts in automobile.

4.2 Automotive Glazing and Global Regulations

Automotive glazing is covered by legal requirements based on window and vehicle type. In the United States, the law requires the glazing system to comply with ANSI/SAE Z26.1 [4-1], and in Europe, it is required to comply with ECER43 [4-2]. Some of the aspects covered in the U.S. standards are abrasion resistance for the back-lite with wiper to withstand 1000 cycles with less than 2% haze and luminous transmittance greater than 70%. According to ECER43, the impact resistance for a head form must be no breakage of a fully separated large piece and head injury criteria (HIC) of less than 1000 [4-3]. In addition to abrasion resistance and light transmittance, it also defines the weathering performance as light transmittance not falling below 95% of the preweathering value. Also, light transmittance must be above 70% for better visibility for the driver with no bubbles or other visible decompositions, discolorations, milkiness, or fine cracks known as crazing. Complete details of the regulation can be obtained from the ANSI/SAE Z26.1 and ECER43 standards.

4.3 Automotive Glazing—Role of Polycarbonate

Automotive designers constantly look for alternate materials to replace glass in order to achieve more styling freedom and weight reduction. One such potential material is PC (polycarbonate) [4-4] because of its transparent nature, good impact strength, and ability to be molded in any complex form.

4.3.1 Weight Reduction

Density of PC ranges from 1.2 to 1.22 g/cm^3, and that of glass ranges from 2.3 to 2.8 g/cm^3. Hence, the use of PC in replacing glass in automotive applications results in weight reduction in the range of 30–40%, depending on the size and thickness of the glass part. Figure 4.2 shows a 3.2-mm tempered glass replacement with 4-mm PC and 2-mm PC/ABS (acrylonitrile butadiene styrene) two-shot system resulting in a 30% weight reduction, and 5-mm glass replacement with 5-mm of PC resulting in a 50% mass reduction, based on the density difference.

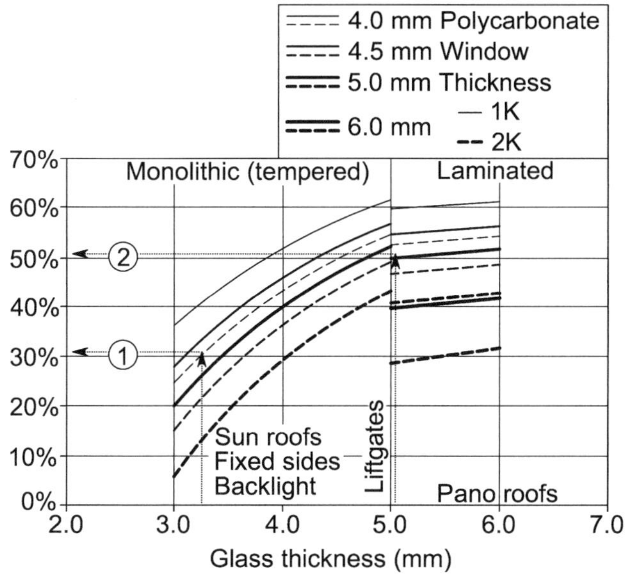

Fig. 4.2 Potential of weight reduction with PC.

4.3.2 Styling and Design Freedom

As PC can be molded in any complex shape compared to glass, it gives designers the freedom to create more stylish designs to improve aesthetics, aerodynamics, and differentiation for their models among the competition. Also, PC glazing helps to integrate the fixation elements, gas struts, and so forth, through second-shot molding, which reduces or eliminates the number of parts in the assembly. Also, integration of several features, such as the center high-mount stop lamp (CHMSL), defroster, and fractal antenna, into PC glazing provides design freedom, which is seldom possible with glass. Some of the styling and design freedom with PC can be highlighted by the possibility of having *wrap-around* styling for front

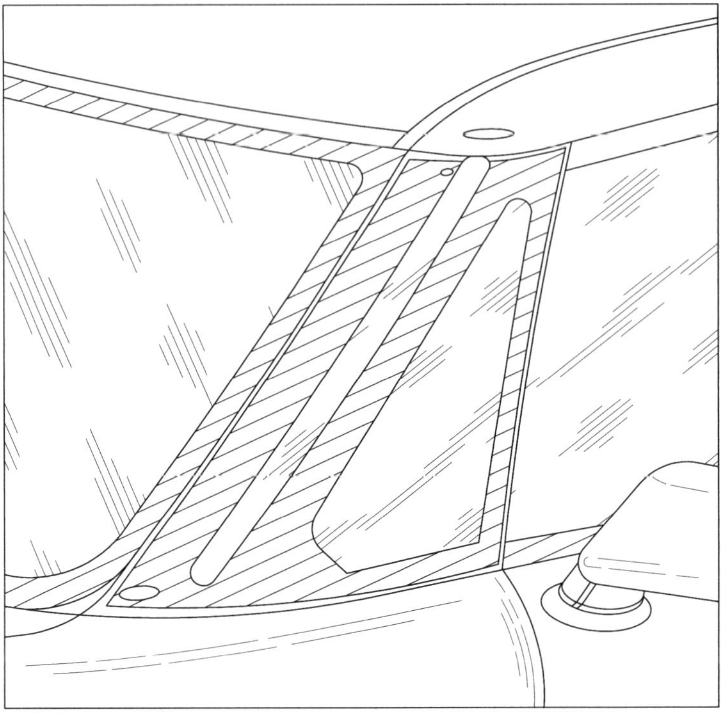

Fig. 4.3 Wrap-around design of the front quarter window.

quarter windows, which increase the visible area (as shown in Fig. 4.3) and integration of the bracket of the mirror housing (as shown in Fig. 4.4).

Similarly, in rear quarter windows the guides for door sliding can be integrated using the two-shot method. Also, the rear quarter windows can be made large enough to include the taillamp lenses to provide signature styling. In the back-lite, several integration opportunities are possible to eliminate or consolidate the components so as to minimize parts for assembly while allowing for better styling. Some examples in tailgate application include integration of handles, taillight lenses, and rear spoilers, as shown in Fig. 4.5.

4.4 Characteristics of a Glazing System

It is essential to understand the characteristics of a glazing system for the successful replacement of glass with PC. Glazing systems can be broadly classified into optical, structural, thermal, acoustic, scratch, and weathering performance requirements. These characteristics emerge from functional and regulatory requirements as well as customer specifications.

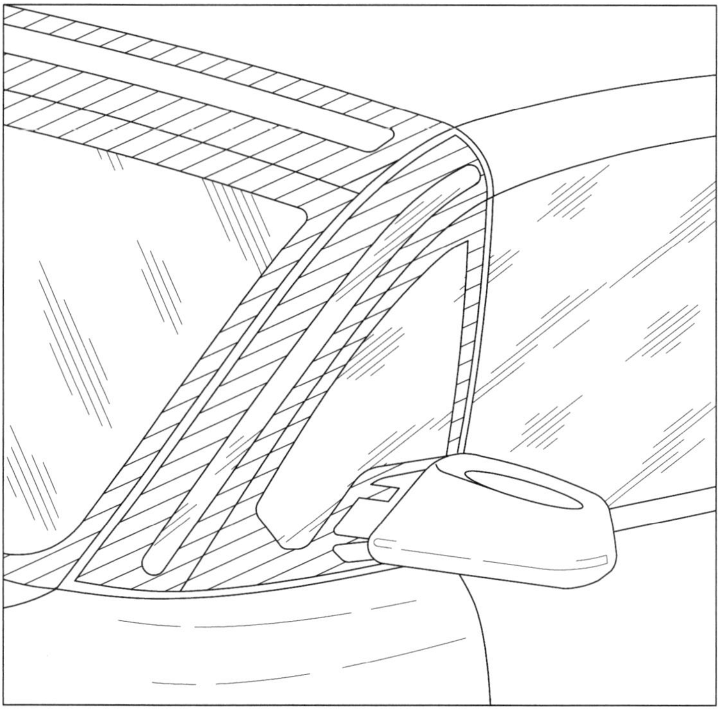

Fig. 4.4 Integration possibility of the mirror housing bracket.

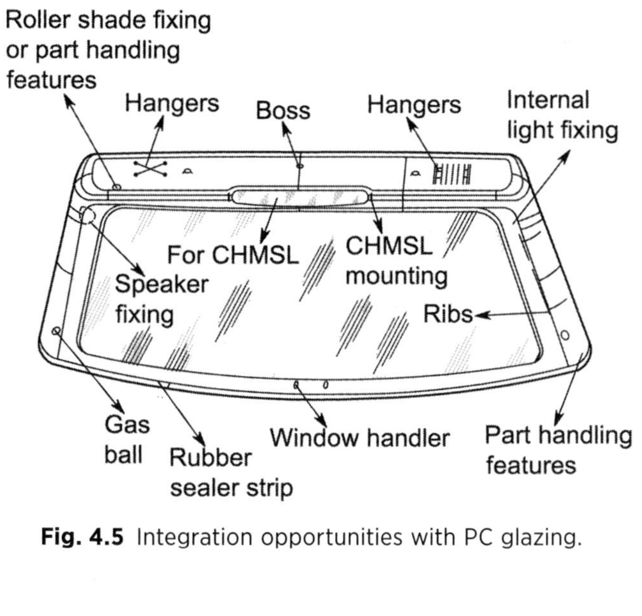

Fig. 4.5 Integration opportunities with PC glazing.

Optical performance of a glazing system is defined by regulations as luminous transmittance greater than 70% [4-1, 4-2]. The molded PC system should be designed and manufactured to meet this regulation for optics.

The structural performance requirement of the glazing system is to be mechanically stiff enough to withstand various loads, such as impact and wind, and not to buckle under its own weight. Another key parameter of significance is the expansion of the part according to the coefficient of linear thermal expansion (CLTE). As PC has a higher CLTE than glass, the management of flush (thickness change of the adhesive) and gap (shear movement) deformation has to be addressed through design parameters such as curvature and thickness of the PC panel and the type, location, and geometry of the adhesive used to attach the panel with body in white (BIW). Also, the glazing system has to meet some of the impact criteria as laid out in the regulations to satisfy the HIC for occupant safety [4-3].

Thermal requirements primarily cover the heat buildup inside the cabin. The heat gain resulting from the use of PC needs to be considered and its effect minimized. Several technologies have been adopted to reduce the cabin heat buildup such as the use of infrared (IR) reflective/absorptive coatings, additives and films.

Another important factor to be considered when replacing glass is acoustics. The replacement should not result in discomfort to the passenger caused by higher noise or vibration. Hence, necessary testing—both experimental and numerical—needs to be carried out to ensure that the proposed PC part performance is better than or comparable to that of glass.

Molded PC is not scratch resistant, whereas glass is naturally scratch proof. This shortcoming can be remedied through coatings. Two types of coating are mainly employed in industry: wet coating and plasma coating. Wet coating is done by applying silicone (polysiloxane) to the surface and curing it by baking [4-5]. Plasma coating offers superior scratch resistance compared to wet coating. Both of these coatings help to improve the weatherability of PC glazing to a large extent; plasma coating offers a weatherability performance of up to ten years [4-6].

Apart from these performance-related requirements, manufacturability of large PC panels also needs to be considered for successful replacement of glass. Advanced molding processes such as two-shot injection compression molding (2K-ICM) is employed to achieve a high level of quality in the manufacture of these parts.

Another important aspect to be considered is the blackout capability. The glazing panel is fixed to the BIW with the help of adhesive, and the blackout at the edges of the glazing panel helps to protect the adhesive from ultraviolet rays, so that it does not become brittle. Hence, a PC glazing should incorporate the blackout feature. This can be achieved in two ways, either by ink printing or through a second shot. The former approach is limited by curvature of the part while the latter offers many opportunities for integration of various components, as explained earlier.

4.5 Structural Performance

Structural stiffness and the dimensional stability of PC glazing panels are of interest for automotive application developers, because any glazing system would be subjected to various loads such as impact, thermomechanical, and wind loads. PC has a high coefficient of thermal expansion (CTE) when compared to glass. Hence, large PC panels, such as those used for panoramic roofs and back-lite, can suffer from unwanted high-thermal deformations, when used without appropriate bonding mechanism. The influence of adhesives connecting the PC panels to the BIW is quite significant in these scenarios and is worthy of critical study. The influence of adhesive size and shape on the thermomechanical behavior of PC panels needs to be studied to minimize thermal deformation. By using a two-phase adhesive with low stiffness and high damping, Krois et al. [4-7] have demonstrated better vibration isolation of glass window panels and subsequent reduction of sound levels in the interior of the car.

4.5.1 Design for Structural Stiffness

Designing PC glazing panels to offer the required structural stiffness to the automobile body is very important, because of the panels' significant effect on acoustic [4-7] and thermomechanical performance. The sensitivity of the thermal deformations of a panel, caused by adhesives acting as rigid end restraints, is a significant factor in the structural design of PC glazing. In addition to the adhesives, the panel curvature and thickness also play important roles in the structural performance.

4.5.2 Role of Restraints

Adhesive stiffness influences the magnitude of restraint offered to the in-plane (flush deformation) panel motion due to the thermal deformation, as well as the transverse displacements of curved panels (gap deformation) [4-8]. It can be inferred from Fig. 4.6 that the influence of end restraints is more significant in flatter panels than in panels with a high curvature.

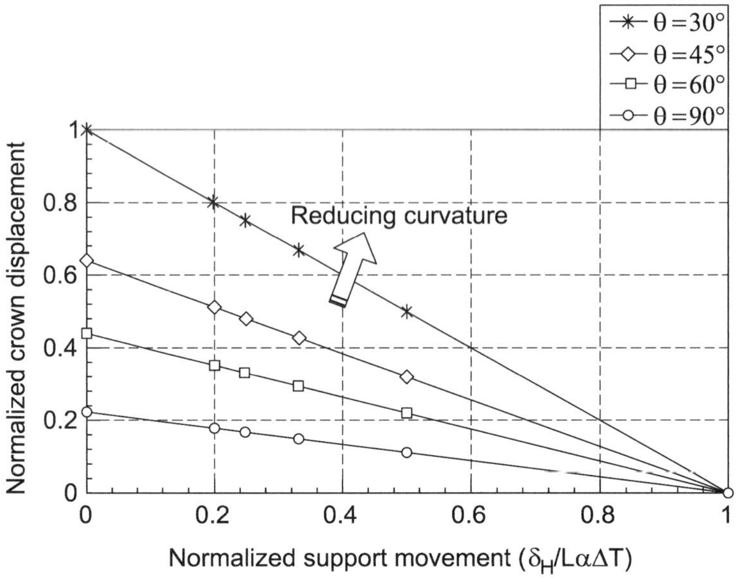

Fig. 4.6 Effect of support movement on crown deflection.

4.5.3 Role of Curvature

Transverse displacements of panels with end restraints are highly sensitive to the curvature of the panels. These transverse displacements are calculated as a function of the ratio of the length of span to radius of curvature (L/R) [4-8]. As can be seen from Fig. 4.7, the crown displacement decreases exponentially as the L/R ratio increases; that is, for a given span, crown displacements show an exponential decrease with higher panel curvature.

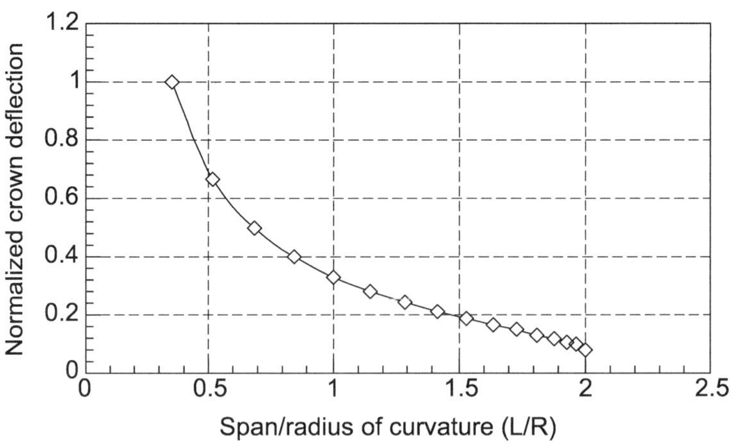

Fig. 4.7 Effect of support movement on crown deflection.

Perfectly flat or zero curvature panels undergo purely axial deformation because of restrained thermal expansion. This situation is rarely attained in practice, as the restraints are not always centered along the panel cross section. The resulting eccentricity would cause the panel to bow and have slight curvatures. Figure 4.8 illustrates the effect of both rigid end restraints and panel curvature [4-8]. Elongation of the panel increases linearly with the span. The corresponding crown displacements occurring due to the restraints exhibit exponential growth.

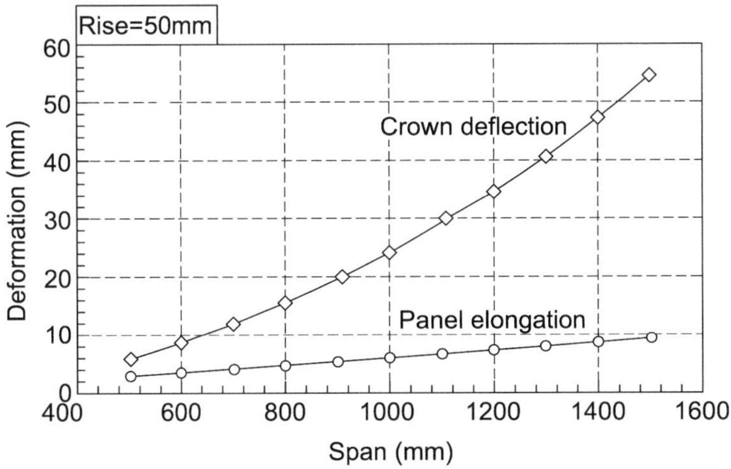

Fig. 4.8 Crown deflection due to restrained thermal deformation.

4.5.4 Role of Thickness

Because of partial restraint against thermal deflections by the adhesive, the glazing panel would deform by way of flexure. But an increase in thickness of the panel will not lead to a reduction in such thermal deformations, because it is naturally expected in bending deflections. In-plane thermal deflections of the panel are independent of its thickness. Thus, increasing panel thickness will have no effect on its thermal deflections. Thicker panels would have higher stress levels for the same thermal deformation. Design of thickness of the glazing panel should be based mainly on the requirement of structural stiffness and its impact performance. Hence, any thermomechanical study of a PC glazing system should include an appropriate model of the supporting mechanism. Adhesives are the restraint, so their role is important.

4.5.5 Importance of Adhesive and Its Characterization

Adhesives are used in the form of thin films bonding two substrates in many applications including PC glazing. Displacements and strains of the adhesive specimens are very low, and hence it is very difficult to measure them accurately. Typically, adhesive in a bonded form can be reasonably assumed to be homogeneous [4-9]. It is also observed that the adhesive behavior in bulk form is same as that in bonded form, as thin films [4-9, 4-10, and 4-11]. This enables characterization of adhesives that are easy to prepare and easy to test bulk samples. Adhesives for the glazing application are chosen based on quality of the bonding and compatibility with PC. Generally, hyperelastic material models are suitable to simulate the behavior of such adhesives.

Numerical studies revealed that strain levels in the adhesive caused by thermal deformations of the automotive window components are lower than 10% [4-10]. A linear elastic assumption was found to be sufficient to model these adhesives. Based on these premises, tests were performed to characterize the adhesives only for elastic modulus. Tests were also performed to characterize and evaluate thermal expansion coefficient over the entire operating temperature range of –40 °C (–40 °F) to 110 °C (230 °F).

Adhesive samples are prepared carefully to eliminate air bubbles [4-11], which is an important step toward accurate characterization. A sheet of adhesive was cast, and specimens per ASTM International standards for plastics [4-12] were punched out from this sheet using standard ASTM punches.

4.5.6 Adhesive Testing—Uniaxial Tension

The adhesive being tested is extremely flexible. Because of this, the standard procedure for tensile testing has few problems with regard to accurate measurements. Clip-on extensometers that are generally used for metals and polymers can cause the adhesive specimen to bend in the gauge length region. Hence, strain measurements would be inaccurate. To overcome this, the standard testing procedure was slightly modified. The test was done without clip-on extensometers, and strains were computed from the crosshead movement itself. But, because of the default positioning of crossheads (115 mm apart), strain measures would be inaccurate [4-8]. Hence, the crossheads had to be moved closer, to have a grip length equal to the gauge length of 65 mm, as shown in Fig. 4.9. Thus, a constant adhesive cross-sectional area is ensured.

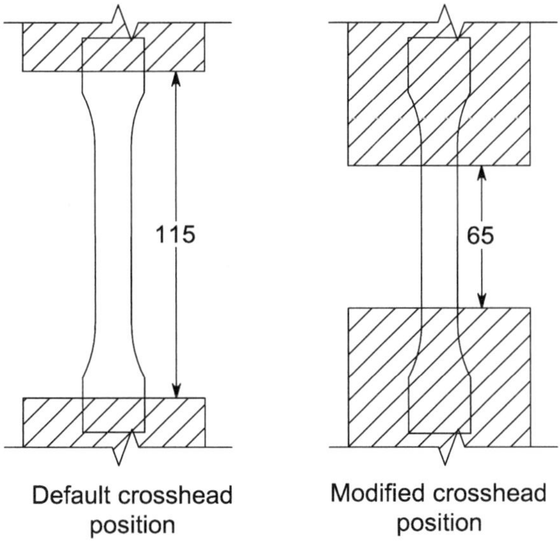

Fig. 4.9 Crosshead positions on adhesive tensile specimens.

Adhesive properties are generally highly sensitive to strain rate. The glazing system is subjected to a gradual thermal loading (a time-independent phenomenon). Hence, tensile testing was conducted at a strain rate of 5% per minute In the modified crosshead position, this translates to a crosshead movement of 3.25 mm/min. Table 4.1 shows the material properties obtained at –40 °C (–40 °F) and 110 °C (230 °F).

Table 4.1 Adhesive Material Properties

	Temperature (°C)	
Adhesive material properties	**–40**	**110**
Elastic modulus (MPa)	5.8	2.5
Thermal expansion co-efficient (/°C)	18.9E-05	189E-05
Poisson's ratio	0.45	0.45

Using a differential scanning calorimetry (DSC) test, the glass transition temperature (*Tg*) for the adhesive was identified to be –70 °C (–94 °F). During the entire operating regime from –40 °C (–40 °F) to 110 °C (230 °F), the adhesive is above its *Tg* and is in a rubbery state. Poisson's ratio was assumed to be 0.45. The CTE of adhesive was characterized by using thermomechanical analyzer. The CTE is found to be constant over the whole range of operating temperature.

4.5.7 Dimensional Stability—Effect of the Coefficient of Thermal Expansion

An actual roof panel with an approximate projected dimension of 1 m^2 was tested to measure the thermal deformations of a PC roof panel bonded to a carbine steel frame with adhesives. The construction of the PC panel, adhesives, and A36 carbine steel frame can be seen in Fig. 4.10.

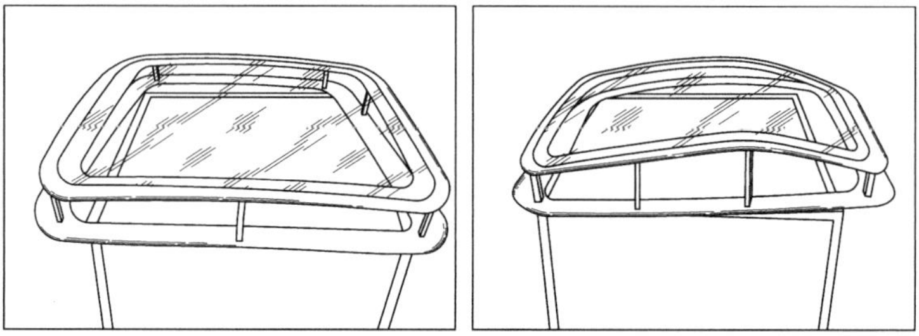

Fig. 4.10 Full-scale PC roof panel bonded onto a steel frame.

Linear voltage displacement transducers (LVDTs) are used to monitor flush and gap displacements on the roof panel at ten locations on each panel. The panel is glued onto the steel frame, and the adhesive is allowed to cure according to specifications. The whole assembly is then placed in a chamber with controlled humidity and temperature levels. A series of heat-generating bulbs were placed on the top surface of the panel to represent the solar load, as shown in Fig. 4.11.

At the beginning of the experiment, the outer surface of the panel has a higher temperature compared to the inner surface. The panel then reaches a steady-state temperature. The temperature of adhesive and the frame will be different from that of the panel. This methodology simulates the condition in which the glazing panels are heated up, for example, when a car is parked in the hot sun.

Generally most OEMs (original equipment manufacturers) specify a 15-min. ramp to a specified temperature followed by a 30-min. soak at the same temperature for the sun-load cycle. A typical sun-load cycle covering three temperature levels (85 °C [185 °F]), 100 °C [212 °F], 110 °C [230 °F]), as shown in in Fig. 4.12, is used for the purpose of the study.

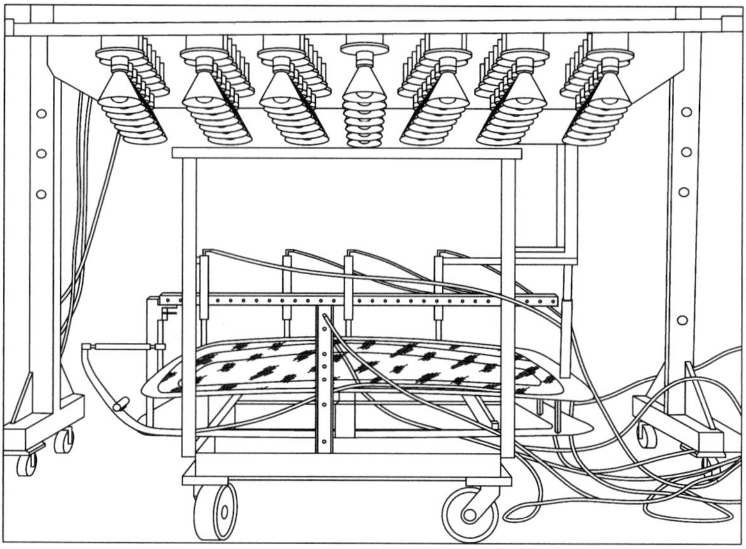

Fig. 4.11 Sun-load test setup.

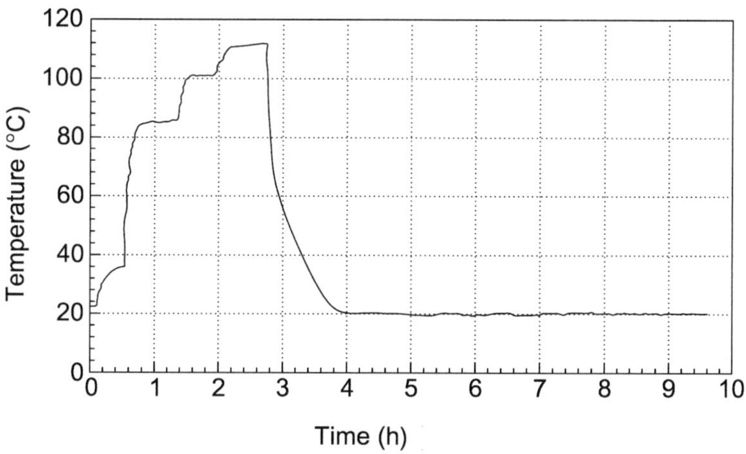

Fig. 4.12 Profile of the sun-load cycle.

4.5.8 Simulations and Experiments

Numerical studies are conducted on the glazing system to establish a methodology for accurate correlation between experiments and simulation in PC glazing panels. The glazing panel was discretized using solid hexahedral elements. Solid elements were used to facilitate the modeling of conductive heat transfer during the sun-load cycle. The steel frame to which the adhesive gets bonded was meshed with shell elements. Figure 4.13 shows the finite element mesh used for the glazing panel setup.

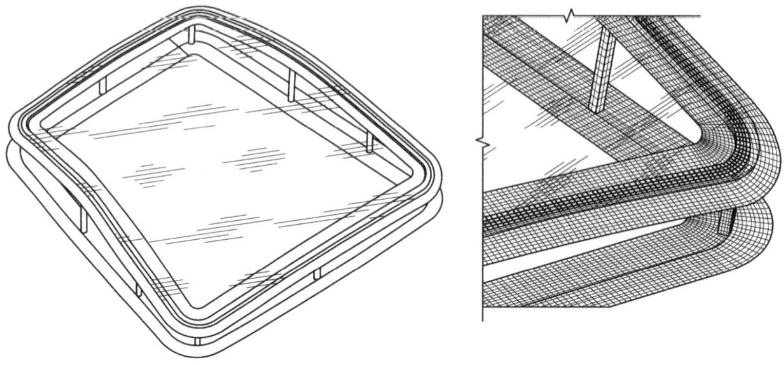

Fig. 4.13 Finite element model of the glazing panel bonded to steel frame with adhesive (strip along the periphery).

The material properties were provided at multiple temperature levels between the minimum and maximum temperature values. The entire system was initially held at room temperature, then gradually ramped up or down to the desired positive or negative temperature. This method simulated the actual thermal cycle as in the experiment.

Absolute deflection values at various locations for the two peak temperatures are shown in Table 4.2.

Very good correlation between experimental and simulation results can be observed for out-of-plane deflections at locations 2–8. Gap and flush values are very sensitive to adhesive dimensions at the location of measurement. Differences between experimental and simulation results for gap and flush deflections are due to the assumption of linear approximation of the actual adhesive width variation. Through this approach, validity of simulation results has been established.

4.5.9 Design of Experiments Approach

Multiple factors influence the thermal deformation of PC panels. A design experiments (DoE) [4-13] approach is used to obtain the qualitative and quantitative measures of influence of each of these factors on the panel response, by using a well-known statistical tool, Design-Expert [4-14]. A panoramic roof panel having a projected dimension of 1.5 m × 1 m was considered for this study. The objective was to arrive at a stiffener and adhesive configuration that would result in out-of-plane and in-plane

Table 4.2 Correlation Results for a Roof Panel

	Flush			Out-of-plane			Gap		
Location	**1**	**5**	**2**	**3**	**4**	**6**	**8**	**9**	**10**
T = -40°C									
Experiment	-0.56	-0.43	-3.81	-3.64	-3.45	-4.74	-5.08	-1.61	-1.85
Simulation	-0.56	-0.25	-3.85	-3.49	-3.51	-4.63	-5.59	-1.13	-1.51
T = 110°C									
Experiment	0.60	0.64	4.69	4.67	4.41	6.30	6.56	2.18	2.86
Simulation	0.64	0.16	4.46	4.14	4.15	5.96	6.34	1.83	2.31
Material properties used for simulation									
At -40°C	$E_{(PC)}$ = 2576 MPa			$E_{(adh)}$ = 5.8 MPa			$CTE_{(PC)}$ = 6.3E-05/°C		
At 110°C	$E_{(PC)}$ = 2075 MPa			$E_{(adh)}$ = 2.5 MPa			$CTE_{(PC)}$ = 7.2E-05/°C		

panel displacements within allowable specification limits. Figure 4.14 shows a schematic of the roof panel along with various parameters used for the DoE. The following are typical performance requirements for the thermomechanical response of automotive glazing panels:

1. A minimum gap of 1 mm to be maintained between the panel edge and the supporting frame
2. Maximum out-of-plane displacements need to be less than 1% of the panel dimension. Three parameters have been studied as below as per Fig. 4.14.
    1. Stiffener width (35–70 mm),
    2. Stiffener thickness (1.5–4 mm),
    3. Adhesive depth (3–6 mm).

A central composite design was built with the three parameters as input variables, resulting in a 15-run setup [4-8]. Main effects plots illustrate the independent impact of each factor on panel responses. The influence of each of the factors can readily be inferred from these plots. It can be seen from Figs. 4.15, plotted through Minitab [4-15], that adhesive depth or stiffness plays a key role in determining the thermal response of a panel, compared to edge stiffeners.

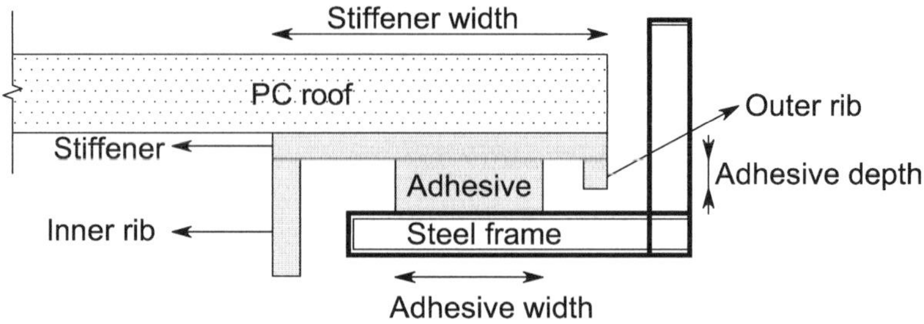

Fig. 4.14 Schematic of the glazing system lay-up.

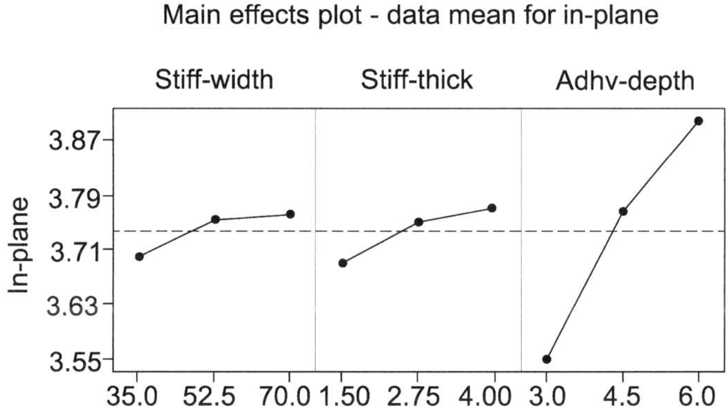

Fig. 4.15 Influence of the parameters on in-plane displacements.

A design optimization is carried out to find the dimensions of stiffener and adhesive to be able to meet the performance objectives using Design-Expert [4-14]. Optimal dimensions of the adhesive and the stiffener were arrived at, to restrict panel thermal deformations within specification limits. The optimum adhesive depth was found to be 5.3 mm, and with corresponding stiffener dimensions, this resulted in a gap of 1.44 mm at 110 °C (230 °F).

Given that the adhesive shape and size directly influence in-plane and out-of-plane deformations, this becomes a critical aspect to be studied in detail. Thermal deformations of PC panels are highly sensitive to adhesive stiffness. The adhesive stiffness is typically controlled both by the choice of the adhesive (material stiffness) and the cross-sectional dimensions of the adhesive lay-up (geometric stiffness). The influence of adhesives is highly significant for low-curvature panels. Stiffeners, all along the periphery, have less influence on thermal deformations of panels. Through this study, a procedure was established for accurate thermomechanical modeling of PC glazing panels, including the role of adhesive as a connecting mechanism.

4.6 Acoustic Performance

The sources of cabin noise can be classified in to airborne noise (e.g., wind and neighboring vehicles), structure-borne noise (e.g., engine vibrations and road noise), and noise caused by a sudden pressure pulse (e.g., a door slam) [4-16]. When glass has been replaced by PC, the latter two sources are dominated by the panel stiffness comprising the material and geometry, and airborne noise is controlled by the panel surface mass. With the help of predictive engineering and experimentation—to evaluate various parameters that influence the comfort of the occupant—PC glazing should either match the current level of glass or, preferably, surpass it.

4.6.1 Transmission Loss

Transmission loss is one of the key parameters used to quantify the effectiveness of glass versus PC. A typical transmission loss spectrum is shown in Fig. 4.16.

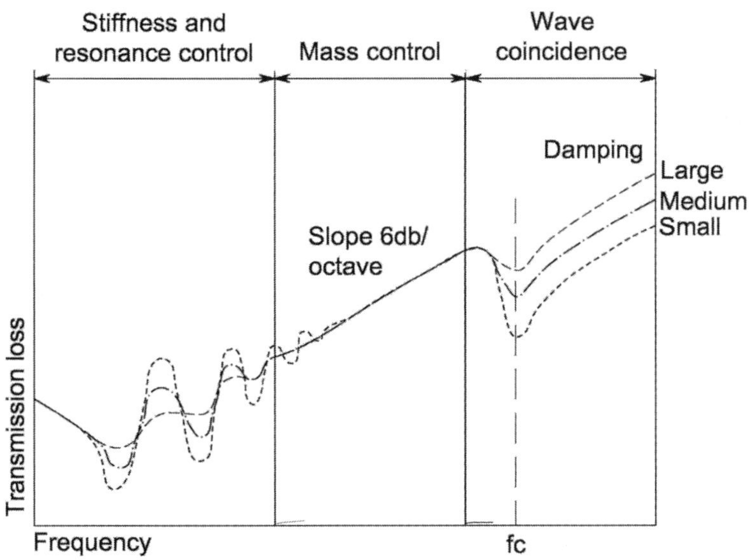

Fig. 4.16 Typical Transmission loss spectrum.

The transmission loss is calculated by [4-17]

$$\text{Transmission Loss (TL)} - 10\log_{10}\left(\frac{W_{src}}{W_{rec}}\right) \tag{4.1}$$

where

W_{src} = Sound energy at source
W_{rec} = Sound energy received

It can be observed from the graph in Fig. 4.16 that the resonance dip occurs when the frequency of incident sound waves matches the natural frequency of the panel, and the coincidence dip occurs when the wavelength of the panel-bending wave matches the incident sound wavelength. The inference is that the higher the transmission loss, better the noise isolation.

Figure 4.17 shows the equal-loudness contours of the human ear.

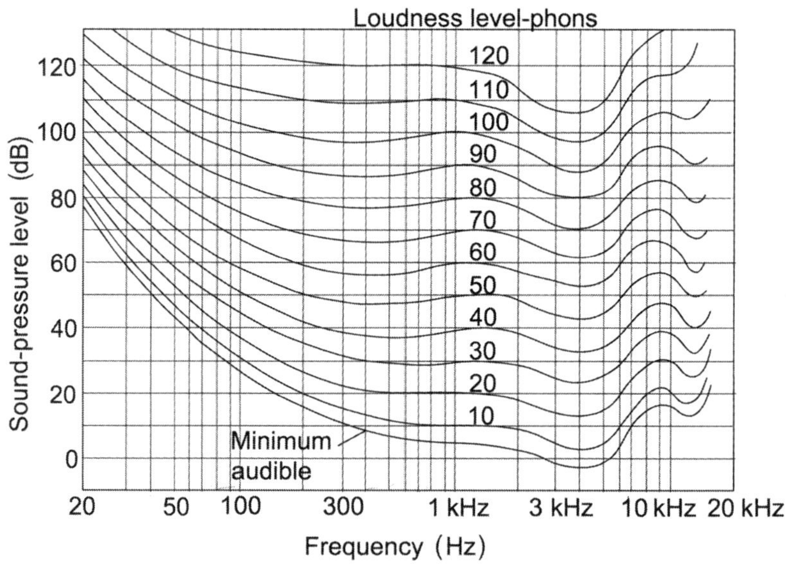

Fig. 4.17 Equal-loudness contours of the human ear.

The frequency response of the human ear can be defined in audible range as 16–20 kHz and the sensitive range as 1–6 kHz. The human ear is extremely sensitive in the 3- to 4-kHz zone [4-18].

4.6.2 Transmission Loss Spectrum: Glass versus Polycarbonate

An independent study was conducted through predictive engineering to evaluate the transmission loss performance of glass and PC. A uniform 4-mm thickness of tempered glass, laminated glass, and PC sheets were taken up for analysis. Airborne noise was taken as the source. The results are shown in Fig. 4.18.

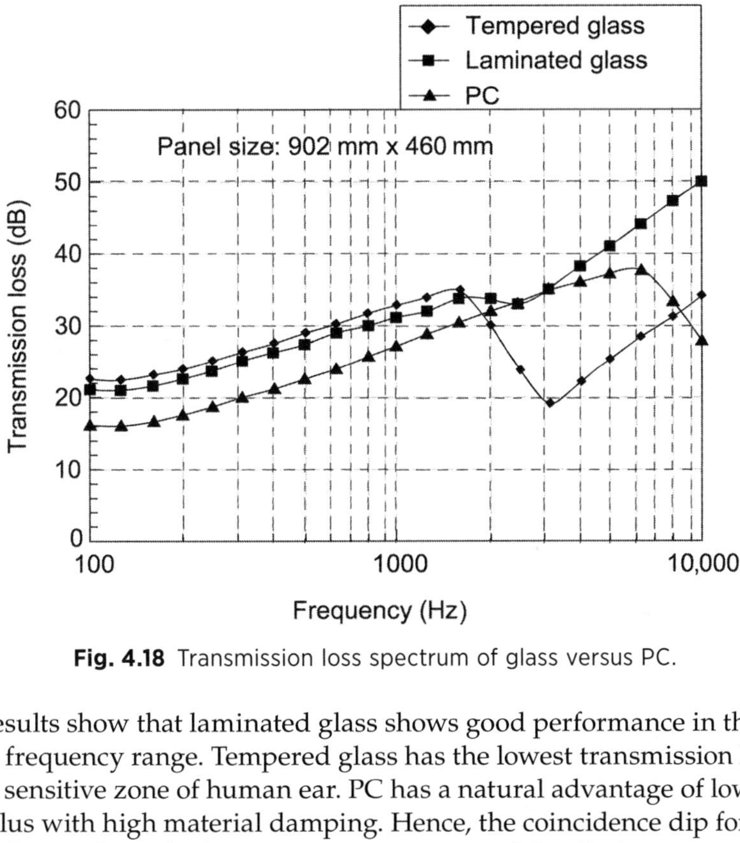

Fig. 4.18 Transmission loss spectrum of glass versus PC.

The results show that laminated glass shows good performance in the entire frequency range. Tempered glass has the lowest transmission loss in the sensitive zone of human ear. PC has a natural advantage of low modulus with high material damping. Hence, the coincidence dip for PC is pushed away from the human sensitive zone, and the dip is not as steep as for laminated glass.

4.6.3 Sound Transmission Loss Performance

Another experiment was conducted to validate the predictive engineering results. The experimental setup to measure the sound transmission loss is shown in Fig. 4.19.

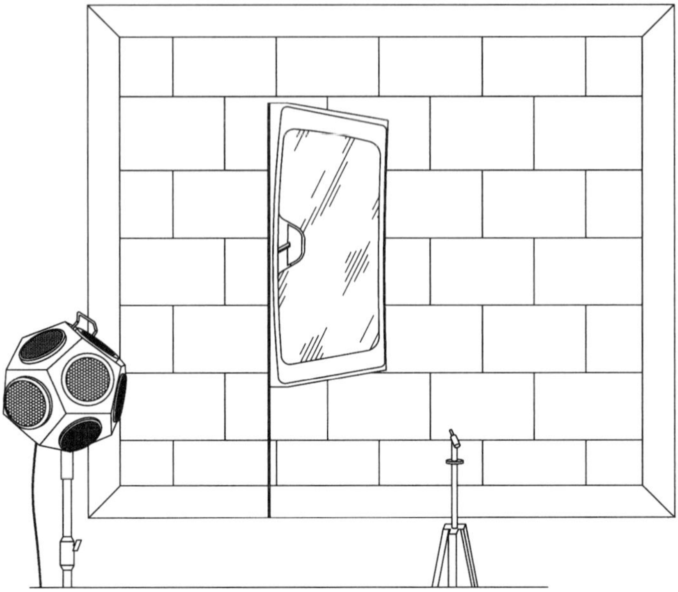

Fig. 4.19 Experimental setup for measurement of sound transmission loss.

The glazing panel is placed between a source and receiver. The glazing panel is insulated on both sides to avoid other sound loss. The experiment is conducted with a 4.2-mm thick glass and PC panel. The results are as shown in Fig. 4.20.

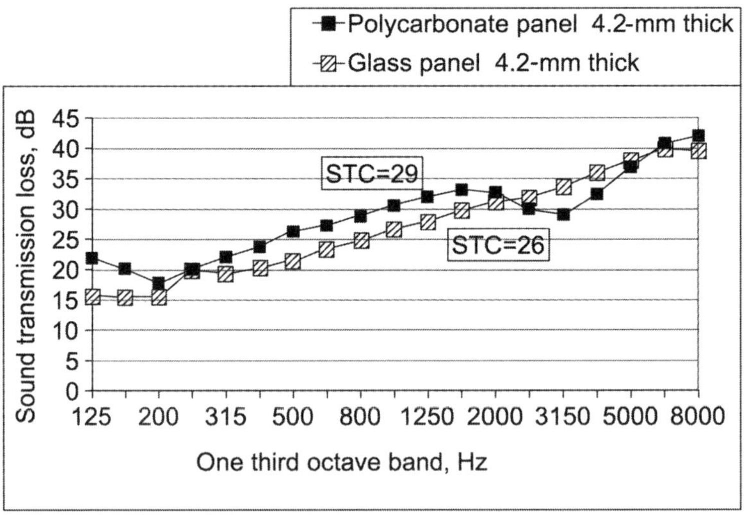

Fig. 4.20 Experimental results comparing sound transmission loss of glass versus PC.

In summary, at low frequencies interior noise is dominated by structure-borne noise, and the PC panel is only a secondary contributor. In the higher frequency range of 2000–8000 Hz, the better performance of PC can be noticed, comparable to that of glass. At lower frequencies, the sound transmission loss of toughened glass is higher than that of PC by around 4 dB. Hence, in the lower frequency range, PC performs better than glass, and in case of higher frequency, the performance is comparable to glass. Overall, the performance of PC will not cause any additional discomfort.

4.7 Thermal Management

Thermal comfort is an important requirement for occupants in an enclosed environment such as the passenger compartment of a vehicle. There has been significant interest on the assessment of air conditioning (A/C) efficiency and its impact on reducing greenhouse gas (GHG) emissions [4-19]. Various studies have concentrated on reducing heating, ventilation, and air conditioning (HVAC) loads through advanced lightweight insulation and window technologies. Through these technologies, fuel economy can be improved substantially without loss of vehicle performance or passenger comfort [4-20]. The heat buildup inside the cabin due to solar irradiation and ambient temperature is one of the factors to be considered during the design of an A/C system. Variation in thermal conductivity and IR transmittance of PC glazing panels compared to glass contribute to the variation of heat buildup inside the cabin. Heat transfer between the ambient air and the air inside a vehicle cabin typically determines the steady-state load on the vehicle's HVAC system. The HVAC load is a significant factor for vehicle efficiency and GHG emissions.

A study highlighting the effect of glazing (i.e., window) thermal conductivity during steady-state was carried out. This study considers the difference in thermal conductivity of a monolithic glass and standard PC [4-21]. Computational fluid dynamics (CFD) simulations were performed on a model car cabin including HVAC vents, interior seating, side windows, windshield, back-lite, and roof-lite. Two locations in United States representing hottest (Phoenix, Arizona) and coolest (Minneapolis, Minnesota) weathers were considered. Average monthly temperature and radiation data for Phoenix and Minneapolis were used to define hot and cold climate scenarios [4-22]. For gasoline-powered vehicles in a hot climate, reduced HVAC load translates to reduced load on the engine. This translates into improved fuel economy and reduced GHG emissions during operation of the HVAC system. As there are a good number of semitransparent glazing materials present, accurate thermal modeling of semitransparent materials becomes very important for accurate thermal predictions.

4.7.1 Thermal Modeling of Semitransparent Materials: Spectral Transmission and Absorption

Radiation through semitransparent materials is typically modeled by using a spectral transmission and absorption approach. Spectral transmission and absorption are calculated for a semitransparent material using Fresnel reflections at normal incidence, the exponential solution for extinction through the transparent layer, and the series solution for multiple internal reflections [4-23]. Fig. 4.21 shows the radiation paths through the first two passes.

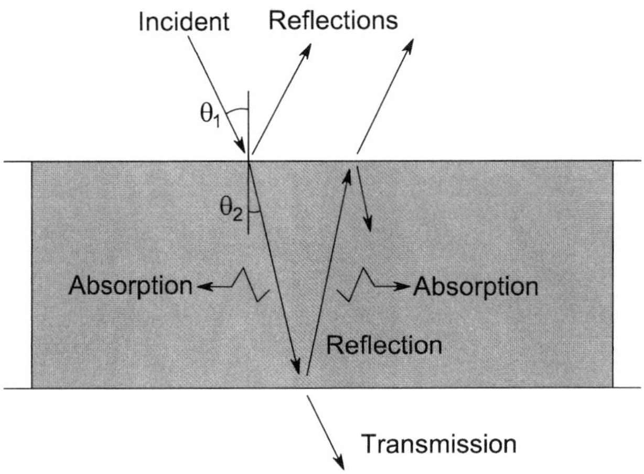

Fig. 4.21 Absorption during two passes of radiation through semitransparent material.

The calculation starts with the given values of index of refraction n, the coefficient of extinction κ, the thickness of the layer d, and the angle of incidence θ_1. Numerals 1 and 2 represent before and after the incidence. The refracted angle θ_2 is found from Snell's law [4-23]:

$$n_1 \sin\theta_1 = n_2 \sin\theta_2 \tag{4.2}$$

The reflectivity for a single reflection at the interface, assuming ideal dielectric medium, is found using Fresnel's formula [4-23]. The reflectivity in the perpendicular and parallel components of polarization, r_1 and r_2, is given by

$$r_1 = \frac{\sin^2(\theta_2 - \theta_1)}{\sin^2(\theta_2 + \theta_1)} \tag{4.3}$$

$$r_2 = \frac{\tan^2(\theta_2 - \theta_1)}{\tan^2(\theta_2 + \theta_1)} \tag{4.4}$$

It is assumed that extinction is through absorption only, without any scattering. The single-pass transmission through the window can be expressed as

$$\tau_a = e^{-\kappa d/\cos\theta_2} \tag{4.5}$$

where τ_a is the single-phase absorptivity.

The overall spectral transmission τ_i, reflection ρ_i, and absorption α_i, including multiple internal reflections, for each polarization (i = 1, 2), are [4-23]

$$\tau_i = \tau_a \frac{1-\rho_i}{1+\rho_i} \frac{1-\rho_i^2}{1+(\rho_i\tau_a)^2} \tag{4.6}$$

$$\rho_i = \rho_i(1 + \tau_a\tau_i) \tag{4.7}$$

$$\alpha_i = (1-\tau_a)\frac{1-\rho_i}{1-\rho_i\tau_a} \tag{4.8}$$

The overall properties are taken to be the average of the values for the two polarization components, assuming unpolarized incident light. To derive the absorption coefficient from the transmission data, equations 4.2 to 4.8 are applied to each wavelength for which data are available and the value of κ that is closer and which yields the smallest sum of errors, relative to the required transmission, reflection and absorption data. Through this process, the absorption coefficients are calculated based on the spectral transmission and refractive index values.

4.7.2 HVAC Load—Advantages of Polycarbonate

A generic geometric model of the car on which the CFD studies were performed is shown in Fig. 4.22. Studies were performed on this generic car model to compare the performance of monolithic glass and general PC as the material for the roof- and back-lite [4-21]. The cabin is enclosed by glazed and nonglazed areas. Heat transfer is assumed to be through convection and radiation for glazed areas and through convection for nonglazed areas. The three seats shown in Fig. 4.22 are modeled thermally as solid plywood with the same thermal conductivity as the nonglazed areas of the cabin envelope.

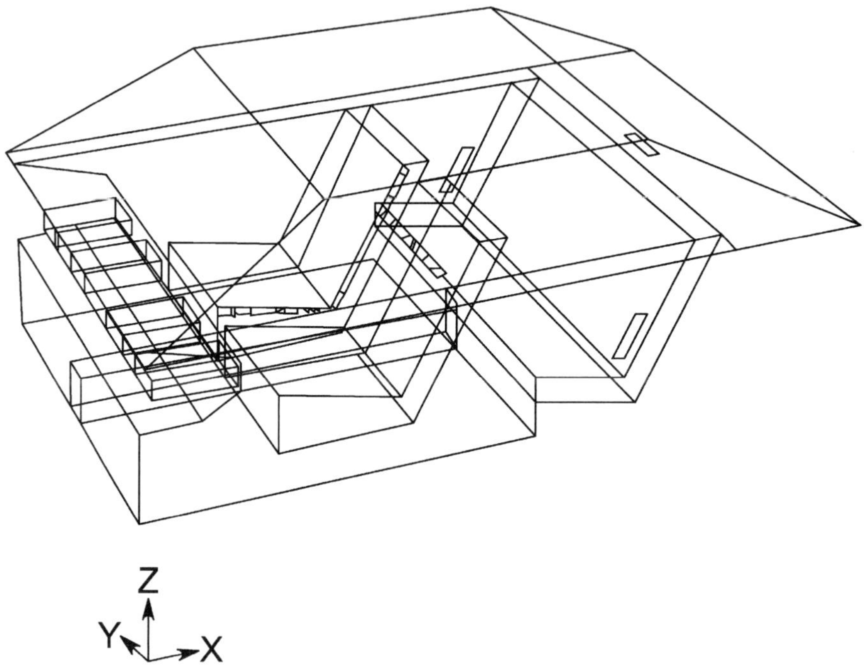

Fig. 4.22 Cabin of a model car.

External heat transfer coefficients used in the study are based on Reference [4-24]. These external heat transfer coefficients are applied to both glazed and nonglazed surfaces. Air from the HVAC system is assumed to enter the cabin through inlet vents facing the front seats and to exit through small holes at the rear of the cabin. The target set value for volume average cabin air temperature is 22 °C (71.6 °F). This temperature is achieved by adjusting the temperature of the air entering the cabin through the inlet vents. For each case, a trial inlet air temperature determines corresponding volume average cabin air temperature by CFD simulations. This average temperature is compared with its target value of 22 °C (71.6 °F), and the inlet air temperature is adjusted accordingly for the next simulation. This trial-and-error process involving successive simulations ultimately yields an inlet air temperature corresponding to the target cabin air temperature of 22 °C (71.6 °F).

Table 4.3 shows the results for total heat transfer through all surfaces of the cabin for different climate scenarios and different weather seasons [4-21].

Table 4.3 Nominal HVAC Load for Car Moving at 100 km/h

City	Phoenix	Minneapolis	Minneapolis
Months	April–October	June–August	October–April
HVAC mode	Cooling	Cooling	Heating
Glass at all locations (baseline)	1342 W	458 W	-1500 W
PC roof-lite and back-lite; glass at all other locations	1257 W	430 W	-1394 W
Magnitude change from baseline	-6.3%	-5.9%	-7.1%

This heat transfer is into the cabin in (positive values) for the first two scenarios. The heat transfer is out of the cabin in the third scenario (negative value). These heat transfers represent nominal cooling and heating loads, respectively, on the HVAC system. Each heat transfer (nominal load) value is an average of the corresponding city and months. The baseline load case from which percentage changes are calculated assumes glass at all glazing locations. In Table 4.4, the reduced loads resulting from the replacement of glass with PC at the back-lite and roof-lite are shown.

The respective percentage changes in magnitude from the baseline load under different climate and state-of-motion scenarios can also be seen in Table 4.4. All these loads reflect the heat transfer through the other glazing and nonglazing surfaces.

The percentage of reductions in nominal HVAC loads is in a narrow range (for either the moving car or the stationary car). The absolute load values vary widely given the range of ambient conditions from severe to mild (e.g., Minneapolis in summer).

Table 4.5 presents the steady-state average inside surface temperature of the roof-lite for each glazing material in two scenarios. This table also contains the respective ambient and average cabin air temperatures. Table

Table 4.4 Nominal HVAC Load for Stationary and Moving Cars

City	Phoenix		Minneapolis			
HVAC mode	Cooling		Heating			
Month	August		January		April	
Car speed	Stationary	100 km/h	Stationary	100 km/h	Stationary	100 km/h
Glass at all locations including roof-lite (baseline)	1361 W	1662 W	–1801 W	–2298 W	–500 W	–662 W
PC roof-lite and back-lite; glass at all other locations	1305 W	1555 W	–1721 W	–2136 W	–476 W	–612 W
Magnitude change from baseline	–4.1%	–6.4%	–4.5%	–7.0%	–4.9%	–7.7%

4.5 shows that in either a hot or cold climate, the average temperature at the inside surface of the glazing is closer to the volume average cabin air temperature for PC than for glass.

Table 4.5 Roof-Lite Average Inside Surface Temperature for a Car Moving at 100 km/h

City	Phoenix	Minneapolis
Month	July	January
HVAC mode	Cooling	Heating
Inside surface temperature—glass	37.6 °C	-0.8 °C
Inside surface temperature—PC	33.8 °C	5.0 °C
Volume average cabin temperature	22.0 °C	22.0 °C
Ambient temperature	40.6 °C	-5.2 °C

Thus, in the hot climate (Phoenix in July), the inside surface of PC glazing is cooler than glass. This translates to a more comfortable subjective experience for passengers. In the cold climate (Minneapolis in January), the inside surface of PC glazing is warmer than glass. This tends to reduce the occurrence for condensation on the inner surface of glazing.

The translation of HVAC load reductions into the lowering of GHG emissions depends on vehicle type (conventional or hybrid) and HVAC efficiency. The A/C efficiency of the car is based on the conventional car considered in ref. [4-25]. The baseline A/C fuel assumes 0.026 L/km (0.0109 gallon/mile). It is assumed that the percentage reduction in A/C fuel use is 0.87 times the percentage reduction in HVAC load. It is also assumed that there are 8,788 grams CO_2 emissions per gallon of fuel consumed [4-26]. Based on these assumptions, a 6.3% reduction in HVAC load in the Phoenix April–October scenario in Table 4.3 translates to an emissions reduction of 5.2 grams CO_2 per mile. (Total baseline emission for this car with A/C operating is 571 grams CO_2 per mile.) This would be the basis for reduced GHG emissions from mobile sources in cold climate due to reduced glazing thermal conductivity. In either a hot or cold climate, the GHG benefits from reduced thermal conductivity are in addition to the GHG benefits from weight reduction (50% lower mass density of PC relative to glass).

4.7.3 Improved Performance of Electric Vehicles

Some of the shortcomings of electric vehicles (EVs) include expensive and toxic batteries with limited life spans and limited driving ranges [4-27]. The range and battery lifetime affect both EV acceptance by vehicle owners and EV impact on the environment. A/C load is one of the biggest auxiliary loads on the battery of an EV [4-28]. The range of an EV can be improved by reducing the auxiliary load on the EV battery. Figure 4.23 shows the model cabin sectioned at its plane of symmetry with steady-state air velocity vectors, and Fig. 4.24 displays the temperature fields.

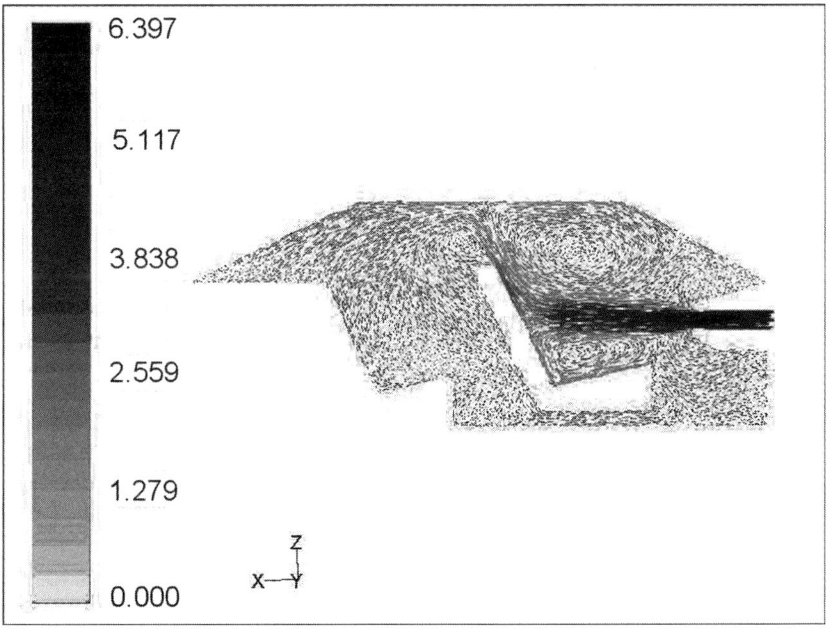

Fig. 4.23 Air velocity vectors in cabin plane of symmetry (m/s).

These profiles are for the model car cabin with a PC back-lite and roof-lite, moving at 100 km/h in Minneapolis in July with a mid-day ambient temperature of 27.28 °C (81.1 °F) [4-29].

Table 4.6 summarizes the results for a moving car under three climate scenarios.

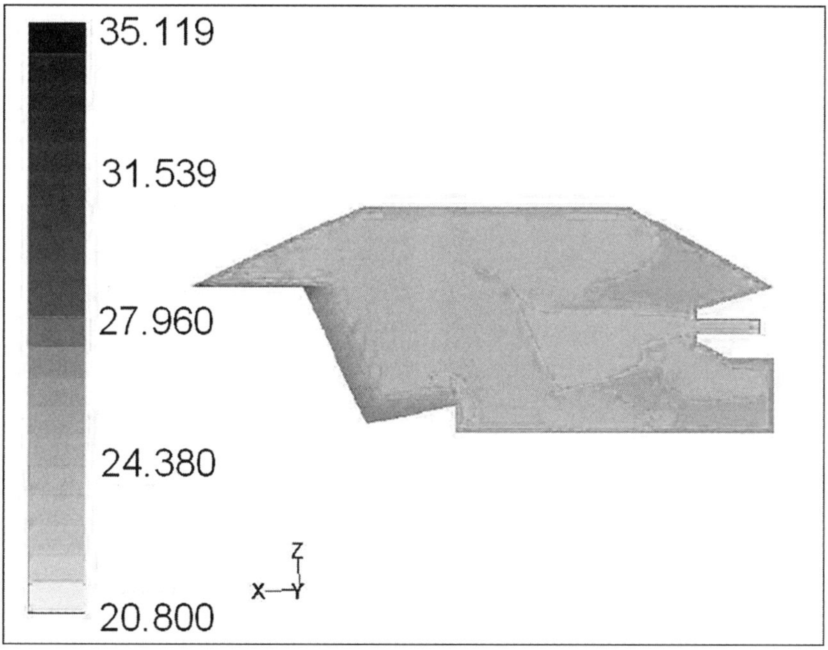

Fig. 4.24 Temperature distribution in cabin plane of symmetry (°C).

Table 4.6 Results for a Car Moving at 100 km/h in Three Climate Scenarios

City	Phoenix	Minneapolis	Minneapolis
Months over which parameters are averaged	**April–October**	**June–August**	**October–April**
HVAC mode	Cooling	Cooling	Heating
Nominal HVAC load (glass)	1342 W	458 W	1500 W
Nominal HVAC load (PC)	1257 W	430 W	1394 W
Change in HVAC load	-6.30%	-5.90%	-7.10%
Effective base load	4000 W	4000 W	4000 W
Change in single-cycle discharge time	2.10%	0.80%	2.60%
Change in battery cycle life	0.05%	0.20%	0.60%
Change in battery lifetime	2.60%	1.00%	3.20%

Table 4.7. compares stationary and moving cars under hot and cold climate scenarios.

Table 4.7 Results for Stationary and Moving Cars in Two Climate Scenarios

City	Phoenix		Minneapolis	
Month	August		January	
HVAC mode	Cooling		Heating	
Car speed	Stationary	100 km/h	Stationary	100 km/h
Nominal HVAC load (glass)	1361 W	1662 W	1801 W	2298 W
Nominal HVAC load (PC)	1305 W	1555 W	1721 W	2136 W
Change in HVAC load	-4.10%	-6.40%	-4.50%	-7.00%
Effective base load	0 W	4000 W	0 W	4000 W
Change in single-cycle discharge time	5.60%	2.50%	6.10%	3.40%

The effective base load is assumed to be at 4000 W for moving cars [4-28]. Change in single-cycle discharge time of 2.1% as a result of the change in HVAC load (–6.3%) is observed. For moving cars, the percent change in EV range is calculated with the percent change in single-cycle discharge time.

Tables 4.6 and 4.7 indicate improvements of using a PC back-lite and roof-lite in measurement of battery performance in all climate and state-of-motion scenarios. Table 4.7 includes only the change in single-cycle discharge time, as the cycle life and lifetime are not practical battery performance measures for stationary vehicles. Single-cycle discharge time for stationary vehicles would be relevant to electric-drive buses. These buses must maintain comfortable cabin temperatures while idling for extended periods. Table 4.7 shows that improvements in single-cycle discharge times are higher for a stationary car than for a moving car. This is mainly due to the absence of load for propulsion and non-HVAC accessories. Results in Tables 4.6 and 4.7 indicate a beneficial role of reduced glazing thermal conductivity in EV battery performance. Replacing a glass back-lite and roof-lite with PC (which has a fivefold lower inherent thermal conductivity) reduces the HVAC load. This translates into a reduced EV battery discharge rate and extended EV range and battery lifetime by around 2–3% in both hot and cold climates. PC's lower mass density relative to glass would extend EV range and lifetime even further.

4.7.4 Soak Performance of Polycarbonate Glazing

During hot weather conditions when a car is parked for an extended period with A/C off and all windows and doors are closed, the car develops elevated air temperature in the cabin compared to the ambient temperature. The settled temperature during these conditions is known as soak temperature. The soak temperature has received attention in the regulatory arena and as a factor to be considered with regard to various glazing technologies [4-25].

Reduced glazing thermal conductivity tends to reduce the HVAC load and hence has a beneficial effect in the steady-state, when the A/C is on [4-21]. At the same time, one needs to consider the possibility of an offsetting effect on soak temperature due to the reduced thermal conductivity. For exterior surfaces of the cabin envelope, the free stream temperatures and heat transfer coefficients are equal to all surfaces directly exposed to ambient air and solar radiation. For these surfaces, the free stream temperature is the prevailing ambient temperature. The heat transfer coefficient is based on a wind speed of 6.44 km/s (4 miles/s) [4-22]. For other exterior cabin surfaces, such as the engine compartment, trunk, and pavement, the free stream temperatures and heat transfer coefficients are generally different. Such differences and the resulting surface parameters also tend to vary among actual cars in their real settings. Thus, various cases are studied to test the sensitivity of CFD results to these parameters [4-30], as shown in Table 4.8

During the soak condition, radiative heat transfer into the cabin is balanced by convective heat transfer out of the cabin. Temperature dependence of air density is modeled, and this enables natural convection of the cabin air. Moisture level in the cabin air is considered low enough that its influence on air transport properties can be neglected. Turbulent flow is assumed to be for all simulations based on Rayleigh number [4-31]. The Discrete Ordinates Radiation model within Fluent [4-32] is applied. The mass average air temperature, along with the temperature at various locations, is monitored.

Table 4.8 Heat Transfer Parameters

Case	A	B	C	D,E,F
Interior, nonglazing elements:				
Emissivity	0.7	0.9	0.5	0.9
Diffuse fraction (0 = specular to 1 = diffuse reaction	0.8	0.9	0.5	0.9
Exterior, free steam temperature (°C):				
Surfaces directly exposed to ambient	39	39	39	39
Surfaces facing pavement	39	39	44	44
Surfaces facing engine compartment	39	39	54	54
Surfaces facing trunk	39	39	49	49
Exterior, Heat transfer coefficient (W/m^2K):				
Surfaces directly exposed to ambient	21	21	21	21
Surfaces facing pavement	21	21	21	21
Surfaces facing engine compartment	21	21	2.6	2.6
Surfaces facing trunk	21	21	3.2	3.2

Figure 4.25 shows the temperature distribution at cabin symmetry section during soak condition.

Figure 4.26 shows the temperature profile at a section between one end and the symmetry section.

The relatively wide temperature scale in Fig. 4.26 reflects the presence of the relatively warm front seat. That seat is absent from Fig. 4.25, allowing a relatively narrow temperature scale and a relatively high contrast between temperature contours. Figure 4.27 shows the airflow field for the same case and plane as in Fig. 4.25.

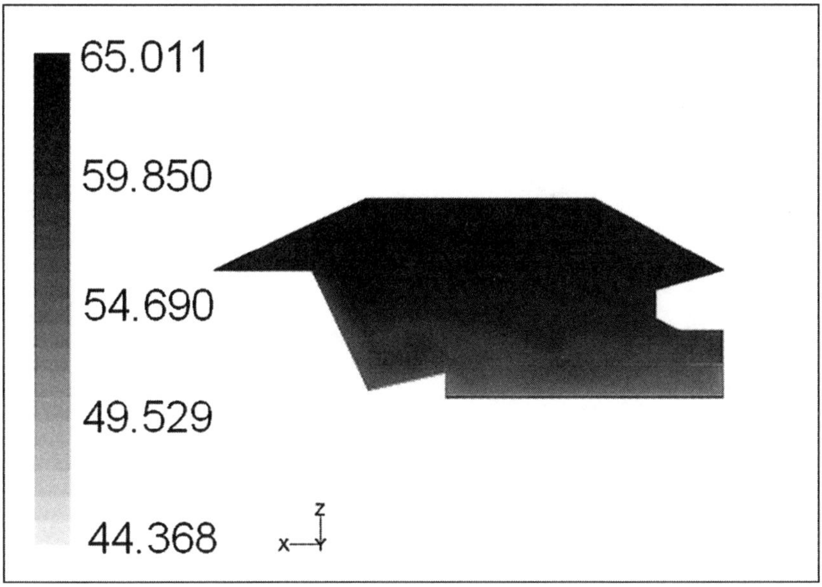

Fig. 4.25 Temperature contours in cabin symmetry plane 4 (°C).

Fig. 4.26 Temperature contours in plane midway between symmetry plane and right side of cabin (°C).

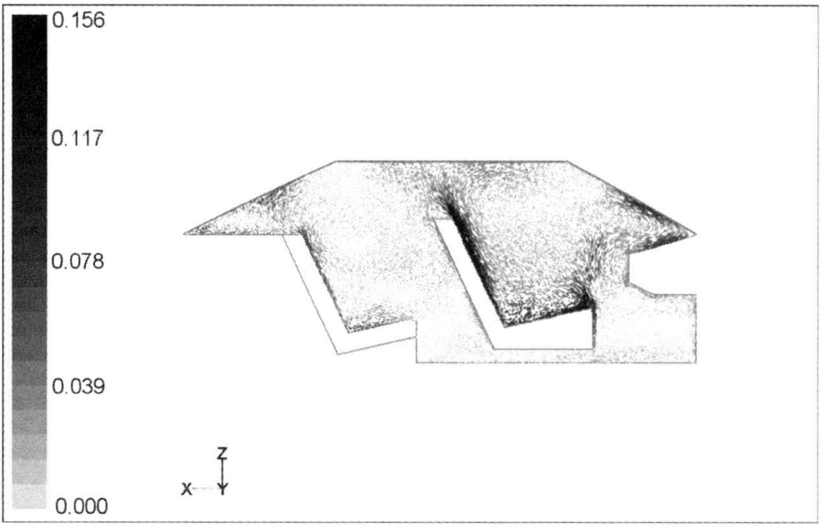

Fig. 4.27 Airflow field in plane midway between symmetry plane and right side of cabin (m/s).

Together, these figures illustrate the role of the relatively warm seats in driving steady natural convection. Recirculation patterns in the front and rear of the cabin appear in Fig. 4.27 to be coupled in the space above the backrest of the front seat.

CFD results for the baseline glazing configuration are presented in Table 4.9.

In addition to the mass average cabin air temperature (soak temperature), results also include mass average air temperatures. Air temperatures representing air-breath and air-foot are also monitored. Radiative heat transfer through the cabin envelope is the sum of the radiative heat transfers through glazing and nonglazing elements of the cabin envelop. The same is applicable for convective heat transfer as well.

For all these cases, the mass average cabin air temperatures remain within a narrow range relative to the temperatures themselves, or even to their elevation over the ambient (39 °C [102.2 °F]). This is most significant for cases C and D, which represent extremes of emissivity and diffuse fraction and combined with common exterior heat transfer parameters. Thus, the sensitivity of mass average cabin air temperature (soak temperature) to the heat transfer parameters in Table 4.9 is very low.

Table 4.9 Results for Baseline Glazing Configuration

Case	A	B	C	D
Back-lite, roof-lite, windshield, side windows	Glass	Glass	Glass	Glass
Mass average cabin air temperature (°C)	55.17	55.83	56.26	58.80
Air-breath temperature, front (°C)	61.9	61.96	62.5	63.2
Air-foot temperature, front (°C)	41.7	41.70	46.2	46.5
Air-breath temperature, rear (°C)	60.9	31.14	61.4	62.5
Air-foot temperature, rear (°C)	44.9	44.36	48.3	50.0
Radiative heat transfer through all glazing (W)	255.87	327.90	145.42	326.34
Radiative heat transfer through all nonglazing (W)	-185.69	-244.32	-101.14	-243.31
Radiative heat transfer through cabin envelope (W)	70.18	83.58	44.28	83.03
Convective heat transfer through all glazing (W)	13.18	-16.77	31.83	-68.87
Convective heat transfer through all nonglazing (W)	-83.36	-66.81	-76.11	-14.16
Convective heat transfer through cabin envelope (W)	-70.18	-83.58	-44.28	-83.03
Roof-lite, average temperature interior (°C)	67.0	67.7	68.0	67.4
Roof-lite, average temperature exterior (°C)	66.6	66.3	67.5	66.8

Table 4.10 indicates a mass average cabin air temperature for case E (reduced glazing thermal conductivity) within 1 °C (33.8 °F) of the comparable temperature for case D (baseline glazing). This temperature difference is smaller than the range in mass average cabin air temperature generated by the variation of heat transfer parameters in Table 4.10.

The practically negligible effect of glazing thermal conductivity on mass average cabin air temperature is attributable to the relatively quiescent air at the interior and exterior glazing surfaces under soak conditions. Hence transfer coefficients are low at those surfaces, which tend to lower the sensitivity of the heat transfer to the glazing conductivity.

The simulated soak states for these two glazing configurations were compared for the interior and exterior surface parameters that maximized convective heat transfer (including conduction) through the glazing. The environment for all simulations was that of Phoenix at midday in June, which represents the maximum solar radiation level at that location. This tends to provide the maximum temperature difference across the glazing (between the cabin and free stream). These conditions were chosen to provide a decisive test of the effect of glazing thermal conductivity. The soak temperature for the reduced thermal conductivity configuration was higher by <1 °C (33.8 °F); air-breath temperatures (front and rear) were higher by about 2 °C (35.6 °F), while air-foot temperatures were practically unchanged.

CFD simulations of a model car cabin showed that the lower thermal conductivity of PC glazing tends to reduce HVAC load, which in turn reduce GHG emissions. This reduction was observed for both hot and cold climate conditions in stationary and moving vehicles. The lower thermal conductivity also translates into a reduced EV battery discharge rate, extending EV range and battery lifetime by 2–3% in both hot and cold climates. The CFD study during soak condition shows that the offsetting effect due to the lower thermal conductivity of PC when compared to glass is also very minimal, with a difference of about 2 °C (35.6 °F).

4.8 Conversion Process

The major advantage of using PC glazing is the ability to mold it into any complex form, as explained earlier. Thermoforming of large transparent PC sheets can be used for prototyping and low-volume production. However, for mass production of a glazing system, either injection compression molding (ICM) or two-shot injection compression molding (2K-ICM) is employed rather than the conventional injection molding

Table 4.10 Results Comparing Glazing Configurations			
Case	**A**	**B**	**C**
Back-lite	Glass	PC	Glass
Roof-lite	Glass	PC	None
Windshield and side windows	Glass	Glass	Glass
Mass average cabin air temperature (°C)	58.82	59.71	52.35
Air-breath temperature, front (°C)	63.2	65.1	51.7
Air-foot temperature, front (°C)	46.5	46.6	45.6
Air-breath temperature, rear (°C)	62.5	64.7	51.7
Air-foot temperature, rear (°C)	50.0	50.0	48.6
Radiative heat transfer through all glazing (W)	326.34	343.40	-
Radiative heat transfer through all nonglazing (W)	−243.31	−254.40	-
Radiative heat transfer through cabin envelope (W)	83.03	88.81	-
Convective heat transfer through all glazing (W)	−68.87	−71.83	-
Convective heat transfer through all nonglazing (W)	−14.16	−16.98	-
Convective heat transfer through cabin envelope (W)	−83.03	−88.81	-
Back-lite, average temperature interior (°C)	58.8	61.6	-
Back-lite, average temperature exterior (°C)	58.1	56.9	-
Back-lite, average heat transfer coefficient, interior (W/m^2K)	3.63	3.85	-
Roof-lite, average temperature exterior (°C)	67.4	70.2	-
Roof-lite, average temperature interior (°C)	66.8	66.5	-
Back-lite, average heat transfer coefficient, interior (W/m^2K)	2.51	2.42	-

process. Given that these panels are large, regular injection molding imparts a high amount of residual stresses, which in turn affects the optical quality, and it requires high clamp force, thus increasing the machine size and cost. These can be overcome by using the ICM or 2K-ICM process.

4.8.1 Two-Shot Injection Compression Molding

Usually, a glazing panel is manufactured using the 2K-ICM process. The process comprises of two shots: The first shot, which is transparent, uses the injection compression process, and the second shot uses the normal injection molding process to mold over the first shot to form the blackout area. The molded glazing panel (tailgate window) using the 2K-ICM process is shown in Fig. 4.28.

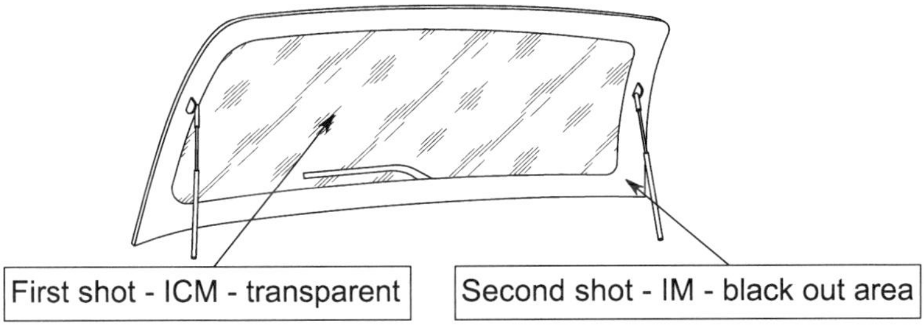

Fig. 4.28 Glazing panel manufactured using the 2K-ICM process.

4.8.2 First-Shot Injection Compression Molding

As these panels are huge, with sizes approximately ranging from 0.5 to 1.5 m^2, conventional injection molding is not the preferred conversion process, as it involves higher pressures and higher clamp forces. Because of the requirement for high optical quality, the internal stresses have to be less in the panel to avoid birefringence patterns when exposed. But injection molding induces higher stress as a result of high pressure, and hence is not a suitable process. Also, given the higher projected area it requires, higher clamp force and hence bigger machine size are needed, resulting in higher cost. All these limitations can be overcome with the injection compression process.

Injection compression process involves two phases as shown in Fig. 4.29. The first phase is the injection phase, during which the tool is kept open to a set limit and a definite volume of material is injected.

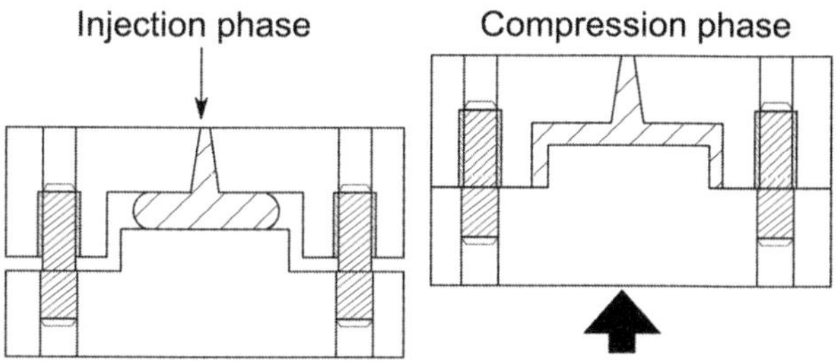

Fig. 4.29 Injection compression process.

If the intended thickness of the final part is 4 mm, and if the tool is open to a set limit of 1 mm, and while flow happens, it assumes the total thickness to be 5 mm, the pressure is reduced compared to 4 mm. During the second phase, the injected material is then compressed to make the tool completely closed, during which the material flows through the rest of part to fill it completely [4-33].

Since the pressure required is significantly lower than the conventional injection molding process, the stresses induced in the part are lower, as shown in Fig. 4.30.

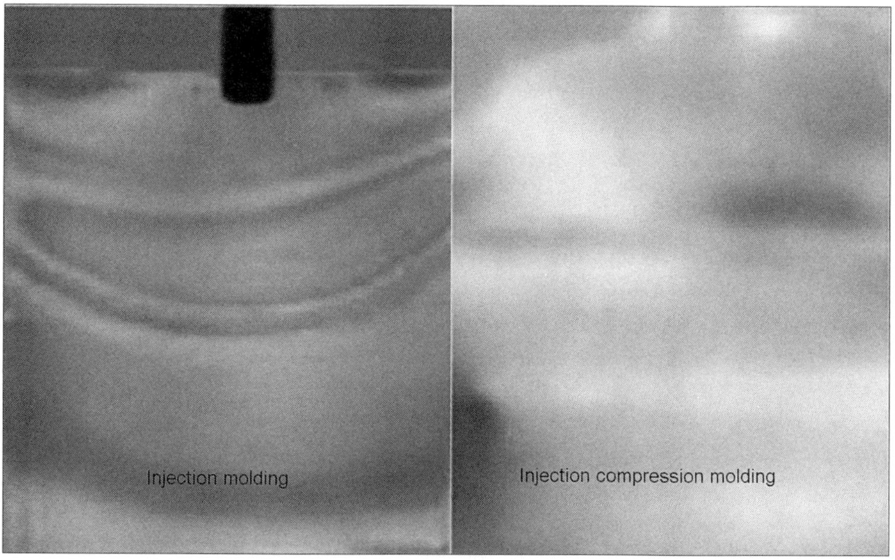

Fig. 4.30 Comparison of residual stresses.

Because the projected area remains the same, the reduction of pressure in turn reduces the required clamp force; hence, these parts can be molded with lower clamp force. A comparison of reduction of clamp force with the use of ICM can be seen from Fig. 4.31.

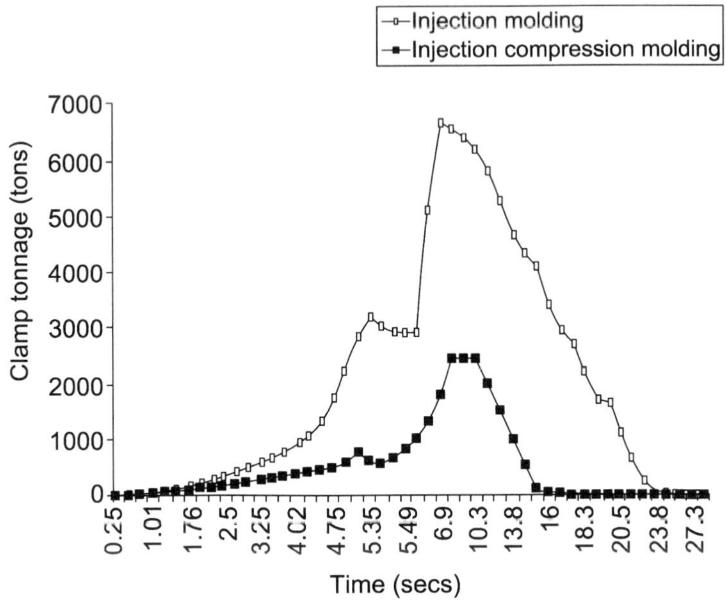

Fig. 4.31 Comparison of clamp force.

There are different types of ICM that are in vogue in the industry. The details are described in the next sections.

4.8.3 Sequential Injection Compression Molding

In the sequential process, the tool is kept open during injection and the compression starts after the injection is complete, as shown in Fig. 4.32.

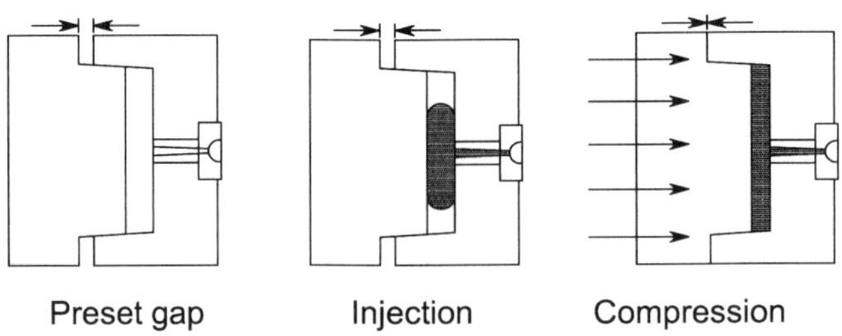

Fig. 4.32 Sequential ICM.

4.8.4 Simultaneous Injection Compression Molding

In the simultaneous process, the tool is kept open during injection and the compression starts during the injection is in progress, as shown in Fig. 4.33.

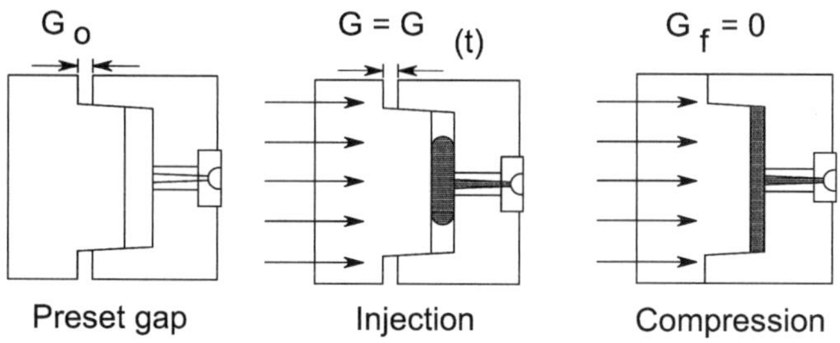

Fig. 4.33 Simultaneous ICM.

4.8.5 Breathing Injection Compression Molding

In the breathing process, the tool is kept closed during injection, but the clamp force is not applied. The tool moves during injection because of the injection pressure of the melt, but it can be controlled by clamp force. After it reaches the set gap, the compression starts to fill the part completely, as shown in Fig. 4.34.

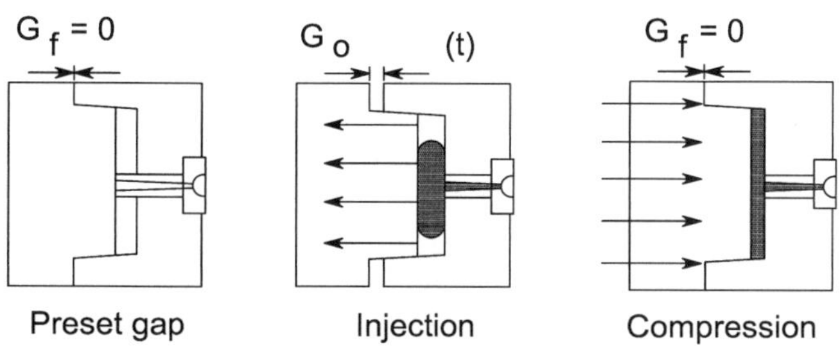

Fig. 4.34 Breathing ICM.

4.8.6 Second-Shot Injection Overmolding Process

The second shot is a normal injection molding process, except that the machine setup is different from the conventional injection molding machine, as shown in Fig. 4.35.

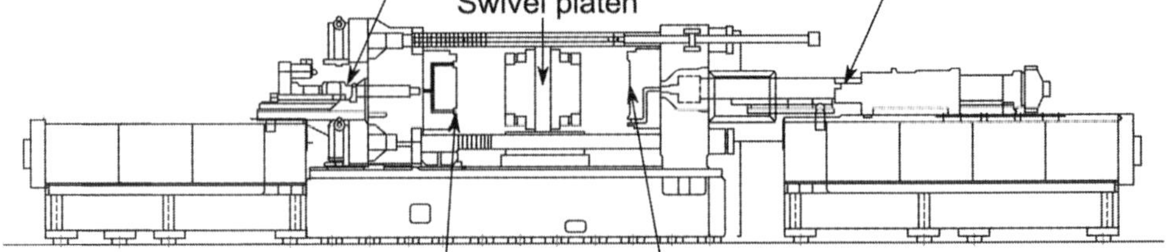

Fig. 4.35 2K-ICM machine setup.

The machine consists of two injection units placed exactly opposite to each other, and the clamping unit has an additional provision for swivel platen, which can rotate over either the horizontal or vertical axis. After the first shot is molded through ICM, the tool is opened and the part is retained in the cavity side of the tool. The swivel platen rotates either in either the vertical axis or horizontal axis. The first shot acts as a core for the second shot, then the tool is closed and the second shot is injected and cooled. During this process, simultaneously the ICM of first shot also happens. The completed two-shot molding is ejected, and the first shot is ready to rotate. In this way the whole cycle is repeated, thus improving productivity. The complete process can be represented as shown in Fig. 4.36.

4.8.7 Prediction Methodology of Two-Shot Injection Compression Molding Process

Prediction of this complex process becomes a key to a priori understanding the design and process-related issues so as to develop a successful glazing part. However, this represents a challenge for part designers and tool designers, since off-the-shelf injection molding simulation capabilities are not now available for these two processes in a combined molding cycle. Commonly available simulation codes [4-34] are capable of either predicting the ICM process or the 2K process in isolation, but not both the processes together for molding a single part, such as an automotive panoramic roof.

4.8.8 Part and Tool Development

A representative part of an panoramic roof, as shown in Fig. 4.37, was designed, and a 2K-ICM tool was manufactured to study the processing and performance parameters.

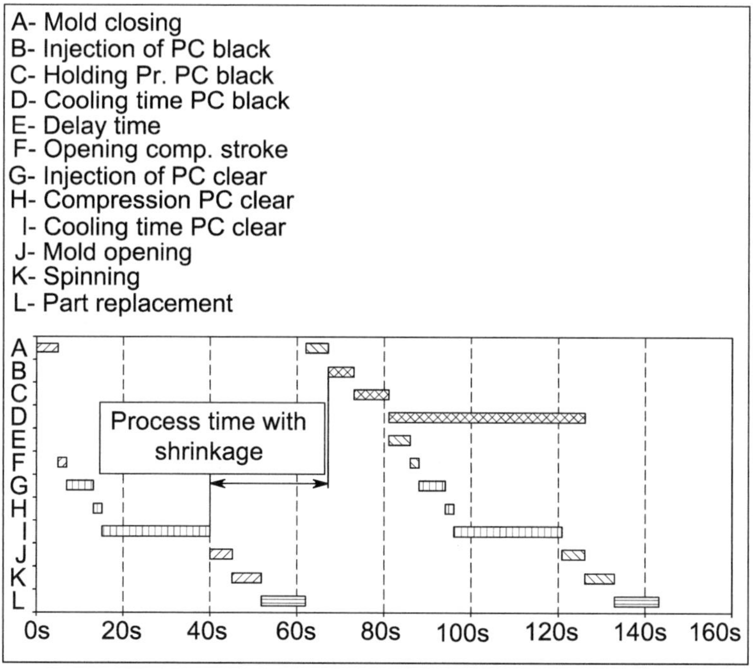

Fig. 4.36 Cycle of 2K-ICM process.

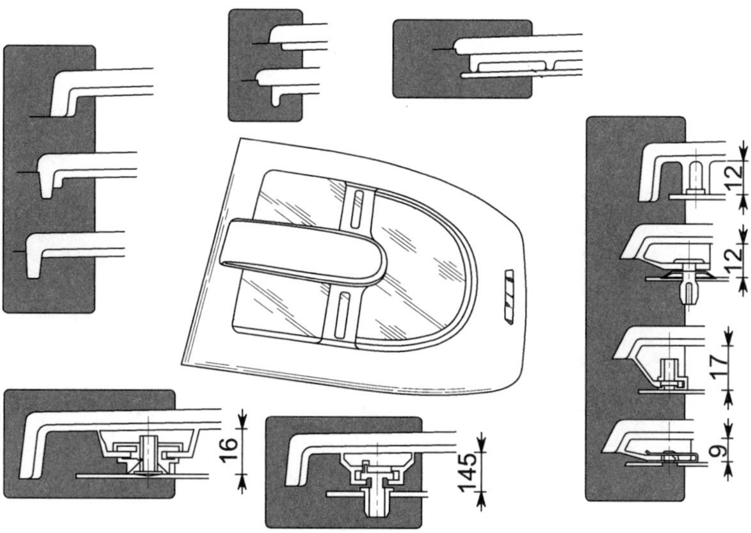

Fig. 4.37 Features of panoramic roof part.

The part has dimensions of half the size of the actual panoramic roof with an approximate area of 0.5 m^2. The part design has multiple edge designs (transition zones) to understand the window effect. Window effect is the distortion of light when it falls at an angle. This effect is particularly noticed in the transition area where the first and second shots merge. These different designs will help to give a suggestion for the best transition design for the customers. Apart from the transition design, the part also has different types of attachment designs, though glazing parts are traditionally attached to the BIW by adhesives or mechanical means, based on the requirements of OEMs. This design feature was also included in the part design, as shown in Fig. 4.37, to study these variations.

4.8.9 Filling Correlation

The approach adopted to predict the combined cycle of 2K-ICM is shown in Fig. 4.38.

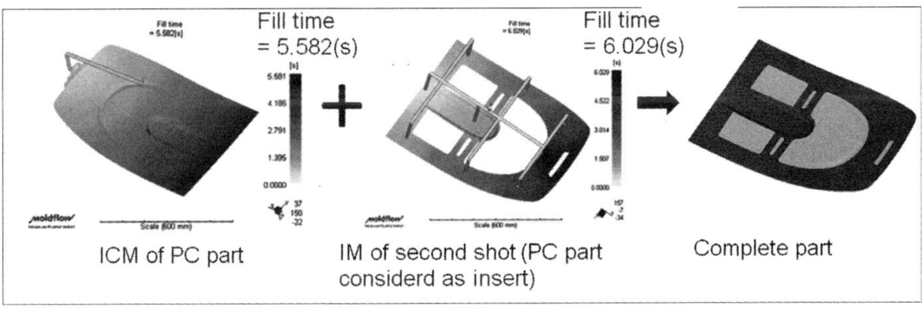

Fig. 4.38 Approach for prediction of 2K-ICM process.

In this approach the two processes are considered as two simultaneous processes. Then the ICM is carried out as a separate process, and the second shot is carried out as an independent injection molding process by considering the first shot as a thermoplastic insert with PC as the material at elevated temperature of 70 °C (158 °F). This approach has been found to be adequate enough to address the filling behavior of the process.

This methodology was used to correlate the filling behavior with real-time molding experiments, and good correlation was achieved, as shown in Fig. 4.39. This helps to build confidence for designing complex parts as well as manufacturing them [4-35].

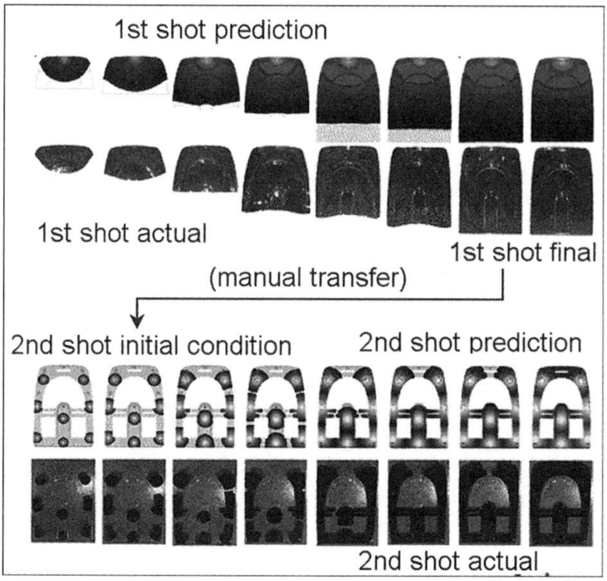

Fig. 4.39 Correlation of 2K-ICM process with actual molding.

4.8.10 Warpage Methodology Development

Warpage is a phenomenon of deflection of the part after molding as a result of the release of internal stresses developed during molding. There are several factors contributing to warpage, such as differential shrinkage, differential orientation, and differential cooling. It is indeed a challenge to predict the combined cycle warpage for such a complex process, in the absence of any established numerical tool. While the filling behavior can be modeled as two separate processes, as explained earlier, the same approach cannot be extended to warpage prediction, as it is a complex phenomenon of combined stresses arising out of both the molding process. Hence, it becomes critical to develop a methodology to predict the combined cycle warpage of the 2K-ICM process to understand the issues upfront before the tool cutting to reduce the rework and cost [4-36].

4.8.11 Measurement Setup

To compare with the predicted warpage values, actual measurement of warpage was done with a 2K-ICM roof part. The parts were molded using PC as both the first-shot and second-shot material. These parts were then measured using the setup shown in Fig. 4.40.

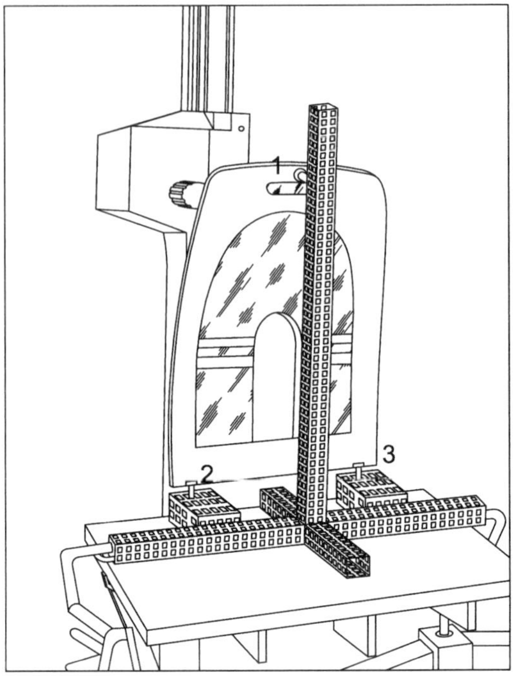

Fig. 4.40 Measurement setup.

The process of measurement was to scan the part to generate point clouds using laser scanning, and these point clouds were meshed to generate a surface. This surface was then superimposed with the original CAD model to measure the difference as shown in Fig. 4.41. The difference was given as a contour plot with minimum and maximum deflections from the original CAD coordinates, as shown in Fig. 4.41.

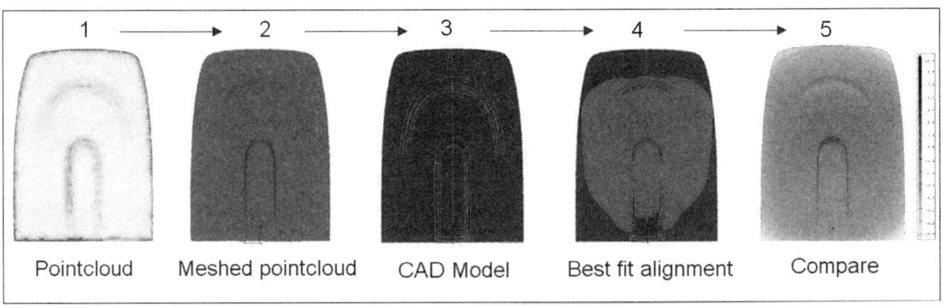

Fig. 4.41 Measurement using CAD tool.

The objective of this warpage prediction methodology was to correlate with the actual measured data, as shown in Fig. 4.42.

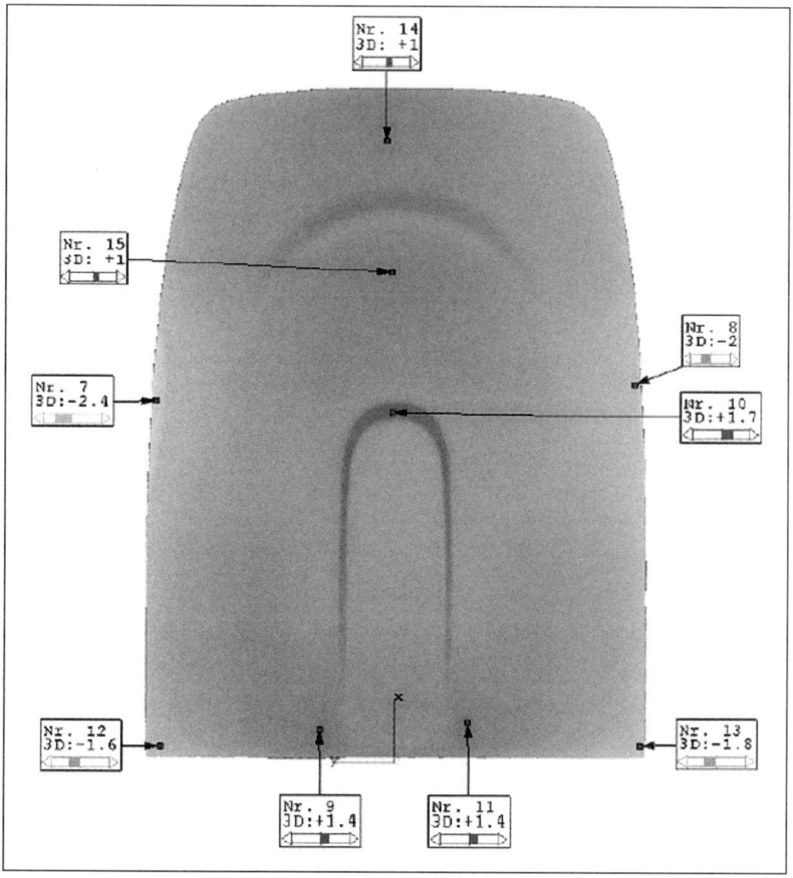

Fig. 4.42 Contour plot of actual measurements.

4.8.12 Approach

The basic concept of the approaches can be summarized as shown in Fig. 4.43.

To balance the software limitations and capture the history of the process, the first-shot molding cycle (e.g., residual stress and shrinkage) has to be combined with the second-shot history to find the combined cycle warpage. Using this methodology, three approaches were found, as shown in Fig. 4.44.

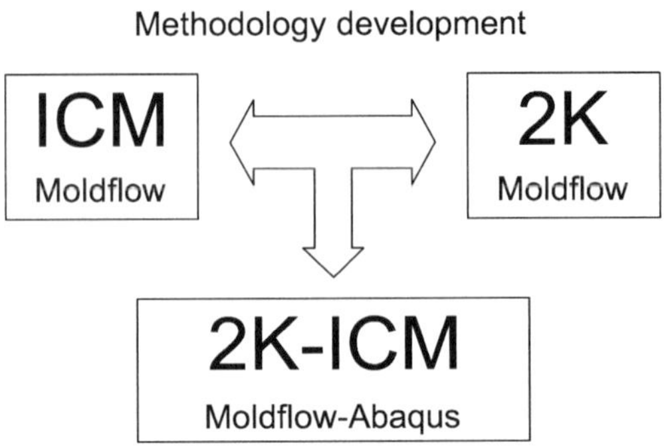

Fig. 4.43 Warpage prediction methodology.

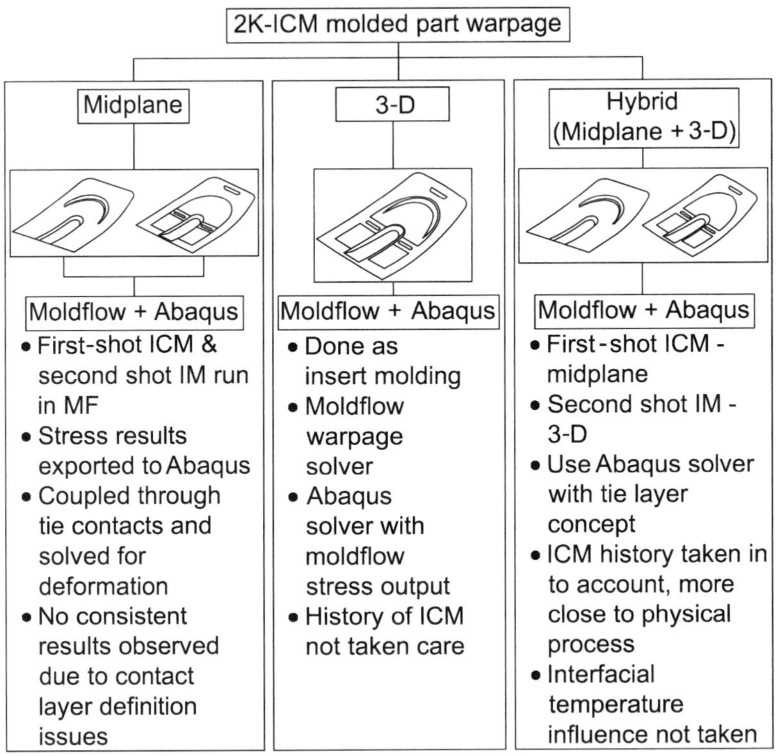

Fig. 4.44 Approaches for 2K-ICM warpage prediction.

The first approach is to combine the results of the first-shot and second-shot runs on the shell mesh inflow solver and solve for deformation in the structural solver by assuming a suitable coupling mechanism. The second approach is to solve using 3-D elements in a flow solver; however, the limitation of this approach is that the history of ICM cannot be accounted for, because of the software limitation that ICM is not available in 3-D mesh. The third approach is to use shell elements for ICM from a flow solver and 3-D elements for the second shot from the flow solver, then couple them in structural solver to calculate deformation.

The results of these three approaches as compared to the actual measurements are shown in Table 4.11.

Table 4.11 Comparison of Results

	Lexan-lexan deflection (mm)		
Analysis method	**Max**	**Min**	**Total**
Actual	1.7	-2.4	4.1
Midplane	-2.14	0.96	3.1
3-D with first shot as insert	-2.21	0.82	3.03
Hybrid-midplane ICM + 3-D second shot	-2.48	0.8	3.28

It can be observed that the hybrid approach of combining the first-shot stress of ICM in shell elements with the second-shot stress in 3-D gives a closer correlation with the actual molding. Hence, it can be concluded that this methodology can be applied to analyze the trend of warpage by varying processing conditions, gating sequences, and the area of the second shot to arrive at an optimum design with less warpage. One important factor—the interfacial temperature effect between the first shot and second shot—is not considered. Hence, while this approach is not complete in capturing the physical phenomena, advances in simulation techniques will help to improve the accuracy in future.

4.9 Summary

PC glazing in the automotive industry provides many benefits such as weight reduction, integration opportunities, and styling freedom as compared to glass. However, glass is a well-proven material that meets the regulatory requirements. Hence, before replacing glass with PC, one should consider important characteristics, such as thermal, structural, and vibrational behavior. Advancements in coating technologies, like plasma coating, help to achieve glasslike abrasion resistance and weatherability. Several advanced CAD and CAE tools are available to evaluate the new designs before they are implemented in the actual automobile. Moreover, advanced manufacturing methods such as the 2K-ICM process are now being used to implement the design.

4.10 Trends

Styling trends and integrated functions will continue to drive innovations in automotive glazing. Currently designers are working to implement several features into the glazing system to add value. One such feature is the use of fractal antennas that can be molded integrally molded with the PC glazing, so that the antennas do not have to protrude out of the car. Another feature is integration of solar films or cells with the glazing to generate electricity. Also, the trend will move toward more visible areas, meaning bigger curved parts, larger panoramic roofs in future automobiles for weight reduction, integration opportunities, and styling freedom as compared to glass [4-37]. PC would also be a material of choice apart from glass where integration opportunities are possible with additional advantage of less mass and has the potential to be a preferred material of choice in future, necessitated by the forthcoming stringent fuel efficiency and reduced GHG emission regulations.

4.11 References

4-1. "ANSI/SAE Z26.1-1996," American National Standard for Safety Glazing Materials for Glazing Motor Vehicles and Motor Vehicle Equipment Operating on Land Highways—Safety Standard, http://books.sae.org/book-z26.1/96. Accessed 22 October 2012.

4-2. "ECE Regulation 43 for Laminated Windscreens," Vehicle Regulations home page, http://www.unece.org/trans/main/welcwp29.html. Accessed March 6, 2012.

4-3. "HIC Criteria," http://www.nhtsa.gov/DOT/NHTSA/NRD/Multimedia/PDFs/Crashworthiness/Air%20Bags/rev_criteria.pdf. Accessed 22 October 2012.

4-4. "Polycarbonate," Wikipedia, http://en.wikipedia.org/wiki/Polycarbonate. Accessed March 6, 2012.

4-5. "Silicone Hard Coating," http://203.158.253.140/media/e-Book/Engineer/Maintenance/Coating%20Technology%20handbook/DK4036ch93.pdf. Accessed March 6, 2012.

4-6. "Plasma Coating," http://files.hanser.de/hanser/docs/20071031_2713115264-32_978-3-446-41120-3_Extract%201.pdf. Accessed March 6, 2012.

4-7. Krois, M., Dilger, K., Bohm, S. and Koch, S. 2003. "Use of Compliant Adhesive Layers in Direct Glazing of Road Vehicles for Improved Sound Attenuation." *International Journal of Adhesion and Adhesives* 23(5): 413–425.

4-8. Anand, E.S., Marur, S., Wittbrodt, J. and King, R. 2007. "Role of Adhesives in the Dimensional Stability of Polycarbonate Structural Panels." *International Journal of Adhesion and Adhesives* 27(4): 338–350.

4-9. Lilleheden, L. 1994. "Mechanical Properties of Adhesives in Situ and in Bulk." *International Journal of Adhesion and Adhesives* 14(1): 31–37.

4-10. Dolev, G. and Ishai, O.J. 1981. "An Effective Stress/Strain Concept in the Mechanical Characterization of Structural Adhesive Bonding." *International Journal of Adhesion and Adhesives* 1(3): 135–140.

4-11. Jeandrau, J.P. 1986. "Intrinsic Mechanical Characterization of Structural Adhesives." *International Journal of Adhesion and Adhesives* 6(4): 229–231.

4-12. Standard Test Method for Tensile Properties of Plastics—D638-03. 2003. ASTM International, West Conshohocken, PA.

4-13. Montgomery, D.C. 2004. *Design and Analysis of Experiments*. Hoboken, NJ: Wiley.

4-14. Design-Expert Software, Version 6.0. 2000. Stat-Ease, Minneapolis, MN.

4-15. MINITAB Release 12.2. 1998. Minitab, State College, PA.

4-16. "Sources of Noise in a Car," Wikipedia, http://en.wikipedia.org/wiki/Noise,_vibration,_and_harshness. Accessed March 6, 2012.

4-17. Fahy, F. and Gardonio, P. 1995. *Sound and Structural Vibration*. Academic Press: Salt Lake City, Utah.

4-18. "Equal Loudness Contour," Wikipedia, http://en.wikipedia.org/wiki/Equal-loudness_contour. Accessed March 6, 2012.

4-19. "Study of Methods for the Assessment of Mobile Air Conditioning Efficiency," http://assets.dft.gov.uk/publications/pgr-roads-environment-research-reportmobileairconditioning-pdf/reportairconditioning2.pdf. Accessed March 6, 2012.

4-20. Türler, D., Hopkins, D. and Goudey, H. 2003. "Reducing Vehicle Auxiliary Loads Using Advanced Thermal Insulation and Window Technologies." SAE Paper No. 2003-01-1076. SAE International, Warrendale, PA.

4-21. Gasworth, S. and Tankala, T. 2011. "Reduced Steady State Heating and Air Conditioning Loads via Reduced Glazing Thermal Conductivity." SAE Paper No. 2011-01-0126. SAE International, Warrendale, PA.

4-22. "Solar Resource Data," NREL Renewable Resource Data Center, http://www.nrel.gov/rredc/solar_data.html. Accessed April 30, 2013.

4-23. Duffie, J.A. and Beckman, W.A. 1991. *Solar Engineering of Thermal Processes.* Hoboken, NJ: Wiley.

4-24. "International Standard for Road Vehicles—Safety Glazing Materials—Method for the Determination of Solar Transmittance." *ISO* 13837:2008, Convention A.

4-25. Rugh, J., Chaney, L., Lustbader, J., Meyer, J., Rustagi, M., Olson, K., and Kogler, R. 2007. "Reduction in Vehicle Temperatures and Fuel Use from Cabin Ventilation, Solar-Reflective Paint, and a New Solar Reflective Glazing." SAE Paper No. 2007-01-1194. SAE International, Warrendale, PA.

4-26. "Emission Facts," http://pbadupws.nrc.gov/docs/ML1204/ML120440122.pdf. Accessed April 30, 2013.

4-27. Becker, T.A. and Sidhu, I. 2009. *Electric Vehicles in the United States—A New Model with Forecasts to 2030.* Berkeley, CA: University of California, Berkeley Center for Entrepreneurship and Technology.

4-28. Farrington, R. and Rugh, J. 2000. "Impact of Vehicle Air-Conditioning on Fuel Economy, Tailpipe Emissions, and Electric Vehicle Range." Earth Technologies Forum, Washington, DC.

4-29. Gasworth, S., Tankala, T., Kancharla, A. and Shuler, S. 2011. "Improved Battery Performance in Electric Vehicles via Reduced Glazing Thermal Conductivity." SAE Paper No. 2011-01-1341. SAE International, Warrendale, PA.

4-30. Gasworth, S. and Tankala, T. 2012. "Effect of Glazing Thermal Conductivity on Cabin Soak Temperature." SAE Paper No. 2012-01-1207. SAE International, Warrendale, PA.

4-31. Holman, J.P. 1997. *Heat Transfer.* New York: McGraw-Hill.

4-32. *User Manual, FLUENT v12.1.* Canonsburg, PA: ANSYS Fluent.

4-33. Krishnamoorthy, N., Nayak, S. and Surisetty, G. 2009. "Polycarbonate Glazing Body Panels For Automotive Applications." SAE Paper No. 079. SAE International, Warrendale, PA. Symposium on International Automotive Technology (SIAT), Automotive Research Association of India (ARAI): Pune, India.

4-34. "Simulation Tools for Injection Molding," Autodesk Simulation, http://www.moldflow.com. Accessed April 30, 2013.

4-35. "Unique Injection-Compression Molding Simulation," Omnexus, http://www.omnexus.com/news/news.aspx?id=20506&q=2k-icm. Accessed April 30, 2013.

4-36. Krishnamoorthy, N., Garg, A. and Kumar, S.M. 2011. "Analytical Methodology for Prediction of Warpage of Two Shot Injection Compression Molded Polycarbonate Glazing." SAE Paper No. 2011-01-0003. SAE International, Warrendale, PA.

4-37. "Market Trends in the Automotive Glazing Industry," Research Analysis Just-Auto, http://www.just-auto.com/analysis/market-trends-in-the-automotive-glazing-industry_id89764.aspx. Accessed March 6, 2012.

Chapter 5
Plastic-Metal Hybrid (PMH) Structures

T.K. Poovanna, P. Arunachala, and Sandeep Kulkarni

5.1 Introduction

Today's stringent emission standards and increasing cost of fossil fuels have put a lot of pressure on automakers to improve the fuel efficiency of their vehicles. Automakers work constantly to make their engines and transmissions more efficient, as well as develop hybrid drives to improve the fuel efficiency. They are also continually striving to reduce the curb weight of their cars. This has led to the extensive use of lightweight materials [5-1] for building cars. Many engineering plastics have already replaced traditional materials in some complex and critical applications. However, the extensive use of plastics in load-bearing structures has been limited because of factors such as cost, safety, and the traditional mind-set about plastics. There is a lot of development in the area of sheet metal, wherein manufacturers are trying out thinner, high-strength materials and lighter alloys of aluminum and magnesium [5-2]. There is a limitation on thinner metal parts that can be used for load-bearing applications; for example, thin metallic shell structures are prone to buckling failure [5-3]. The geometric imperfections that are dent induced—either during manufacture of the part itself, while handling, or during the use phase of the part—can further contribute to the buckling strength of thin shells dramatically. The welding or bonding of a stiffener may not be very practical in thin structures, because of the cost and assembly complexity of the structure where plastics can play a major role.

Traditionally, metals and plastics in the building of automobiles compete with each other; however, a new synergy between these two materials has been emerging recently. The hybrid technology combines the advantages of the inherent material properties of both materials and the manufacturing flexibilities of plastic. The plastic-metal hybrid (PMH) structure offers more lightweight construction potential than single-material propositions. Compared to either only metal or only plastic construction, PMH construction offers cost and mass advantages. The cost advantage can be further enhanced by the integration of additional functionalities into the hybrid structures.

5.2 Why Hybrid Designs?

In a load-bearing application, just by replacing the metal with thermoplastics, it is almost impossible to meet the structural requirements, unless the metal part was highly overdesigned. Many a times, design modifications are required to achieve equivalent performance in a thermoplastic part design. Although reinforced thermoplastics exhibit lower strength and stiffness than metals, it is possible to design a plastic part to provide equivalent performance to that of metal parts, if suitable changes in cross section and/or thickness are allowed. However, if the design solutions required are highly demanding from a structural perspective; and/or there is a constant load on the part, which can lead to creep-related problems; and/or the service temperature is high, then there are only few possible cost-effective all-plastic solutions. In such scenarios, making the best use of the inherent properties of metal and plastic leads to a hybrid design, in which most of the load carrying is accomplished by a thin formed shell of metal, reinforced by plastic ribbing. The resulting structure offers lighter weight, more compact structure, better dimensional properties, and multifunctional parts. These hybrid structures enhance productivity, improve the structural properties for the given system weight, and reduce the overall system cost.

Historically, semistructural applications are designed using composite materials that are reinforced using fibers. Some of the well-established fibers that are widely used in this kind of application are glass fibers, carbon fibers, chemically treated natural fibers [5-4], and polymeric fibers. Long fibers, with a length-to-diameter ratio (known as the aspect ratio) of more than 100 are becoming popular in structural parts manufactured through injection molding. Such filled materials exhibit good a stiffness-to-weight ratio. However, to meet the intense structural requirements, only a composite approach would disturb the cost-versus-benefit balance. In such cases, the PMHs offer a good value proposition. The choice of plastic, metal, and the method to combine the two will have great influence on the cost, performance, and weight of the overall system. In the best designed hybrid systems, the contribution of each component is maximized and a fine balance is struck, which yields optimal structural performance, lightweight structure, and lowest cost.

5.3 Types of Hybrids

The simplest and most common PMH is made by insert molding of the metal member in the mold, wherein the metal insert is completely submerged in plastic. This method is popular for small parts where the local loading is taken care by the metal insert. However, this may not be the most optimal way of producing large parts, where the much thinner metal members are normally used. Conventional insert molding of thinner formed profiles is not effective in providing structural stiffness, and manufacturing of them is difficult.

In structural and semistructural hybrid applications, two types of reinforcement structures are popular: open structure and closed structure. Based on the type of metal profiles and ways to attach them to the plastic body, hybrids can be broadly classified as shown in Fig. 5.1.

In open-structure reinforcement, typically a C-section thin steel or aluminum profile is held with the plastic structure. If the assembly of steel and plastic takes place in the mold, then sometimes it is called in-mold assembly. The bond between metals and plastics can be achieved either through physical, chemical, or mechanical means (e.g., mechanical locks). Some of the key methods of joining metal and plastics are summarized in the next sections.

5.3.1 Overmolding

A perforated metal profile is used as reinforcing member, which is joined through the *overmolding* of plastic. During molding, the plastic flows through the perforation and forms buttons that hold the profile in place and effectively distribute the load. Plastic ribbing inside the profile provides the required support to the profile member, which will resist local buckling. The process uses a standard metal forming and injection-molding process. The reinforcement can be planned locally based on the structural requirements. Through this process, multiple profiles can

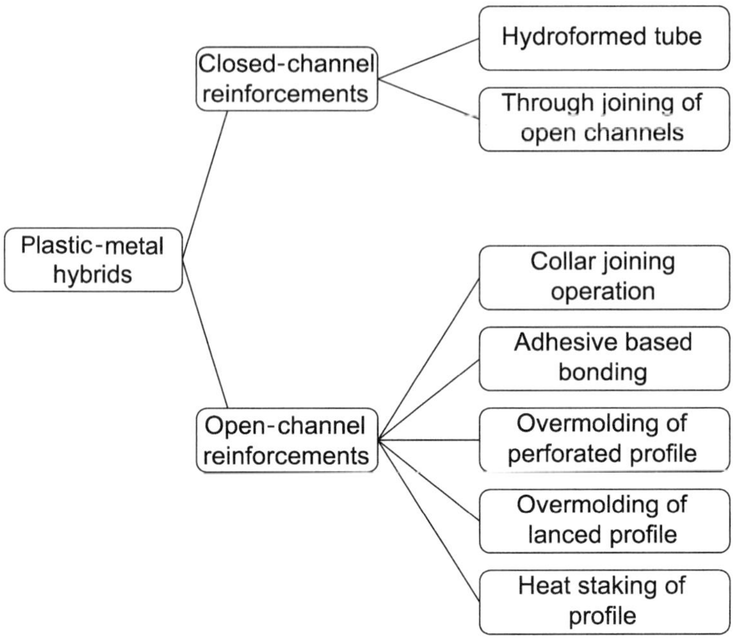

Fig. 5.1 Classification of hybrids based on type of profile and joining method.

also be joined together, while molding the body without requiring any additional steps [5-5].

5.3.2 Adhesive Bonding

In the method of *adhesive bonding*, a closed or open section is achieved by bonding two open sections, one metal and the other plastic. The main advantage of using the adhesive joining method is to achieve a closed profile, which is otherwise not possible through conventional overmolding techniques of open steel profile. Through the adhesive technology, a continuous bond can be achieved between the molded plastic member and the metal profile. The closed profile achieved through this method gives maximum stiffness for the given weight. The special adhesive helps in bonding materials that have low surface energy without pretreatment in most of the cases. One of the main advantages of using this process is the possibility of adding local reinforcement without major changes to the injection-molding part. Since the bonding is independent of the primary molding or sheet metal forming process, the tight tolerance requirement, as in some of the other hybrids, is not necessary [5-6].

5.3.3 Collar Joining

In the *collar joining* method, the plastic part and the stamped part are produced separately. Then both of them are joined together through a secondary process known as collar joining [5-7]. The assembly is a three-step process. First, small holes are punched in the sheet metal profile. Second, the flange is produced through a collaring process by drawing material inwards. Third, this profile is cold-pressed directly onto the wall of molded part. The metal profile gets locked onto the plastic because of the special profile of the previously formed collar. The collar can also be further engineered to have special features such as groves on the flange to enhance the lock strength. The flexibility of sheet metalworking can be utilized to get undercuts in the flange, which can also enhance the lock strength. The main advantage of this process is that it does not need accurate positioning of metal part with respect to the plastic body [5-7].

5.3.4 Polymer Injection Forming

As the name *polymer injection* suggests, the forming and bonding of metal takes place in one step. The metal blank is coated with a thin layer of polymer and placed in an injection-molding tool. During the closing of the mold, the basic forming of the sheet takes place with the help of a modified molding tool. The final forming takes place with the help of the hydrostatic pressure of injected material, as in the hydroforming process. The method is more economical as it reduces number of steps in the manufacturing process. Some of the challenges of the process include trimming of metal and the difficulty in forming complex and deep drawn profile shapes [5-8].

5.3.5 Direct Metal Deposition

With *direct metal deposition*, a very thin layer of metal is deposited over the plastic body. The ability to deposit metal onto the surface of plastic parts gives the flexibility of getting metal into complex shapes, which is very difficult to accomplish with the traditional metal-forming process. Various methods of deposition can be used to deposit metal over the plastic body. When the grain size of the metal is very small, the strength of the metal goes up considerably; a similar principle is utilized here to achieve very-high-strength skin on the plastic surface by coating materials with a grain size several orders smaller than those of conventional metals. While the main focus of the PMH is to get better stiffness and strength properties, the primary use of the coating method is to achieve better creep resistance, electrical conductivity, EMI (electromagnetic interference) shielding, thermal management, and improved permeability of the part. The other major advantage of the process is that the coating can be tailor made; the thickness and location can be adjusted as required by the structure to make it more cost-effective [5-9].

5.3.6 Mechanical Fasteners

The traditional way is to use *mechanical fasteners* such as rivets, screws, and bolts to join metal and plastic together. This calls for an additional assembly step during manufacturing. The assembly process can also be automated, if required, based on the production volume [5-10].

5.3.7 Heat Staking

In the *head staking* method, the plastic member is molded with studs or bosses protruding from the body. The reinforcing profile is made with corresponding holes in the same positions. During the assembly process, the profile member is inserted into the plastic body, the protrusions are softened by a heat gun, and a mechanical force is applied to form a head, which is larger than the opening in the profile member. These rivet heads thus formed to prevent the profile member from getting separated from the plastic body. Frictional welding techniques can also be used to get the rivet head, depending upon the material used for the plastic body.

5.4 Reinforcing Structure

Depending upon the load requirements and manufacturing method adopted, the reinforcing structure can be selected. When the structure sees more of torsional loading, a closed reinforcing member can be used. When the structure experiences more of bending load, a simple open channel can be used as reinforcement.

5.4.1 Closed-Channel Hybrid Structures

In a closed-channel hybrid structure (CCHS), a closed profile is achieved through either hydroforming [5-11] or through conventional sheet metal forming process. In hydroforming, high-pressure fluid (typically water) is used to form the part in closed die, which helps in forming the part with uniform wall thickness. A hydroformed steel tube [5-12] can be used as the backbone of the hybrid system in CCHS, which is used for structures requiring high load-bearing capacity. The hydroformed steel used as a backbone is the main load-carrying member. The plastic functions as a part integrator and a load-transferring member. While designing CCHS, the steel is positioned at the main load path for maximum efficiency. In the example shown in Fig. 5.2, the hydroformed steel spans the entire width of the front-end module (FEM).

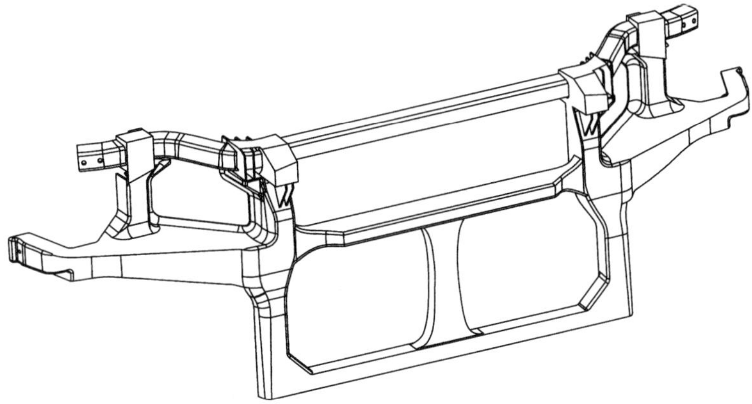

Fig. 5.2 A closed channel hybrid front-end system.

Secondary load paths are achieved by welding hydroformed tubes onto the main tube, before molding, to give rail support. The entire tube is placed in the mold, and plastic is injected to form the rest of the part geometry. Plastic helps mainly in integrating various parts and acts as the load-transferring member. Making use of the feasibility of molding complex shapes and integrating various functions into a single part with plastics, various elements such as attachment mechanisms, mounting guides, wiring conduits, and cable guides are created, in addition to the local load-bearing member.

Key to this technology is the lock [5-13], a design feature that enables the steel tube and plastic to be securely held together. Completely submerging the steel tube in a polymer is not only expensive but also leads to lot of manufacturing issues like tube collapse caused by injection pressure. An innovative joining feature known as SmartLok [5-12] enables a secure attachment between steel and plastic. A cut section of the molded lock is shown in Fig. 5.3.

When it comes to getting a closed profile, there are several options available. The most common and easy way is to use a hydroformed steel tube. Some of the widely used tubular hydroforming methods are low-pressure hydroforming, pressure sequence hydroforming (PSH) [5-14] and high-pressure hydroforming. In this CCHS, the closed profile member is produced through PSH, which is known to produce tight, reproducible tolerances [5-15] and limited controlled thinning of the walls. The process

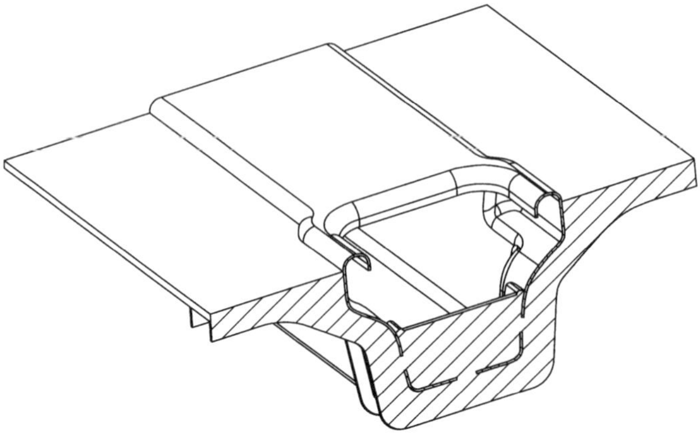

Fig. 5.3 Section of the lock feature.

starts with CNC (computer numerical control) bending of the tube to a near-finished profile, with a length of 1–3 m, depending on the length of the reinforcement required. The main advantage of hydroforming is that the section changes can be significant [5-14] to integrating various features. Areas can be locally modified, for example, a flat face to create a mating surface for attachment features. The other advantage is in-mold hole piercing for further consolidation of manufacturing steps. Sections can be engineered to get maximum strength to structures. The opening for the lock is punched while forming the tube itself.

The hydroformed tube has special features that aid the molding of the lock: the three flow channels in the form of holes, one at the bottom side and two on side walls, allow plastic to flow from outside to inside of the tube. On the top side, a large opening is created whose edge is folded back to compensate for stiffness loss and to support the tube while molding. These special features on the lock area can be seen in Fig. 5.4.

The main reason for the typical geometry of the lock is to allow the feature to be molded without a die lock. Normally, near the lock area the tube would slightly be bulged to allow some of the key features of lock to be molded. The core insert from the mold provides structural support at this location to prevent local tube collapse and keep plastic melt from entering the rest of the tube. The inner lock feature is formed from the tool insert, which helps to create a band that serves as a locking feature as well as a side wall stiffener. During the filling of the part, these bands act as a flow bridge from one side of the tube to other side. These bands help in reducing the pressure drop of the melt from one side of the tube to other side.

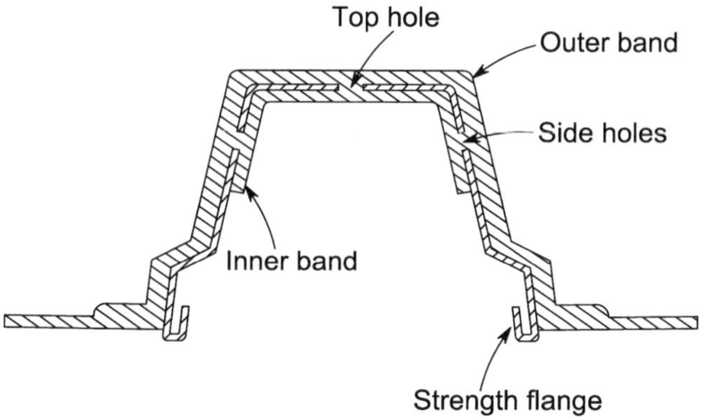

Fig. 5.4 Anatomy of a SmartLok.

5.4.2 Open-Channel Hybrid Structures

There are various advantages as well as limitations associated with hybrid structures with closed and open profile reinforcements. Closed profile members have excellent strength and good torsional rigidity. On the other hand, open profile members allow excellent design flexibility and can easily accommodate cross-sectional changes along the length of profile members. One of the challenges of open-channel hybrid structures (OCHSs) is that despite having a reinforcement member, they can have limited torsional and load-carrying capacity. This calls for stiffening the profile member suitably, without increasing the injection-molding complexity and keeping the metal forming simpler. An example of stiffening with cross ribs is shown in Fig. 5.5.

The reinforcement member is formed through the conventional forming process. The additional feature, which is used for locking plastic with steel, is achieved through a lancing operation. The locking feature in the open channel design is achieved through the special lanced feature [5-16] in the steel channel, as shown in Fig. 5.6, which improves material flow and forms a continuous lock between steel and plastic.

In the lancing operation, a depression is formed by shearing the side walls though a slitting operation or directly while forming the grove. Care has to be taken to avoid excessive thinning of the lock areas, as the forming is mainly achieved through local thinning. The loss of stiffness due to thinning is compensated by the geometric stiffness induced. To produce a PMH component, a preformed, thin-walled metal profile with alternate lancing is located in the injection-molding tool and molded with engineering thermoplastic. While filling the cavity, the plastic

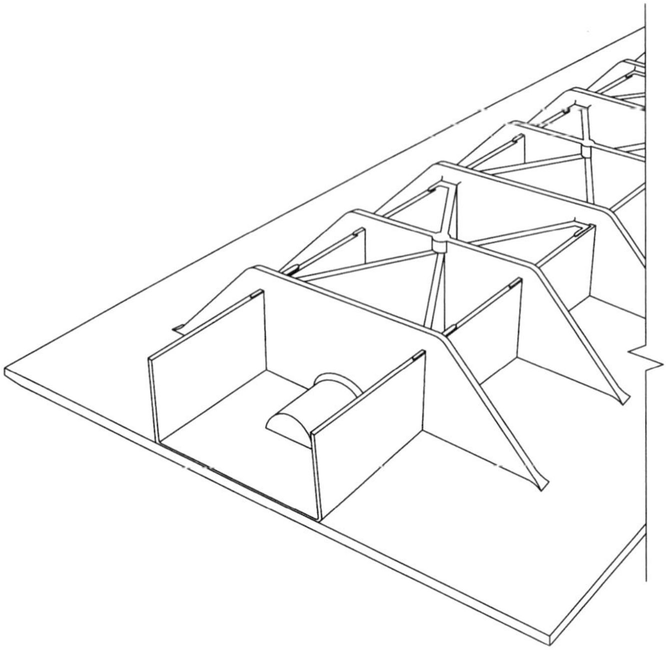

Fig. 5.5 An open channel is reinforced through plastic cross ribbing.

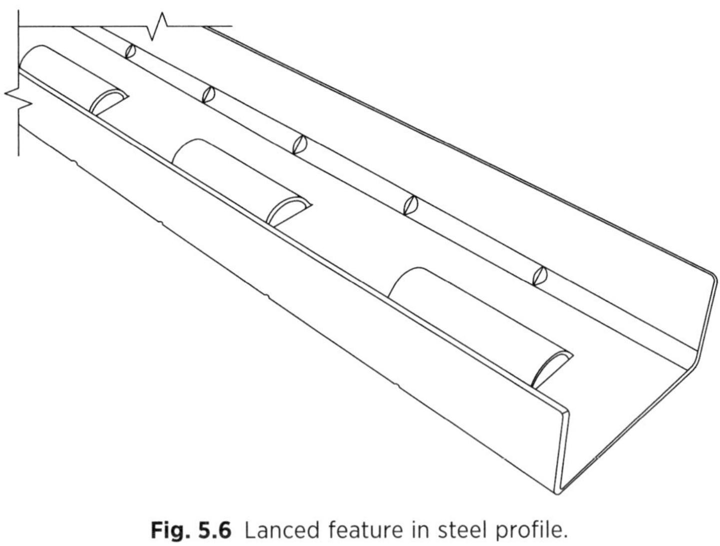

Fig. 5.6 Lanced feature in steel profile.

flows through the runner channel formed by the lancing operation on the reinforcing member. The runner channel also acts as a flow leader and main carrier of melt throughout the cavity, thus enhancing the moldability. Solidification of plastic creates a mechanical, interlocked connection between the materials, producing a high-strength, unified component. This novel locking concept [5-17] enables ease of fabrication of OCHSs at a much lower cost. Also, this concept enables one to use thin sheet metal and complex shapes to make hybrid components with lower weight.

Multiple configurations [5-17] for the lock are possible based on the structural needs of the geometry. The size and cross section of the lock geometry depends on the overall size of the part, required load-carrying capacity, and type of material being used for molding. Normally, it is easy to create circular or elliptical profiles, which would also help in creating excellent flow leaders during molding.

The thickness reduction while forming the lancing feature should be taken into account to ensure that no local tearing of metal occurs. Quick metal-forming simulations can actually help in determining the optimum grove depth for the given profile. Figure 5.7 shows the thickness reduction while forming a lanced feature in the profile member.

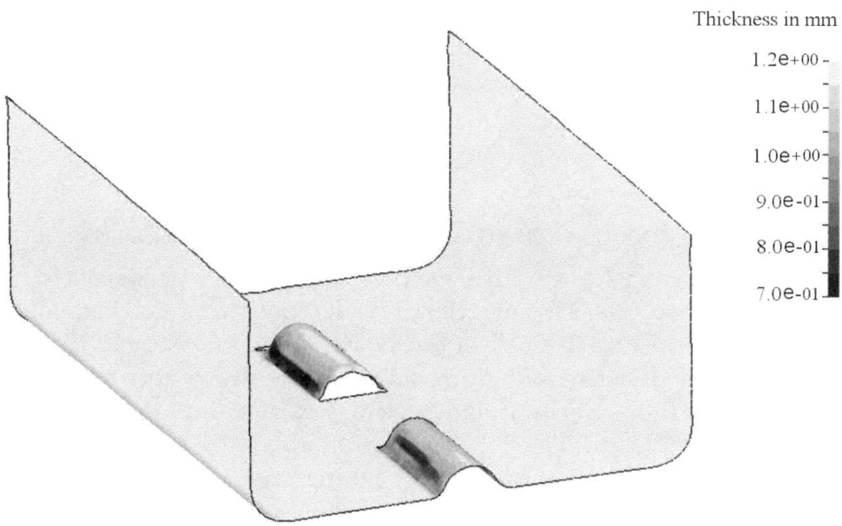

Fig. 5.7 Thinning at the lancing area.

Using the forming limit diagram [5-18], if one observes the possible failure of the material, then multiple smaller locks can be designed. Sometimes multiple locks are required when the profile width is larger. Enough care should be taken to minimize the sink mark potential on the outer surface because of the thick cross section of the plastic profile. Side beaks can be provided to locally set the sheet metal, which would further enhance the dimensional stability of the profile member. When the sheet thickness becomes higher, a planned draft for the side wall will help easy release of the part from the mold. The profile can be stiffened by ribbing or though gussets while molding. The actual ribbing pattern can be achieved through optimization studies taking the different load cases that the structure would be subjected to into account.

5.5 Processing of Hybrids

Conventional injection molds are used in the production of hybrid components. The additional step for molding hybrids is positioning of the profile in the mold before the material is injected. This calls for additional features in the mold to locate and hold the profile. Sometimes, there may be a need to preheat the insert also, close to the mold temperature to prevent the freezing of material as soon as it comes in contact with the cold metal insert. The other challenge is to produce a reasonably high-quality profile, as it is necessary to match provisions in the injection mold precisely. The tolerance requirement from the metal forming would be also slightly stringent compared to general-purpose sheet metal forming. A computer-aided simulation would help in optimizing the processing design and reduce the rework, once the tool is built.

5.5.1 Processing of Closed-Channel Hybrid Structures

When it comes to molding CCHS, the processing plays a major role in producing a good part. The integrity of the hollow section during molding is a key consideration. This calls for extensive simulation studies to optimize the feed system and processing window. The proper modeling of the hybrid structure is important to achieve a meaningful outcome in mold-filling simulations [5-19]. The steel profile actually acts as a flow barrier creating many local islands, depending upon the layout of the tube. The art is to fill each isolated zone without disturbing the tube integrity. The pressures experienced while molding such large structures with materials that have a high melt viscosity can be quite high (up to 120

MPa) in some areas. This pressure can easily crush the hollow steel profile member inside the mold. The whole objective of the filling simulation is to lay out the gating in such a way that low cavity pressure is experienced during molding; moreover, the tubes see very little cavity pressure in critical areas.

The modeling of steel tubes in filling analysis helps to predict the chilling effect of melt caused by a cold tube while molding. The pressure drop across the lock can be optimized by choosing the required diameter of side holes and the connecting band dimensions. Gating on both sides of the lock, if permissible, would help to reduce the pressure experienced by the tube. Figure 5.8 shows the pressure loss in the lock in the closed channel.

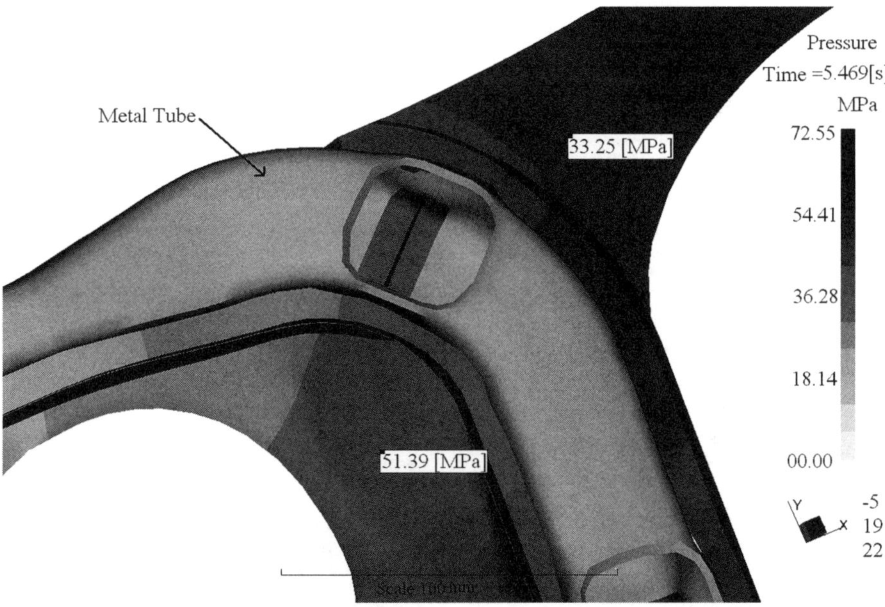

Fig. 5.8 Pressure difference across the hydroformed tube in the mold cavity.

It is necessary to identify the high-pressure areas of the tube, which can be achieved by tabulating the transient variation of pressure. In this example, pressure is measured at the points on the tube, as shown in Fig. 5.9.

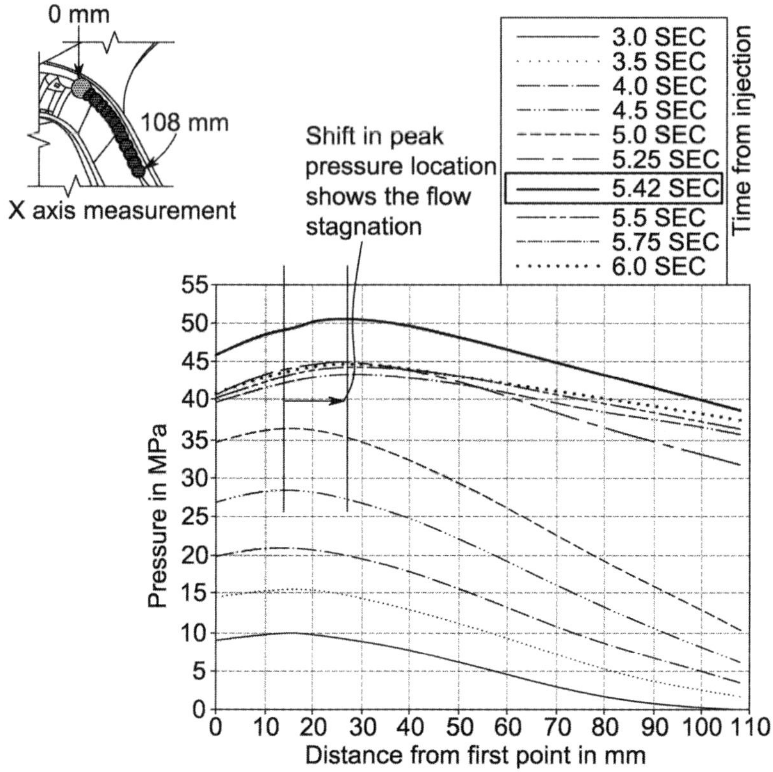

Fig. 5.9 Pressure curves showing flow stagnation.

The graph depicts a pressure gradient. The peak pressure corresponds to the stagnation point. This is where the flow velocity is zero and the pressure peaks. Identifying this area will help to find ways to reduce pressure through the manifold, by pressure relief wells or process conditions. It is not necessary to have metal-to-plastic contact all along the tube; if required, the areas can also be relieved where we see high pressure.

5.5.2 Processing of Open-Channel Hybrid Structures

The major manufacturing challenges, while molding an OCHS, are registering the profile in the mold and preventing the profile shift during molding. Before injecting the plastic, the metal profile needs to be placed in the specified location in the mold. This calls for tight dimensional tolerance and good control on the spring back [5-20] of the formed profile. In particular, slender profiles would pose challenges during the handling of the part and registering it on the mold. The mold is built in such a way that there are few registering or locating elements, which would help in holding the profile in place when the mold is in open condition. The entire process of placing the profile in the mold can be automated using industrial pick-and-place robots. Depending on the end-use conditions, the profile may require pretreatment like anticorrosion coating or painting before molding. Sometimes these profiles may also need preheating to enhance the quality of the part. When the flow is not balanced in the cavity, the shifting or deformation of the profile can be observed. Careful feed system optimization and mold-filling simulations would help in reducing or eliminating the molding defects of OCHS.

5.5.3 Mold Design

The mold for hybrids is designed and built virtually the same way as any other injection-molding tool. Die draw must be maintained or movable slides must be added to account for die lock conditions that may sometime arise because of some of the profile requirements of the tube. Effective cooling lines are required for effective heat removal; sometimes expansion of the tube can also create ejection problems. The mold material should be selected based on the wear properties so that it can withstand the metal profile rubbing the sidewalls during ejection and placement of insert in the mold. With CCHS, fitting of the hydroform insert to the mold is the major distinctive feature of the mold. The projected core inserts, as shown in Fig. 5.10, hold and align the tube during molding and form the lock inner band features.

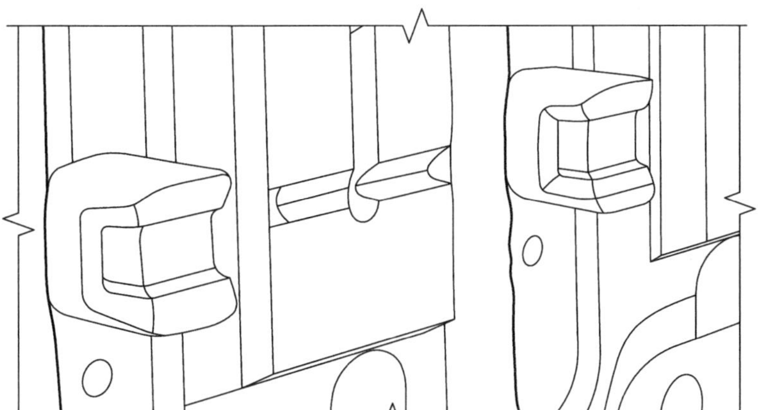

Fig. 5.10 The insert feature that forms locks.

The tolerance in the tool must be considered based on the dimension of the tube in cold and hot conditions. It is even more critical to consider the expansion of the tube when high-melt-temperature materials are used. Sometimes the high-melt temperature also calls for preheating of the tube considering the temperature gradient it sees.

5.6 Performance of Hybrid Structures

Typically, an open- or closed-steel section is used for the main structural part in an automotive. The PMH structure should give performance equivalent to that of an existing steel structure. To compare the performance of a PMH structure, simple beam geometry is chosen, which is subjected to bending and axial load. Displacement is applied on the beam, and the reaction force is measured. Table 5.1 shows the performance results, which establish that the hybrid system can demonstrate better performance and can be used effectively for structural application without compromising on the stiffness requirements.

Another important aspect for hybrid application development is the need for an accurate predictive methodology. Predicting part performance using finite element (FE) simulations necessitates accurate material modeling and geometric modeling. With CCHS, the modeling of the metal–plastic intersection plays a very important role. Thus it is critical to validate the simulation technique before employing it for the bigger part design.

Two test conditions are used to validate the FE technique used for modeling CCHS. See Ref. [5-21] for details about the material modeling and FE modeling techniques. The axial and shear tests are performed to evaluate stiffness of the hybrid system (Fig. 5.11) and the strength of the metal–plastic joining.

Break load and displacement at peak load predicted in simulation matches closely with tests for both axial and shear loading. The robustness of the plastic steel attachment was also evident in the mechanical performance, as the average peak loads are in the range of 3600 N for axial loading and 9200 N for the shear loading condition with an efficient attachment strategy used in closed-channel hybrid technology.

Table 5.1 Performance Comparison of Hybrid and Stamped-Steel Sections

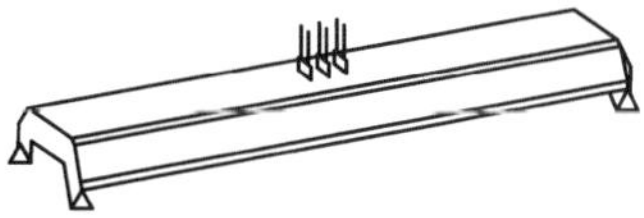

Bending

	Mass (gm)	Disp (mm)	Force (kN)	Normalized stiffness/Wt
Open-channel steel	150	3	0.43	1
Closed-channel steel	260	3	0.75	1
Open-channel Hybrid	210	3	1.87	3.1

Axial

	Mass (gm)	Disp (mm)	Force (kN)	Normalized stiffness/Wt
Open-channel steel	150	3	0.96	1
Closed-channel steel	260	3	1.72	1
Open-channel Hybrid	210	3	3.0	2.22

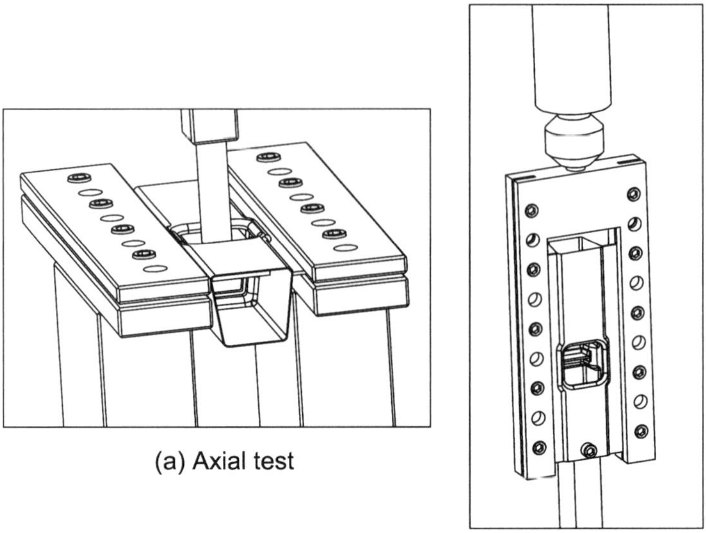

(a) Axial test

(b) Shear test

Fig. 5.11 Test setup of the axial and shear loading on a hydroform hybrid.

Table 5.2 shows the comparison of the test and FE study.

Table 5.2 Test and FE Comparison of CCHS

	Maximum load (N)		
Load	**Test**	**FEA**	**% Difference**
Axial	3643	3470	4.75
Shear	9243	9475	2.51

5.7 Application of Plastic-Metal Hybrids

Better part consolidation and production efficiencies of the PMH process open up new opportunities for designers to develop structures that meet performance demands at a lower weight and cost. Hybrid structures are now being explored for many structural and semistructural applications in automotive industry. Some of the examples using hybrid technology include front-end carriers (FECs), tailgates, door structures, cross-car beams, seating systems, floor panels, and roof structures. An integrated automotive roof module concept using PMH is discussed in Ref. [5-22]. Original equipment manufacturers (OEMs) are also exploring the possibilities of positioning PMH solutions in load-bearing body-in-white (BIW) components [5-23]. Fig. 5.12 shows examples of some of the potential applications using PMH.

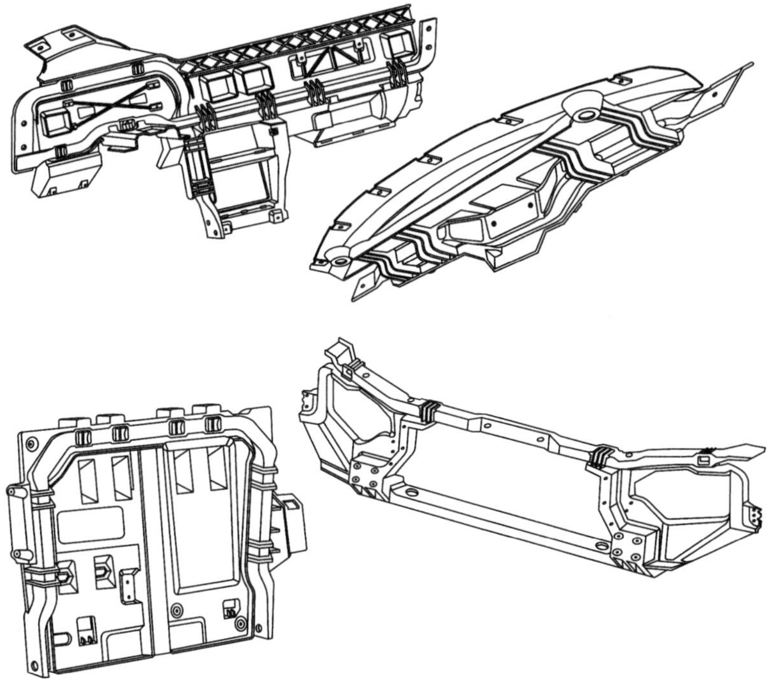

Fig. 5.12 Potential application of hybrid technology.

In most cases, PMH is developed with an injection-molding process, but the other manufacturing processes are under evaluation [5-24]. Some of these applications have already reached the production phase, and some are still in the concept development stage. Hybrid systems can be further used in other industries such as appliances, sports and exercise equipment, railway coach interiors, and aircraft interiors. The current study explores the use of hybrid technology for FEC application.

5.7.1 Front-End Module Application Development

FEC refers to a part fixed on the front of a car supporting most of the cooling package (radiator, A/C heat exchanger, turbo cooler), headlights, latch, fascia, and many other components. The part consists of high-level integration and has to fulfill multiple functions. FEC also connects the upper and the lower longitudinal rails and plays a role in the global and local stiffness of the car. Fig. 5.13 shows the exploded front view of a vehicle, with fascia, bumper beam, FEM, latch, headlamp, and radiator assembly. Latch, headlamp, and radiator assembly are mounted on the FEC structure.

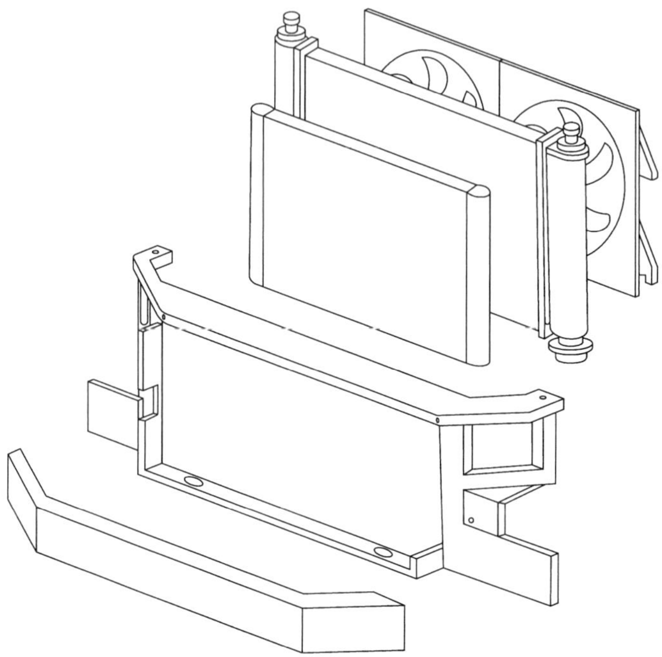

Fig. 5.13 Exploded view of vehicle front showing FEC and other components.

Thus the FEM has to broadly fulfill the module-carrying and structural functions. It carries several parts completely such as the radiator, headlights, bonnet lock, sensors, and pedestrian-safety lower-leg support beam. It also has to support (rather partially) other parts such as the fascia, splash shield, and fenders.

The FEC fulfills several structural functions including stiffness for hood slam or safety hook, stiffness for headlight mountings, pedestrian-safety lower-leg support, and protection of the package in the FEM.

Traditionally, the FEC was constructed by welding the sheet metal assembly. Typically, a steel FEC consists of 15–20 stamped-steel parts welded or bolted together. Figure 5.14 shows one example of an FEC that is made of more than 30 individual steel parts.

Such a design will obviously need higher assembly time and will be heavier in weight. Again, as industry is moving more toward sustainable

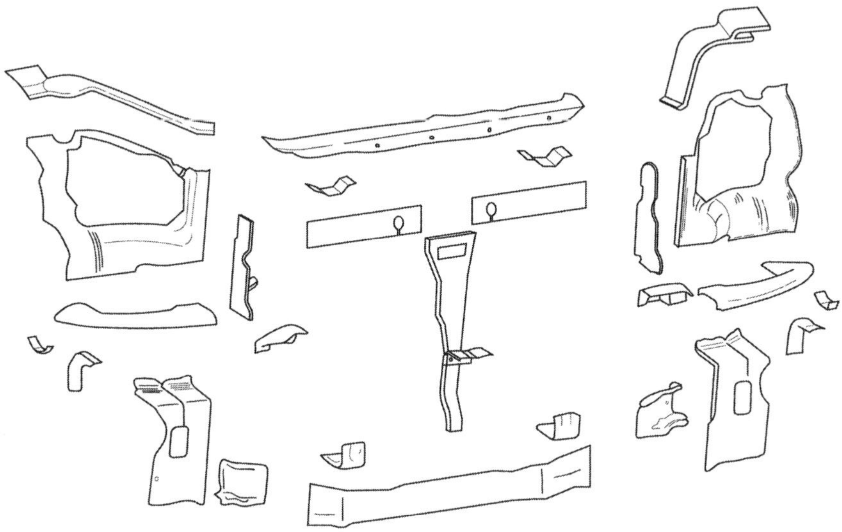

Fig. 5.14 Exploded view of the steel FEC.

solutions, one can imagine the amount of energy consumed in the production of one steel FEC.

With the emerging focus toward a lightweight system for fuel efficiency and a modular system for easy assembly, OEMs are looking into alternate solutions for FECs that can meet the performance of steel while offering higher design freedom. A detailed discussion on the new trend in FECs has been reported in Ref. [5-25], which discusses modular-build benefits, materials options to displace metal, and the like. In recent years, the trend in the automotive industry has swung toward the use of PMH systems [5-26]. PMH FECs can offer weight reduction and modularity. The modular system refers to an FEC that is preassembled with all the surrounding components such as headlamp, radiator, latch, and sensors. Modular FECs offer benefits to carmakers in terms of reduced number of operations on their assembly lines, simplified supply chains, reduced modular weight, and improvement in the ergonomic aspects of the assembly process. The complete assembled system is then referred as FEM. Additionally, the use of FEM improves fit-and-finish on the vehicle's front end; accurately locates headlamps, bumpers, fascia, and grilles in a single package; and, in some cases, provides styling opportunities that would not be possible with conventional builds. Finally, dimensional accuracy is much easier to control on a large, highly integrated composite part than on an assembly of multiple stamped-metal parts, where errors multiply via stack tolerances. Figure 5.15 shows an example of a hybrid FEC fitted with all the components. The complete module is then transferred to the assembly line at the OEM site and fitted on the BIW structure.

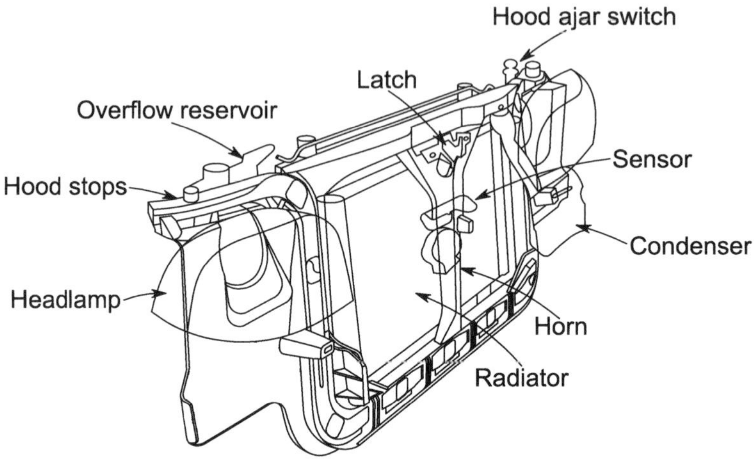

Fig. 5.15 FEM with complete assembly.

The main purpose of using a hybrid structure for FEM application is to use metal for most of the mechanical performance and plastic for the functional integration and creation of complex shapes. The key advantage of plastic is its ability to mold complex shapes and integrate various functions of FEM into a single part, eliminating a multitude of parts and process steps.

The hybrid structure can be designed with either of the two methods mentioned earlier, CCHS or OCHS. Use of CCHS for an FEM application is discussed in detail in this study.

CCHS is ideal choice for the load-bearing applications such as FEM, because CCHS parts have excellent structural characteristics, dimensional accuracy and part integration [5-27]. Engineering thermoplastics (ETP) is used in PMH for automotive front-end structures such as grille opening retainers, headlamp brackets, and FEC, as it offers excellent part integration and dimensional accuracy with moderate structural performance. A CCHS can be achieved by combining hydroformed steel with an injection-molded ETP to create a structure that has enhanced properties [5-28]. The hydroformed steel functions as a backbone that is the main load-carrying member. The plastic functions as a part integrator and a load-transferring structure.

5.7.2 Design Methodology

CCHS technology is discussed here for the FEM application [5-29, 5-30]. Proper hydroformed plastic structure design will position the steel for the maximum load-carrying function. The primary load path is the route the tube will follow. Major load paths are generated from hood slam, road vibration, and natural frequency. Recently introduced pedestrian-protection regulations also demand stiffer design of the lower part of FEM. The plastic panel is designed to integrate automotive hardware and to complement the strength of the hydroform tube. The plastic panel is designed to create attachment features for headlamps, the radiator, washer bottles, and so forth. Gusset and rib features are included to reinforce the entire structure. Locally, the plastic can be thickened to reinforce high-stress areas.

The functional and structural requirements of the FEM structure vary from OEM to OEM. Understanding the OEM specifications is an important first step of the FEM design. Along with the mechanical requirements, the assembly sequence of the FEM and the surrounding components also need to be considered during the design phase. The designer is responsible for the efficiency of the design to achieve minimized cost, mass, and part count.

The present study gives details about the hydroform hybrid FEM development for a sports utility vehicle (SUV) platform. Current production radiator support is a stamped-steel assembly that supports the radiator, headlamps, and hood latch (Fig. 5.16). The FEM is made with a total of 17 stamped-steel parts.

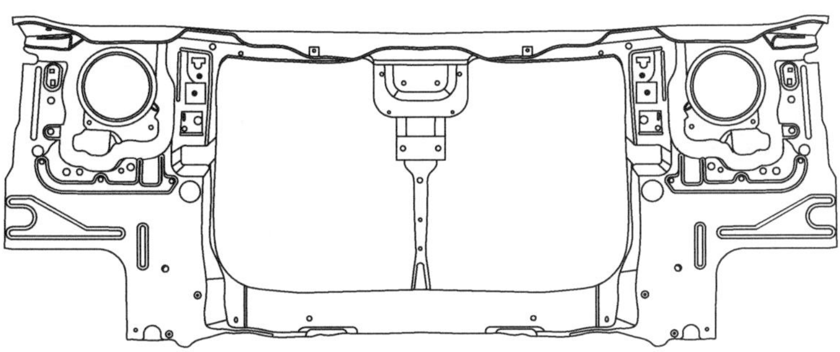

Fig. 5.16 Stamped-steel FEC.

The design concept for hybrid FEM was created by identifying major load paths and hardware integration opportunities. A topology optimization technique has been used to arrive at the first skeleton design of the FEM within the available design space to meet the mechanical loading requirements. Topology optimization also indicates the important load path of the structure that will take the maximum load. Based on the initial skeleton design, the designer can further improve the efficiency of the design by routing hydroform steel along the major load paths and using plastic for secondary load paths and parts integration. For the current FEM, with many possibilities of paths, a U-shaped backbone is the most efficient approach. The tube starts at the fender attachment and extends inboard beyond the headlamp, where it turns vertically downward. The tube then turns inboard to support the radiator and acts as a lower tie bar. The tube continues until it comes to the end of the radiator, where it turns upward and creates the same shape as that on the other side of the vehicle. When the routing of the tube is complete, the next step is to create a plastic bolster. The design of the bolster starts at the location of the attachments on the hydroform tube. From there, the plastic structure fans out to support the hardware mounted on FEM. The function of the bolster is to integrate the module hardware and provide supplementary load management. Thirteen locking features are used in the present design to connect metal with plastic. The designer can further optimize the number and location of these locking points to achieve a proper load-transfer mechanism between metal and plastic. Fig. 5.17 shows the initial design concept of the hybrid FEM design.

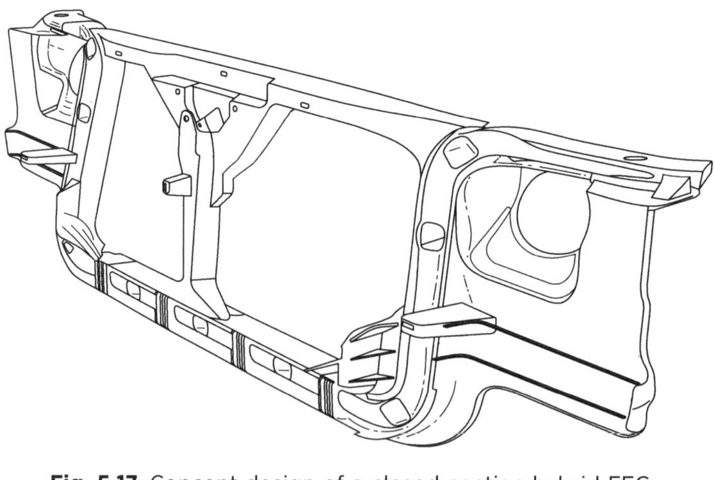

Fig. 5.17 Concept design of a closed-section hybrid FEC.

The industrial design concept is used as a roadmap to guide CAD. The hydroform tube is designed first. It is designed to follow the path dictated by the design concept. Next, the design features that enable the overmolding are set up in key areas on the tube. Finally, the attachment features to the fender and rail are created. Upon completion of the hydroform design, the plastic panel is designed using the hydroform tube as a constraint. The plastic goes from the attachment locations to create the panel that integrates parts and offers the structural rigidity. Further, the open C-section plastic upper beam and vertical support is stiffened with ribs to achieve optimum load transfer from the loading point to the constraint area. Ribs play a very important role in the stiffening of the open C-section. In fact, it can be said that stiffness and mass optimization of the part largely depends on the rib design. Figure 5.18 shows the CAD design of an FEM that is ready for the simulation study.

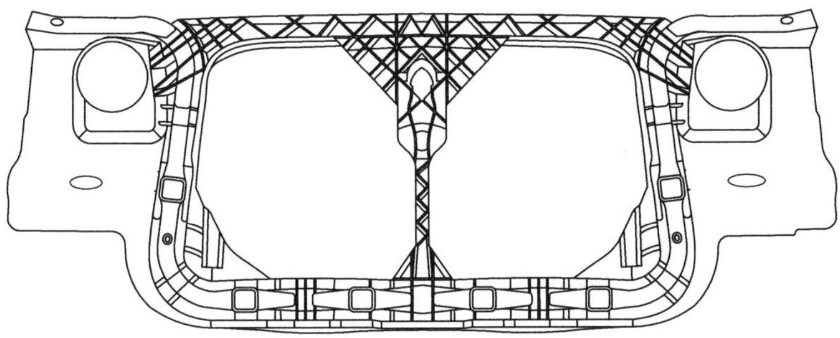

Fig. 5.18 Engineering design of closed-section hybrid FEC design.

5.7.3 Performance Evaluation

FE techniques can be used effectively to get insight into the performance of the hybrid concept and are further extended for the design, optimization, and improvement of the design. FEM is evaluated for various mechanical loads such as hood lift, hood slam, headlamp connection, torsion, and NVH.

To evaluate an initial design, one can adopt a frequency mode evaluation, which will give a good idea about the global structural stiffness of the part and potential connection issues. With normal mode studies, many structural issues can be sorted out before analyzing for the mechanical load cases. For the present FEM design, both the stamped-steel baseline and the hybrid plastic structure (HPS) had a first mode that oscillated fore–aft of the car. The HPS had a stiffer response of 63.5 Hz versus the baseline steel performance of 52 Hz (Fig. 5.19).

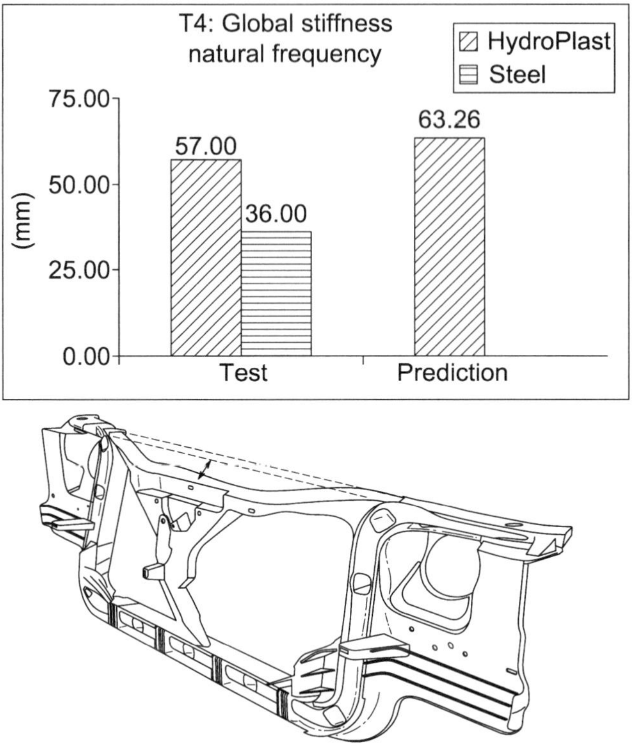

Fig. 5.19 First model frequency evaluation and comparison.

In NVH physical testing, the same trends as seen in the simulation were observed. First frequency from the test as 63 Hz, which is close to the simulation results.

The FEM is further tested for other load cases such as hood lift test, torsion test, and lower tie-bar cross-car stiffness test. In the hood lift test, an upward vertical load is applied to the hood latch mechanism. Torsional stiffness measures the resistance of the structure to twist. On the road this can be equated to potholes. Lower tie-bar stiffness measures the resistance to the vehicle rail spread. A transverse load is input to a lower attachment, while the other body attachments remain fixed.

The lower tie-bar stiffness test setup is used to measure the deformation at the load point. The FE study is correlated with the experimental test study. Figure 5.20 shows the force-deformation comparison of the test and FE study, which confirms the very good correlation between the two.

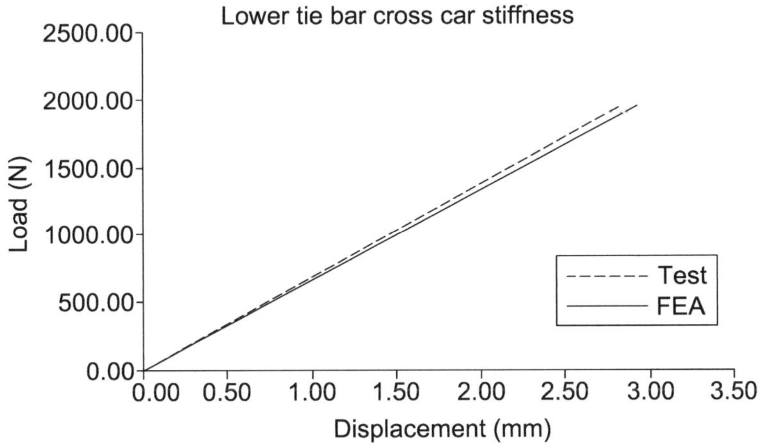

Fig. 5.20 Force-deformation comparison for lower tie-bar cross-car stiffness.

Results of the correlation studies for other tests are tabulated in Table 5.3. Simulation results compared well with the test results, within a 10% variation [5-30].

Table 5.3 Performance of Hybrid Front-End Module Under Different Loading

		Closed-Channel Hybrid	
Measurement	**Unit**	**Simulation**	**Test**
Natural frequency	Hz	63	57
Cross car stiffness	N/mm	980	1086
Torsional stiffness	N/mm	29	30
Hood latch strength	N	3563	3559

Another important test, the durability test, is done to evaluate the performance of FEM in the vehicle [5-30]. A physical simulation of 10 years' durability is conducted using a two-post durability test (Fig. 5.21), where the input acceleration is given, matching the one that the vehicle would experience while driving through various courses of a proving ground.

The acceleration data are repeated 4000 times for a total of 150 hours. Temperature is cycled from −40 °C (−40 °F) for 4 hours to 23 °C (73.4 °F) for 4 hours to 85 °C (185 °F) for 4 hours. The PMH FEM shows no cracks at the end of test. The steel tube provided a stable foundation. The plastic panel showed no sign of fatigue.

Fig. 5.21 Test setup for the durability test.

5.8 Summary

PMH technology offers weight and cost savings for the automotive industry by joining the steel structure with glass-filled engineering thermoplastics, through a novel approach. The main idea behind the hybrid system is to make the best use of the inherent properties of metal and plastic, in which most of the load bearing is achieved by a thin formed shell of metal and reinforced by plastic ribbing, while the part integration and additional functionality of the design are accomplished by the plastic. The resulting structure offers lighter weight, more compact structure, better dimensional properties, and multifunctional parts. These hybrid structures enhance productivity, offer better structural properties for a given system weight, and reduce the overall system cost. Adoption of hybrid structures is on the rise in the automotive, aerospace, appliances, sports equipment, and other such industries.

Two hybrid systems were discussed in this chapter: OCHS and CCHS. Each technology has its own pros and cons, as shown in Table 5.4. It is the designer's responsibility to choose the appropriate one, to obtain maximum system mass and cost advantages for a given application.

Table 5.4 Comparison of OCHS and CCHS

OCHS	CCHS
1. Excellent design flexibility.	1. Good torsional rigidity.
2. Cross section can be changed easily.	2. Excellent structural stiffness.
3. Easy to process metal structure.	3. Limitation on the metal forming.
4. Thin metal structure can be used.	4. While molding, extra care is required to prevent tube collapse.
5. Molding of hybrid is relatively easier.	5. Limited joining methods.
6. Moderate structural stiffness.	

A CCHS for the FEM application was explored in this chapter. The hybrid carrier will help in providing greater styling freedom, up to a 30% mass reduction and 70% part count reduction and packaging space reduction, which would enable aggressive lighting design. The strength of a hydro-formed backbone creates a robust foundation. The plastic panel forms the load-transferring mechanism and integration of geometries to consolidate hardware. Hybrid FEM meets the structural stiffness requirement and offers a modular assembly option to OEMs. The same hybrid can also be developed with open-channel steel reinforcement. It depends largely upon application requirements, processability constraints, and design space availability.

While FEM is one of the examples discussed in detail in this chapter, many other applications such as tailgate, door structures, cross-car beam, seating system, floor panel, and running board offer potential opportunities for hybrid structures.

With stringent regulations on emissions and tougher pedestrian-standards being put into action in more regions, one can expect to see more novel uses of metals and plastics in BIW, FEC, cross-car beams, door modules, seating systems, bumper beams, and tailgates. The constant pressure on suppliers to cut back on the cost while shedding more kilograms will push the modular-build concept to the next level. The stringent targets set for recycling for end-of-life vehicles (ELVs) puts additional pressure on PMH, as there is a consistent demand to reduce the quantity of shredder residue. As the PMH falls in the multicomponent category, material selection and joining methods play an important role, given the way components get recycled. The design of the PMH must take into account the requirements and target quantities for dismantling, recycling, and reuse of ELVs.

5.9 Trends

The traditional PMH components are usually 20–30% lighter than their pure steel counterparts, while offering equal performance efficiency in many applications. In view of the market trend toward lightweight electric cars and lighter, fuel-saving gasoline-motor vehicles, automotive industry is working further toward improving the performance to weight ratio of the hybrid structures. Moving forward, alternate materials could be used for structural applications, replacing metal materials. Continuous carbon-fiber composite (CCFC) and continuous glass-fiber composite (CGFC) two such promising materials, which could replace metal to create an all-plastic-hybrid system. In this technology, a composite sheet reinforced with continuous fibers, shaped in accordance with hybrid technology, is injected with engineering thermoplastic. This new structure would retain all the benefits of established plastic or metal ones, such as lightweight combined with high strength and stiffness, while doing away with the metal. Implementation of composite hybrid is already showcased for some automotive applications such as FEM [5-31] and seat backs consisting of woven-glass or nylon sheet inserts overmolded with 35% glass-filled nylon 6, which provides 20% less weight than standard seat backs [5-32].

On the other hand, it is also important to develop a process to design an application using a composite hybrid structure. Efforts are under way for developing a fast and highly automated process so that a composite hybrid could be used for mass production. This process mainly combines injection molding with thermoforming of composite sheets. Composite fabric made of endless fibers in a thermoplastic matrix is heated, shaped in the injection mold, and then back-injected. This process could be used to add ribs for extra stiffness and to integrate other functions [5-32].

Such all-plastic-hybrid structures can offer higher design freedom (layer stack-up and orientation based on loading) and greater weight saving compared to traditional metal hybrid components. We expect to see an increase in such applications in near future.

5.10 References

5-1. Martin, G., Marc, S., Lukas, R., Gundolf, K. and Elmar, B. 2009. "Super Light Car—Lightweight Construction Thanks to a Multimaterial

Design and Function Integration." *European Transport Research Review* 1(1): 5–10.

5-2. Friedrich, H.E., Beeh, E., Lawson, T. and Zaffaina, L. 2008. "Cost Attractive Lightweight Solutions through New Mg Concepts for the Vehicle Structure Technologies." *Proceedings of the 65th Annual World Magnesium Conference*. Warsaw International Magnesium Association, Wauconda, IL.

5-3. Jeom, K.P., Jae, M.L. and Dong, H.L. 2003. "Ultimate Strength of Dented Steel Plates Under Axial Compressive Loads." *International Journal of Mechanical Sciences* 45(3): 433–448.

5-4. Jay, A., Paulo, A.S., João, C.G. and Glauco, M. 2008. "Natural Fibers Plastic Composites for Automotive Applications." SPE Automotive Composites Conference and Exhibition, Troy, MI.

5-5. Hubert, G. and Boris, K. 1993. "Structural Shell with Reinforcing Ribs Connected Via Perforations." Patent Publication No. US 51908033.

5-6. Teli, S., Naughton, P., Droste, A. and Kotnis, A. 2005. "Bonded Hybrid Automotive Front End Carriers." 5th Annual SPE Automotive Composite Conference, Troy, MI.

5-7. Endemann, U., Glaser, S. and Völker, M. 2002. "Strong Joint between Plastic and Metal." *Kunststoffe* 92: 110–113.

5-8. Van Der Aa, H.C.E., Verdier, A.S. and De Wolf, H.G. J. 2005. "Method And Device For Producing a Composite Product, and Composite Product Produced Therewith." Patent Publication No. WO2003057446.

5-9. Palumbo, G., McCrea, J., Tomantschger, K., Brooks, I., Jeong, D., Limoges, D., Panagiotopoulos, K., Erb, U. and Wang, A. 2006 "Article Comprising a Fine-Grained Metallic Material and a Polymeric Material." Patent Application No. 20060135282.

5-10. Bauhoff, M.J. 2002. "Integral Structures of Metal and Plastic with Fastening Means." Patent Application No. 20020160145.

5-11. Mizumura, M., Honda, O., Yoshida, T., Iguchi, K. and Kuriyama, Y. 2004. "Development of Hydroforming Technology." Nippon Steel Technical Report No. 90.

5-12. Goral, T. 2004. "For Cars, a New Style of Plastic Metal Hybrid." *Plastics Technology*, November, p. 11.

5-13. Staargaard, J.E., Klanges, J., Longhouse, B., Kowal, E., Butterfield, L.D., Goral, T.M., Brown, M., Panter, K.R., Nardone, T.E. and Gaida, D. 2008. "Method of Attaching Plastic to a Metal Section and Part Made Thereby." Patent Publication No. US 7449137 B2.

5-14. Longhouse, B. 2004. "Hydroforming High and Ultra-High Strength Steel Tubular Automotive Parts." Great Designs in Steel Seminar, Autosteel. Livonia, MI.

5-15. Morphy, G. 1998. "Tube Hydroforming: Dimensional Capability Analysis of a High Volume Automotive Structural Component Production Process." SAE paper No. 980450. SAE International, Warrendale, PA.

5-16. Groover, M.P. 2010. *Fundamentals of Modern Manufacturing: Materials, Processes, and Systems* (4th Edition). Hoboken, NJ: Wiley.

5-17. Mooijman, F.R., Prasad, P.S. and Poovanna, T.K. 2008. "Hybrid Structure and Method." Patent Publication No. US 2008/0138586 A1.

5-18. Marciniak, Z., Duncan, J.L. and Hu, J. 1992. *Mechanics of Sheet Metal Forming* (2nd edition). London: Wiley Butterworth-Heinemann.

5-19. Bozzelli, J. 2011. "Mold Filling Simulation: What, When, Why, How." *Plastics Technology*, July, http://www.ptonline.com/articles/mold-filling-simulation-what-when-why-how. Accessed January 15, 2012.

5-20. Gomes, C., Onipede, O., Lovell, M. 2005. "Investigation of Springback in High Strength Anisotropic Steels." *Journal of Materials Processing Technology* 159(1): 91–98.

5-21. Maity, R., Prasad, P., and Goral, T. 2006. "Material Modeling and Finite Element Analysis of Hydroform—Short Glass Fiber Filled Thermoplastic Front-End Structures." SAE Paper No. 2006-01-0824. SAE International, Warrendale, PA.

5-22. Kirson, C. and Stratton, D. 2005. "An Integrated Automotive Roof Module Concept: Plastic-Metal Hybrid and Polyurethane Composite Technology." 5th Annual SPE Automotive Composite Conference, Troy, MI.

5-23. Grujicic, M., Sellappana, V., Omara, M.A., Seyrb, N., Obieglob, A., Erdmannc, M. and Holzleitnerc, J. 2008. "An Overview of the Polymer-to-Metal Direct-Adhesion Hybrid Technologies for Load-Bearing Automotive Components." *Journal of Materials Processing Technology* 197: 363–373.

5-24. Leaversuch, R.D. 2008. "Plastic-Metal Hybrids Are Now Blow Molded." *Plastics Technology*, June, http://www.ptonline.com/articles/plastic-metal-hybrids-are-now-blow-molded. Accessed January 15, 2012.

5-25. Malnati, P. 2008. "What's New in Automotive Front-End Modules?," *Composites Technology*, http://www.compositesworld.com/articles/

what39s-new-in-automotive-front-end-modules. Accessed January 15, 2012.

5-26. Kuvin, B.F. 2005. "Marrying Plastic and Metal for Front-End Modules." *METALFORMING*, PMA Services.

5-27. Naitove, M. 2006. "SPE Automotive Awards Highlight Materials and Processing Advances." *Plastics Technology*, January, http://www.ptonline.com/articles/spe-automotive-awards-highlight-materials-and-processing-advances. Accessed January 15, 2012.

5-28. Mooijman, F.R. and Prasad, P.S. 2008. "Closed Hybrid Structure and Method." Patent Publication No. US 2008/0317988 A1.

5-29. Staargaard, J.E., Kowal, E., Goral, T.M., Brown, M., Panter, K.R., Butterfield, L.D., Longhouse, B., Klanges, J., Nardone, T.E. and Gaida, D. 2004. "Front End Module." Patent Publication No. US 2004/0160088 A1.

5-30. Goral, T., Prasad, P., Brown, M., Panter, K., Klanges, J. and Longhouse, B. 2005. "Automotive Front End Structures Constructed by Over Molding Hydroform Metal Tubes to Engineering Thermoplastic Structures." SAE Paper No. 2005-01-1680. SAE International, Warrendale, PA.

5-31. Defosse, M. 2010. "In Automotive Engineering, Lenxess Sees Bright Future for New Hybrid System." *Plastics Today*, http://www.plasticstoday.com/mpw/articles/lanxess-sees-bright-future-for-new-hybrid-systems. Accessed January 15, 2012.

5-32. Naitove, M. 2010. "Organic Hybrid Composites Highlighted at K Show." *Plastics Technology*, December, http://www.ptonline.com/articles/organic-hybrid-composites-highlighted-at-k-show. Accessed January 15, 2012.

Chapter 6
Headlamp Applications

Triloka Tankala, N. Krishnamoorthy, and Sudhakar R. Marur

6.1 Automotive Lighting Overview

The lighting system of an automotive vehicle consists of lighting and signaling units mounted on to the front, sides, and rear of the vehicle. The purpose of the lighting system is to provide illumination for the driver to operate the vehicle safely during dark conditions. Automotive lighting also needs to show the vehicle's presence, position, size, direction, and speed of travel to oncoming drivers and to the pedestrians.

Forward lighting is generally provided by low beam (LB; dipped beam) and high beam (HB; main beam). Some auxiliary lamps include front fog lights, cornering lamps, and spotlights. Front position lamps, daytime running lamps, parking lamps, and turning lamps constitute the signal and indication lamps [6-1]. A representative automotive headlamp is shown in Fig. 6.1.

The automotive headlamp assembly consists of the reflector, lens, bulb, shield, bezel, and housing, as shown in Fig. 6.2.

Many of these components are being replaced with plastics for better performance, system integration, and weight reduction. All the parts inside the headlamp get heated up, as a result of the thermal loads from the bulbs, the primary source of heat and light.

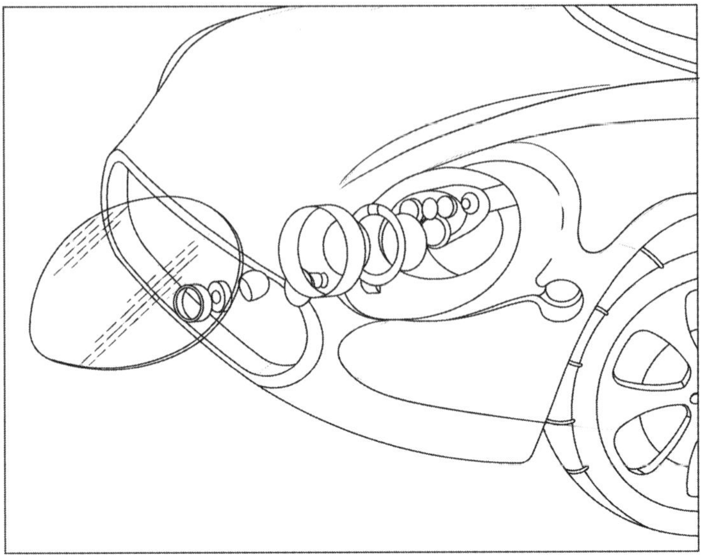

Fig. 6.1 A typical automotive headlamp.

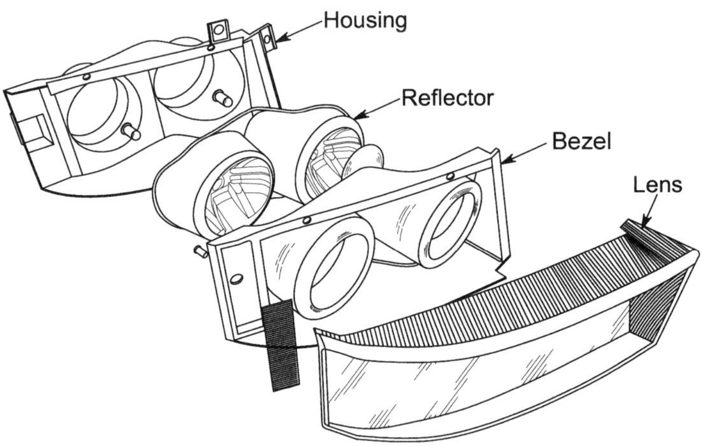

Fig. 6.2 Various components of an automotive headlamp.

In general, headlamps are electrically operated and positioned in the front of the vehicle. A headlamp assembly is required to produce a low and a high beam. This can be achieved either by an individual lamp for each function or through a single lamp with both the features [6-2]. HB is used when driving on poorly lit roads where there is no visible traffic. HBs project most of their light straight ahead and maximize the seeing distance. HBs can also cause glare to the drivers of the vehicles on opposite side of the road. HBs do not have any control of upward light. LB is used when there are other vehicles approaching or when you are following another vehicle. LBs have stricter control of upward light and can direct most of their light downward and rightward or leftward (based on the direction of the traffic in the country). In this process, they provide safe forward visibility without any glare.

Common sources of headlamp light are halogen, xenon, neon tubes, and light-emitting diodes (LEDs). Halogen light bulbs with a tungsten filament are the most prevalent light source for headlamps and other forward illumination functions. Headlamps with a xenon light source are known as xenon headlamps or high-intensity discharge (HID) lamps. Because LED lights offer very long service life and extreme vibration resistance, they are becoming increasingly popular with new generation automotive.

6.2 Automotive Lighting Global Regulations

LB headlamps are normally designed to provide distribution of light that provides adequate forward and lateral illumination. Also, there are regulations that govern the intensity of light directed toward the eyes of oncoming drivers, to control glare. LB is intended for use whenever other vehicles are present in the opposite direction. The international Economic Commission for Europe (ECE) Regulations for filament headlamps [6-3] and for HID headlamps specify a beam with a sharp, asymmetric cutoff preventing significant amounts of light from being cast into the eyes of drivers of preceding or oncoming cars, as shown in Fig. 6.3.

Control of glare is less strict in the North American SAE International beam standard contained in FMVSS/CMVSS 108 [6-4]. ECE regulations permit higher-intensity HB headlamps than do North American regulations [6-4].

6.3 Automotive Lighting—Role of Thermoplastics

Many of the headlamp components are being replaced with plastics for better performance, weight reduction, and styling requirements. One good example is the lens of the headlamp system. Designers have more freedom when designing a plastic lens compared to glass. A variety of shapes and features can be put in place in a plastic lens. Polycarbonate (PC) lenses are very popular in the lighting industry today because of the easy molding of complex shapes and their versatility.

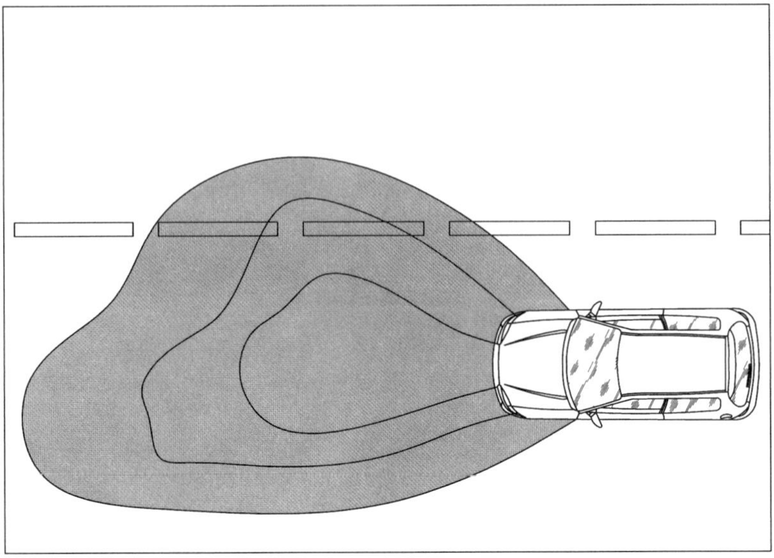

Fig. 6.3 LB light intensity and beam illumination of road surface.

As headlamp shapes become more complex, their inner components, such as bezels, also become more complex. Plastics are the ideal fit in these applications because of the ease of manufacturing complex bezels. Thermoplastics offer a good material fit for the reflector when compared to stamped steel or bulk molding compounds (BMCs). Polyetherimide (PEI) material [6-5] offers various advantages in terms of reduction of various steps involved during molding, such as deflashing, washing, drying, curing, and scrap disposal, as well as elimination of the base coat. Also, PEI scrap can be recycled and directly metallized. Through this process, lean manufacturing procedures can be established that lead to lower system costs.

6.4 Headlamp Reflectors

6.4.1 Material Replacement

Reflectors are functional parts of the lighting system that focus the light beam to a required pattern, to provide clear vision and to meet the regulations. Typically, thermoset-based BMC is the material of choice for this application because of its ability to withstand high heat and its low coefficient of linear thermal expansion. BMCs are also moldable to any shape, given the fact that modern reflectors can be complex in shape and style with zero shrinkage [6-6]. However, the molded parts using this material have to undergo several operations before they could be used in the actual assembly. This can be explained as shown in Fig. 6.4. After the BMC part is molded, it has to be deflashed to remove excess material and then thoroughly cleaned. All these reflectors need to be metalized to achieve a highly reflective surface. To prepare the surface for better adhesion, a base coat is applied, then cured. The part is then metalized and ready for assembly. A better alternative material is PEI (polyetherimide), which has good thermal performance capability. The onset haze temperature [6-7] can vary between 200 °C (392 °F) and 260 °C (500 °F) based on the grade used, which makes an ideal fit for the reflector application. Also, as shown in Fig. 6.4, after injection molding, it can be directly metallized, eliminating the intermediate operations.

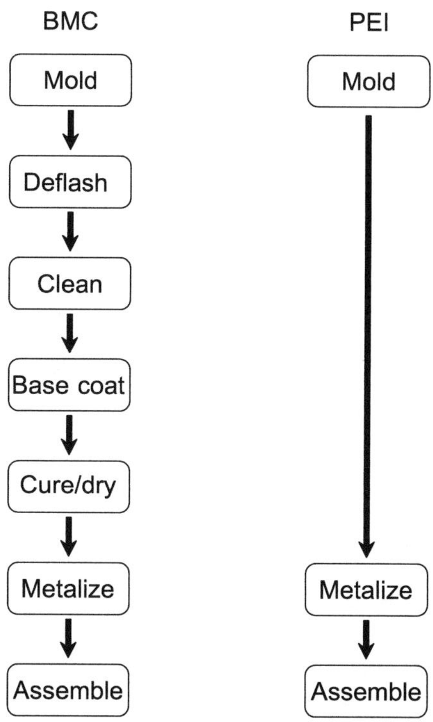

Fig. 6.4 BMC versus PEI process steps.

The challenge is to have a good product design for conversion from BMC to PEI. Generally BMC results in higher wall thickness, as it is easy to flow and process. The thickness distribution of a BMC reflector is as shown in Fig. 6.5.

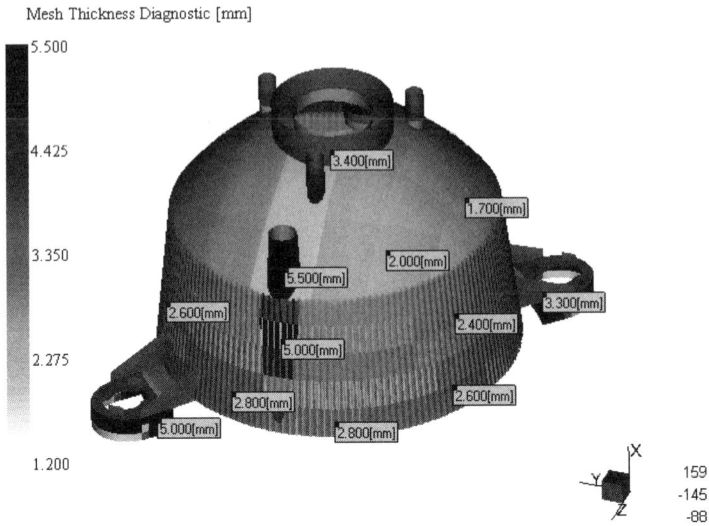

Fig. 6.5 Typical BMC reflector thickness distribution.

As can be observed, the thickness distribution is not uniform and very much higher than for a normal thermoplastic design.

A direct conversion of a BMC design to a PEI design may not be advisable; instead a complete redesign with thermoplastic material, exploiting the weight advantage that it offers, would be a better alternative. The specific gravity of BMC ranges from 1.85 to 1.95 [6-8], and the specific gravity of PEI is 1.27 [6-9]. A PEI reflector can have a generic wall thickness ranging from 1.8 to 3 mm and can be lowered to a minimum of 1.5 mm for smaller parts. The same part can be redesigned with a more uniform wall thickness, as shown in Fig. 6.6, while still meeting the functional requirements. This difference in density and lower wall thickness helps in reducing weight.

It is a good practice to follow the general thermoplastic design guidelines to design a PEI reflector [6-10, 6-11]. Most importantly, the thickness distribution has to be uniform for ease of filling the part and to avoid differential shrinkage resulting in warpage.

The flow of the top and bottom surface of the reflector should not be designed as shown on in the left illustration in Fig. 6.7, where the shift between the front surface and rear surface is not uniform. This results in nonuniform thickness distribution, which can cause several issues during molding, which in turn can drastically affect the performance.

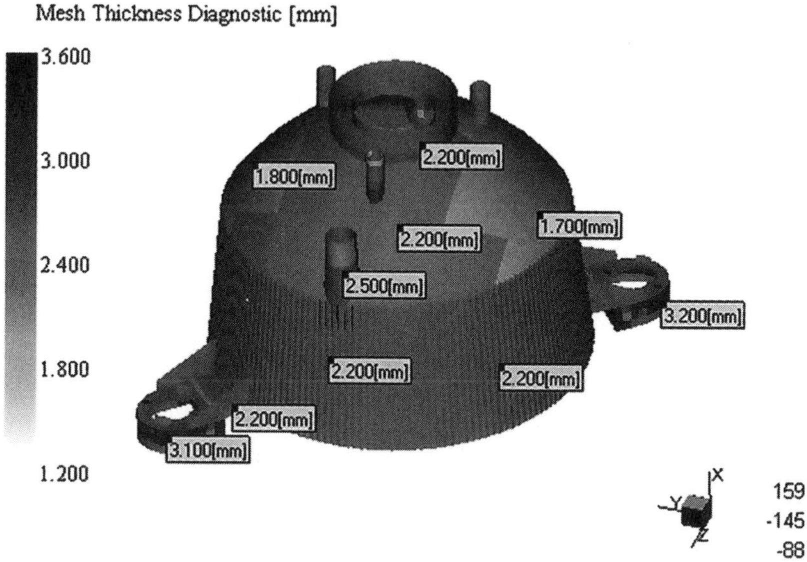

Fig. 6.6 PEI reflector thickness distribution.

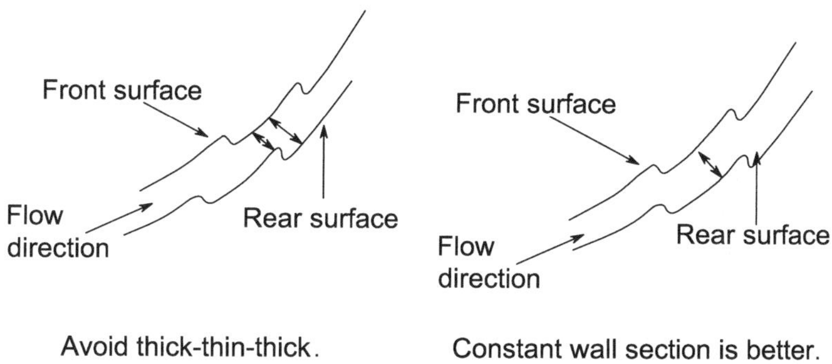

Fig. 6.7 Reflector section design—best practice.

The best practice is to follow design on the right in Fig. 6.7, which has uniform shift in the wall thickness. This helps to give uniform shrinkage of the part, resulting in a good-quality reflective surface after metallizing.

Also, the ribs and bosses have to be designed in such a way as to avoid sink marks, which adversely affect the performance of the reflector [6-10, 6-11]. In general, various methods of coring out can be applied to reduce the effect of the sink marks. The thickness at the base of the ribs or bosses should not be more than 60% of the base wall thickness to avoid sink marks [6-10, 6-11], as shown in Fig. 6.8.

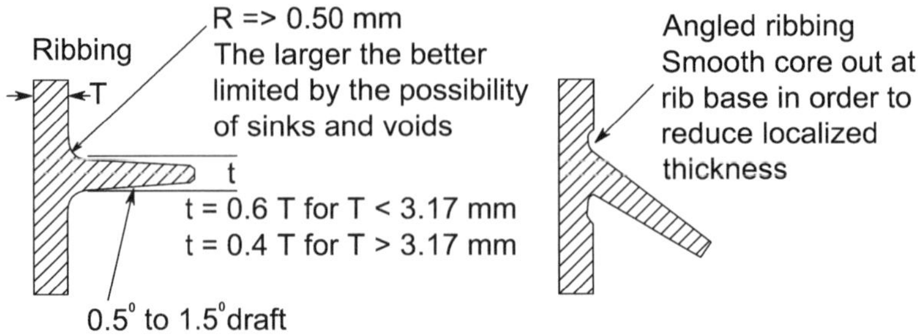

Fig. 6.8 Rib design to avoid sink marks.

6.4.2 Thermal Management

Thermal management of a headlamp reflector is essential as the maximum temperature of the reflector needs to be less than the heat-deflection temperature (HDT) of a particular plastic material [6-12]. The design cycle of a headlamp reflector starts with a given design of the reflector based on the optical requirement from the lighting company, as shown in Fig. 6.9.

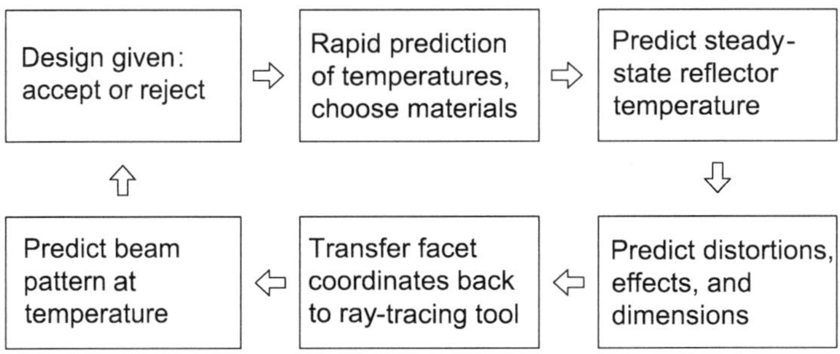

Fig. 6.9 Design cycle of a headlamp reflector.

The next step includes quick prediction of the reflector temperature through hand calculations and quick methodology. Thus, one can get a quick idea of the temperature level in the headlamp assembly. The next step is to calculate the steady-state temperature profile of the reflector, when all the bulbs are switched on. The steady-state temperature profile is transferred to a finite element (FE) solver to predict the thermal distortion based on the constraints imposed by mounting of the reflector. Then this deformed geometry of the reflector is passed on to a ray-tracing tool to estimate the resulting beam pattern. The next step is to check whether the beam pattern is meeting all the regulations. When all the regulations regarding the beam pattern have been met, a given design is accepted.

All the parts inside the headlamp get heated up as a result of the thermal loads from the bulbs, the primary source of heat and light. Thermal management of automotive headlamps becomes critical as well as challenging, given that the need for optical output keeps increasing while the design space for the headlamp remains the same or keeps decreasing because of the styling constraints. A comprehensive thermal analysis can give an idea of the temperatures on the components inside the headlamp assembly. The customer can choose the grade of the plastic based on the thermal requirement of the particular component.

Computer-aided engineering (CAE) analysis is important to estimate the order of temperatures and thermal deformations on the components inside the headlamp assembly. It also gives the flexibility to choose the material based on the thermal requirement of a particular component. In general, a computational fluid dynamics (CFD) study is performed on the headlamp, considering the natural convection inside the assembly and modeling the heat transfer including thermal radiation to find the temperature of various components inside the headlamp assembly.

The objective is to determine the maximum temperature of the reflector in the headlamp assembly in the steady-state condition and to predict the thermal deformations based on the temperature profile of the reflector. Figure 6.10 shows temperature profile of a typical reflector at the steady-state condition.

Along with the temperature profile, the airflow pattern inside the headlamp is also studied at various locations to observe the flowfield inside the headlamp. Figure 6.11 shows the velocity vectors at a cross section of a headlamp.

High velocities can be observed near the bulb shield, and the high temperature on the reflector is caused by the flow of air (natural convection). In natural convection, as the temperature of the air increases, its density decreases and becomes lighter. This causes the air to travel toward the upper side of the reflector, resulting in higher temperatures on the upper side of the reflector, as shown in Fig. 6.10.

The flow is mainly driven by natural convection. It can also be observed, from Fig. 6.10, that the hotspot location is a small percentage of total area. Temperature decreases quickly as we move away from the hot spot. Hotspots are concentrated at typically less than 15% of the total surface area. Multiple methods exist that enable reduced temperature and low-cost thermoplastic reflector solutions. Infrared (IR) transparent resins provide a total temperature reduction of up to ~12 °C (53.6 °F). Novel bulb shield

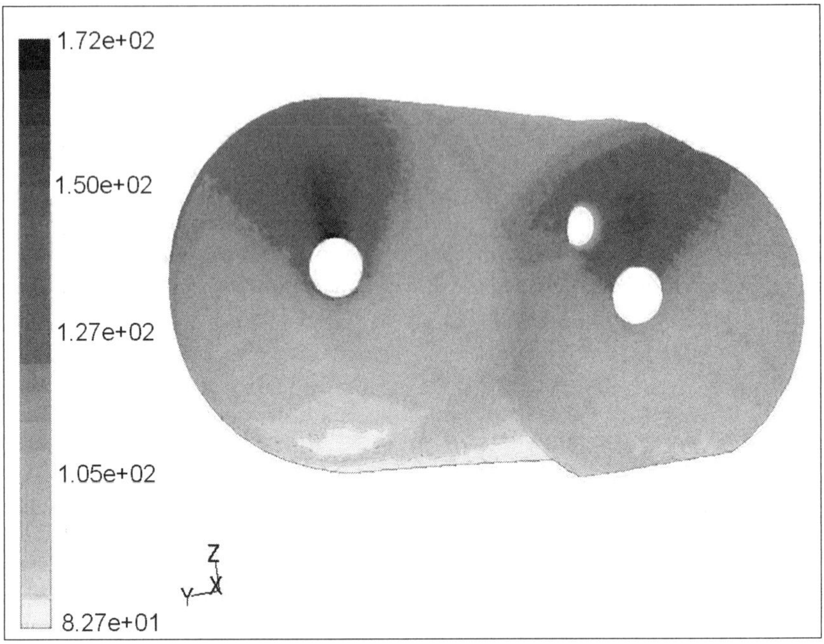

Fig. 6.10 Temperature contour of a reflector.

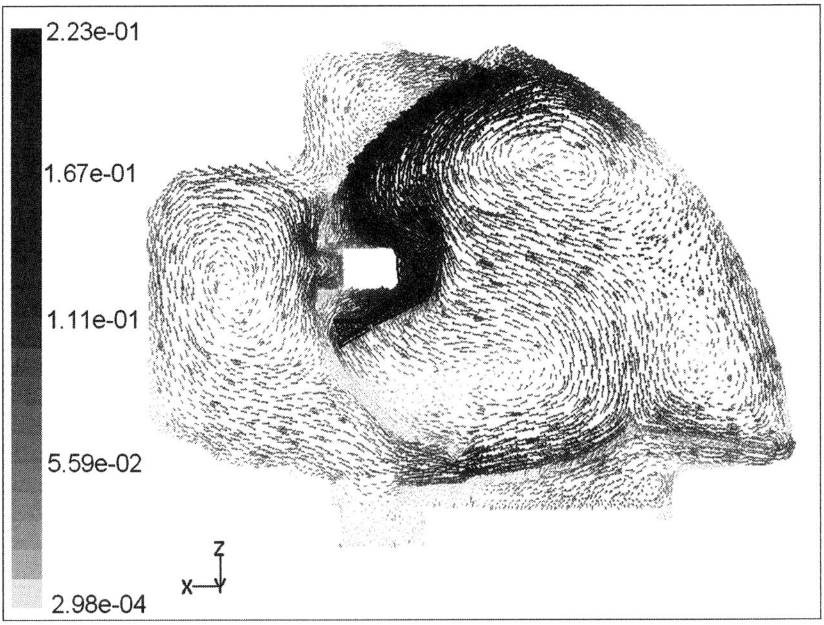

Fig. 6.11 Velocity vectors at a cross section of a reflector.

technologies can provide an additional ~25–30 °C (77–86 °F) temperature reduction based on the design of the headlamp assembly. Attention to heat management enables the lowest cost material selection and robust thermal performance.

Quick tools are also available [6-13] based on the representative geometrical parameters of a headlamp assembly to predict the temperature on the reflector and lens. Thermal predictions by these quick tools are developed with parametric transfer functions using a design of experiments (DoE) approach [6-14]. This tool predicts the maximum temperature of the reflector and lens based on certain geometrical dimensions of the headlamp assembly [6-15]. The geometric features of HB and LB reflectors, along with all other components of the headlamp assembly, are considered, and their influence on the reflector and lens temperature is studied thoroughly. Eighteen parameters are identified that may affect the temperature of the HB and LB reflectors. Through screening a process was established to decouple the whole headlamp assembly into separate LB and HB reflectors.

Further, two configurations for the headlamp assembly were considered: horizontal and vertical. In the horizontal configuration of the headlamp, LB and HB are generally positioned side by side. In the vertical configuration, the LB reflector is typically on the upper side of the HB reflector in the headlamp assembly. In horizontal configuration, it was observed that the geometric parameters of the HB reflector were not influencing the LB reflector temperature. It was also observed that the influence of HB parameters on LB reflector temperature was significant in the vertical configuration of the headlamp. Based on these observations, an individual transfer function for each reflector [6-15] can be developed. To establish the link between HB and LB reflectors for the vertical configuration, a correction factor has been derived to take into the account the influence of HB geometric parameters on LB reflector temperature.

Three different transfer functions have been established for LB temperature, HB temperature, and a correction factor, as shown Tables 6.1, 6.2, and 6.3. Through these transfer functions, one can predict the temperature of LB and HB reflectors based on the geometric parameters of the headlamp assembly and the configuration of the headlamp (horizontal or vertical). The temperature results are estimated to be within ±10% for hot spot. This tool evaluates design changes instantly. Also, it evaluates the optimum location for the bulb shield and the best way to design a bulb shield [6-16]. Figure 6.12 shows the user interface of this tool that predicts the temperature of the reflector and lens based on the geometrical parameters of the headlamp assembly. The model design space is also shown in Fig. 6.12.

Model design space			
High beam parabola			
Parameter	Min	Max	Input
Reflector depth (mm)	30	110	
Reflector height (mm)	70	340	
Wattage of the bulb (watts)	35	75	
Low beam parabola			
Reflector depth (mm)	30	110	
Reflector height (mm)	70	340	
Reflector width (mm)	56	400	
Diameter shield (mm)	11	65	
Spacing of shield (mm)	0	38	
Wattage of the bulb (watts)	35	75	
Calculate			

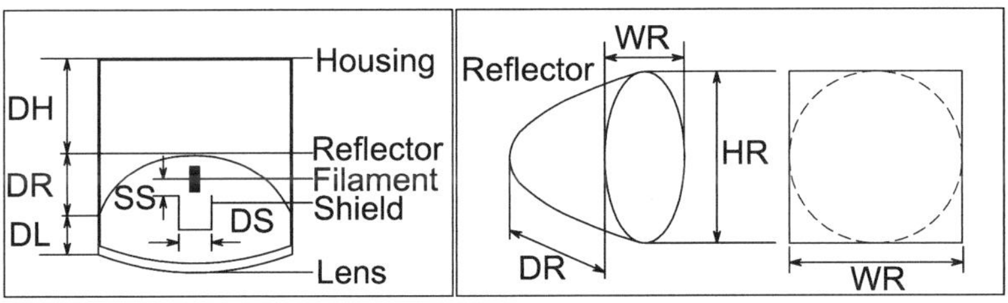

Fig. 6.12 Tool to predict the hot-spot temperature of the reflector and lens.

Table 6.1 Transfer Function for the LB Reflector Temperature [6-14]

Results for Low_Beam_Temperature

Variable	Coefficient	*p*	Interpretation
Constant	1034.929608	7.56338E-70	
reflD_LB	-7.506954413	1.24723E-54	Significant
fact_focal_LB	-1033.185699	5.74187E-34	Significant
fac_width_LB	-68.67085701	9.53978E-07	Significant
fact_dia_shield	-10.18787879	0.000322131	Significant
fact_dist_shield	-539.9381127	1.86892E-27	Significant
Wattage	4.034241241	4.80466E-25	Significant

Table 6.1 Continued

Results for Low_Beam_Temperature

Variable	Coefficient	*p*	Interpretation
reflD_LB*fact_focal_LB	4.28875	9.44534E-23	Significant
reflD_LB*fac_width_LB	0.322265625	0.061562327	Insignificant
reflD_LB*fact_dist_shield	1.476339286	2.44691E-10	Significant
reflD_LB*Wattage	−0.014800781	3.36644E-12	Significant
fact_focal_LB*fac_width_LB	62.375	0.253616548	Insignificant
fact_focal_LB*fact_dist_shield	469.8571429	2.87418E-10	Significant
fact_focal_LB*Wattage	−2.21875	0.000129619	Significant
fac_width_LB*Wattage	−0.3640625	0.286026607	Insignificant
fact_dist_shield*Wattage	−2.298214286	1.66122E-07	Significant
reflD_LB^2	0.026041912	1.7464E-08	Significant
fact_focal_LB^2	452.2917647	0.272099075	Insignificant
fact_dist_shield^2	540.9651861	0.011827449	Significant

Statistics

Std Err	10.819
R-sq	99.05%
R-sq(adj)	98.76%
R-sq(pred)	98.20%

Table 6.2 Transfer Function for the HB Reflector Temperature [6-14]

Results for High_Beam_Temperature			
Variable	**Coefficient**	***p***	**Interpretation**
Constant	919.2699523	1.66259E-45	
DR	-5.800734892	8.36974E-33	Significant
HR	-687.9549164	1.65907E-20	Significant
WR	-233.2068779	0.000343617	Significant
LS	-0.121414216	2.06902E-06	Significant
Wattage	1.332585784	2.35509E-13	Significant
DR*HR	2.9875	1.18956E-15	Significant
DR*WR	0.21875	0.070515113	Insignificant
DR*LS	0.002620833	0.000229949	Significant
DR*Wattage	-0.0095	6.87804E-09	Significant
HR*LS	-0.28	0.169497619	Insignificant
LS*Wattage	-0.001316667	0.297757877	Insignificant
DR^2	0.021181216	1.57255E-11	Significant
HR^2	335.3565684	0.104835758	Insignificant
WR^2	99.74865952	0.212326741	Insignificant
Statistics			
Std Err	5.265		
R-sq	0.99561		
R-sq(adj)	0.99342		
R-sq(pred)	0.98827		

The tool for the HB and LB parabolas predicts the corresponding hot temperature by taking the influence of bulb shield geometry into account. The purpose of a bulb shield in headlamps is to eliminate glare (unwanted reflections) on the eyes of oncoming drivers. The bulb shield plays an important role on the hot-spot temperature of the reflector, making the position and size of bulb shield as critical parameters in design. Such a

Table 6.3 Transfer Function for the Correction Factor [6-14]

Results for LB_Temp

Variable	Coefficient	*p*	Interpretation
Constant	776.2205035	1.11675E-54	
LB_Depth	-3.385632031	2.53202E-38	Significant
LB_Fact_Focal	-334.2249375	1.50639E-25	Significant
HB_Depth	-0.393611523	1.66878E-14	Significant
HB_Fact_Focal	-14.43125	0.330097503	Insignificant
HB_wattage	0.208695313	0.000159764	Significant
LB_Depth*LB_Fact_Focal	2.327375	7.41874E-20	Significant
LB_Depth*HB_Depth	0.003566797	2.38882E-13	Significant
HB_Depth*HB_Fact_Focal	0.164375	0.03560943	Significant
HB_Depth*HB_wattage	-0.001789844	0.000768652	Significant
Statistics			
Std Err	2.077		
R-sq	99.96%		
R-sq(adj)	99.94%		
R-sq(pred)	99.91%		

predictive tool in hand would save a lot of time and cost involved in manufacturing and testing each new bulb shield.

6.4.3 Structural Performance

A headlamp reflector is subjected to structural loads that result from mounting of the reflector to the housing inside the headlamp assembly. Thermal load on the reflector is caused by the temperature of the reflector. The reflector has to withstand these thermomechanical loads. Typically, a finite element analysis (FEA) is performed to study the thermal deformations of the headlamp reflector. Typical mounting locations of a reflector are horizontal aiming, vertical aiming, and at base point (Fig. 6.13). The modern-day aiming system generally consists of a reflector that is mounted at three points, as shown in Fig. 6.13.

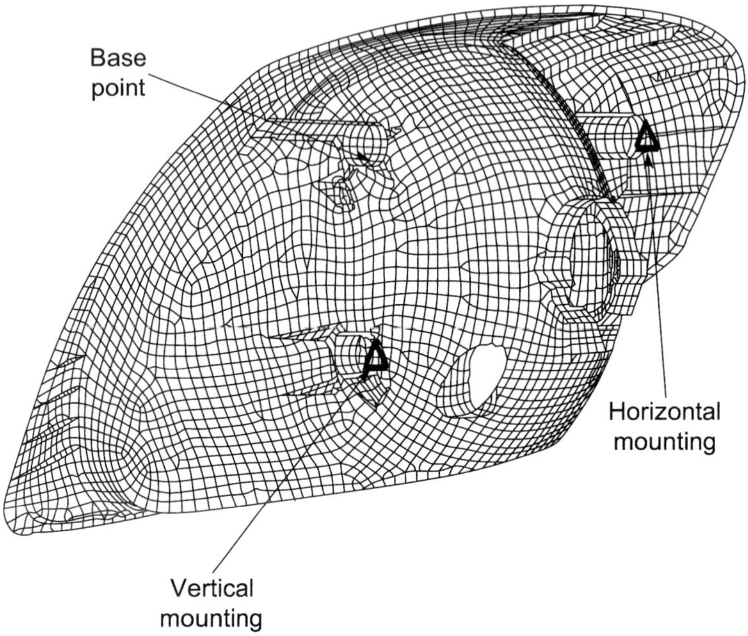

Fig. 6.13 Mounting locations of a typical reflector.

The reflector mounting scheme is generally L-shaped in configuration and provides a three-point suspension for the reflector, in which two of the points are adjustable for horizontal and vertical aiming purposes. The mounting features at these two points also allow for translational motion between the reflector and the adjustment screw. The two adjustment points are located on opposed legs of the "L." The third point is fixed in position and is located at the intersection of the two legs. Traditionally, these mounting provisions are made on the back side of the reflector through ribs and bosses.

A CFD study [6-16] is used to calculate the temperature distribution on the reflector, and an FE solver [6-17] is used to study the mechanical behavior of the reflector subjected to thermal load. A sequentially coupled approach is used to convert the temperature data on the CFD mesh. An FE mesh is generated and imported into the CFD solver. For each node on the FE mesh, corresponding elements of the CFD solver are located, and the temperature at the node is interpolated and written as the temperature output for each of the nodes of the FE mesh. Temperature output from the CFD solver is used as input to carry out thermomechanical analysis [6-18]. Figure 6.14 shows the thermal deformation of a typical reflector subjected to structural and thermal loads.

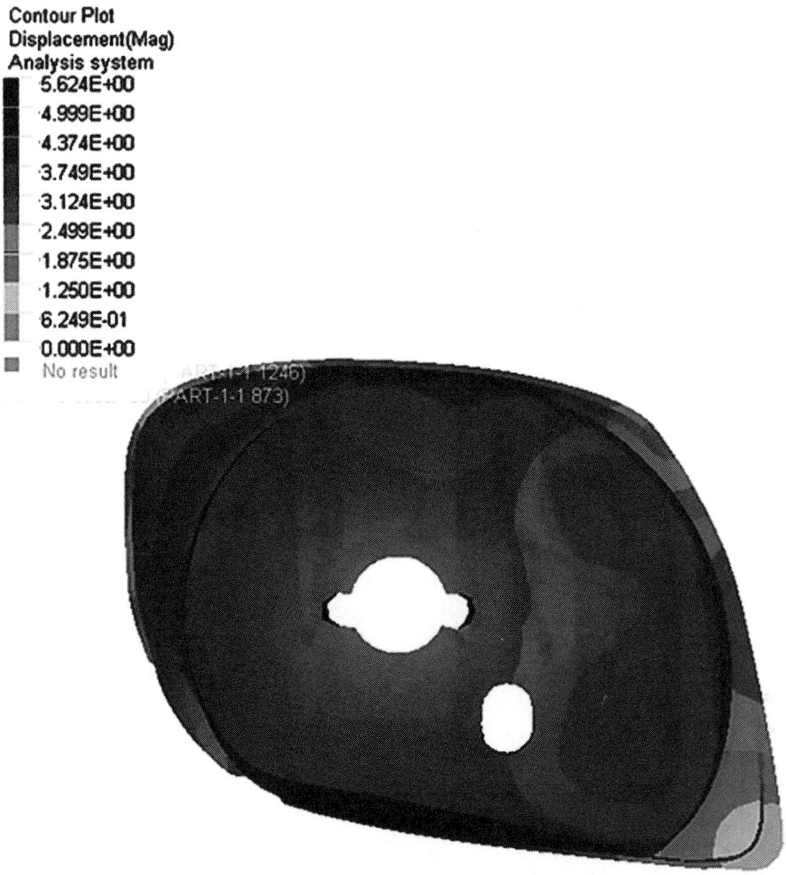

Fig. 6.14 Thermal deformation pattern of a typical reflector.

It can be observed that the deformation of the reflector toward the end side increases as the distance from mounting location increases. Through CFD and FE studies, the deformation of the reflector and stresses on the reflector can be predicted.

6.4.4 Beam Pattern and Optical Performance

One of the functions of a headlamp reflector is to generate the beam pattern, according to ECE R31 regulations. This beam pattern will be deflected as the thermal deformation of the reflector increases. An example illustrating a beam pattern and its movement is described below. A reflector beam is also modeled in an FE model, using 1000-mm rigid elements (Fig. 6.15).

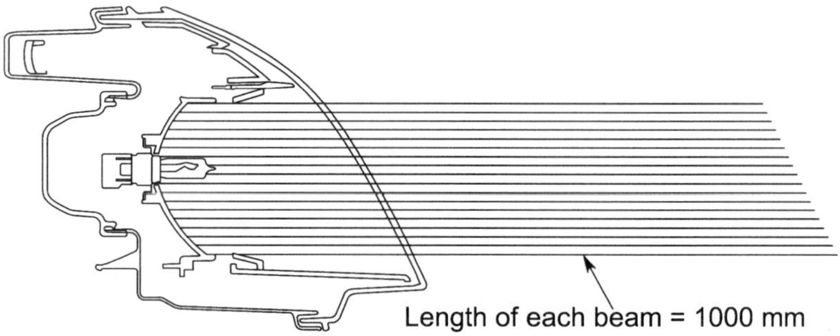

Fig. 6.15 FE mesh with representative beam model.

The massless 1-D horizontal beam is monitored for its angular deformations in horizontal and vertical directions, which represent the changes in beam angle values in both these directions.

To identify the beam angle change pattern at different zones of reflector, the complete reflector is segregated into six different areas, as shown in Fig. 6.16.

Through this process, one can gain a good understanding of the beam angle changes in different areas of the reflector. Figure 6.17 shows the nodal locations for the big area A.

All the nodes displayed in Fig. 6.17 are monitored through massless 1-D horizontal beams attached at these nodes. The angle changes for big zone A, each small area contained therein, and the average change of big area A in the horizontal and vertical directions can be seen in Table 6.4. Through this process, the computing of a beam angle change due to thermal deformation of the reflector has been enabled.

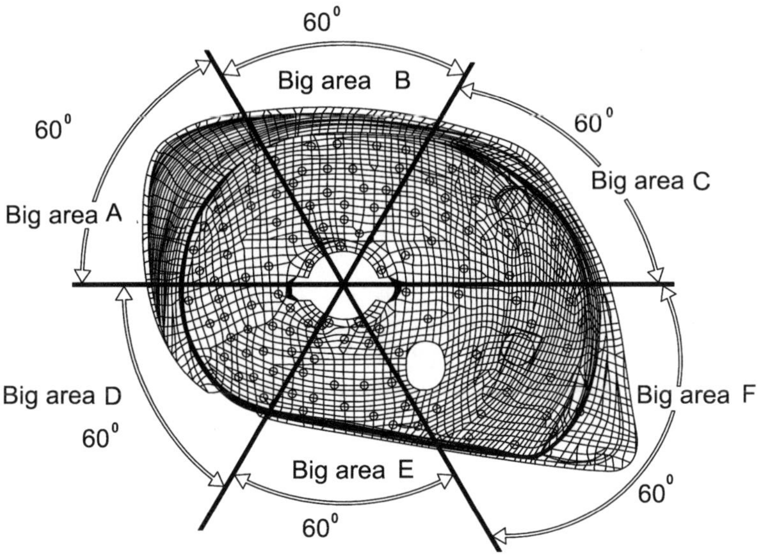

Fig. 6.16 Zonal distribution of the reflector face.

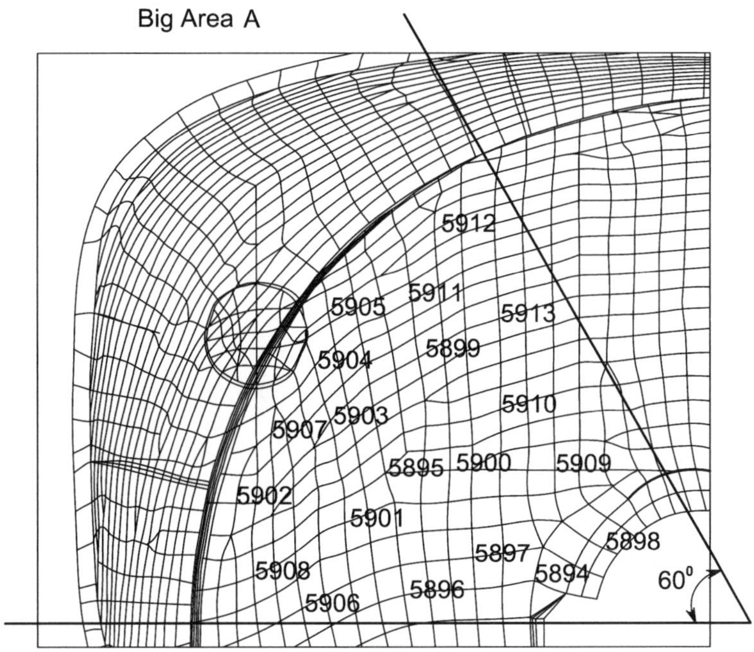

Fig. 6.17 Nodal locations for big area A.

Table 6.4 Beam Angle Change for Big Area A

		Beam End Displacement		Angle Change		Average Displacement		Average Angle Change	
Big Area	Small Area	Y-Disp. (mm)	Z-Disp. (mm)	About Y (in Deg.)	About Z (in Deg.)	Y-Disp. (mm)	Z-Disp. (mm)	About Y (in Deg.)	About Z (in Deg.)
A	5894	0.658	0.190	0.048	0.011	0.897	0.876	0.060	0.050
	5895	0.383	1.327	0.033	0.076				
	5896	0.868	0.692	0.061	0.038				
	5897	1.526	1.080	0.098	0.062				
	5898	0.645	1.395	0.046	0.079				
	5899	0.861	0.439	0.060	0.027				
	5900	1.228	1.282	0.081	0.074				
	5901	0.385	1.372	0.034	0.077				
	5902	1.980	1.047	0.101	0.057				
	5903	0.967	1.366	0.066	0.078				
	5904	0.741	0.844	0.054	0.048				
	5905	0.054	0.033	0.015	0.001				

5906	0.242	1.119	0.026	0.061
5907	0.673	0.977	0.027	0.054
5908	0.286	1.622	0.029	0.089
5909	2.444	1.247	0.149	0.074
5910	1.463	0.605	0.093	0.037
5911	0.711	0.307	0.052	0.020
5912	0.596	0.546	0.045	0.036
5913	1.233	0.036	0.080	0.002

6.4.5 Stress-Free Reflector through Reflector Bracket

The primary optical performance requirement for forward lighting is that the headlamp must conform to one of the established SAE International or ECE luminous intensity standards [6-3]. To meet these regulations, the headlamp beam pattern requires a sharp cutoff line, as shown in Fig. 6.18.

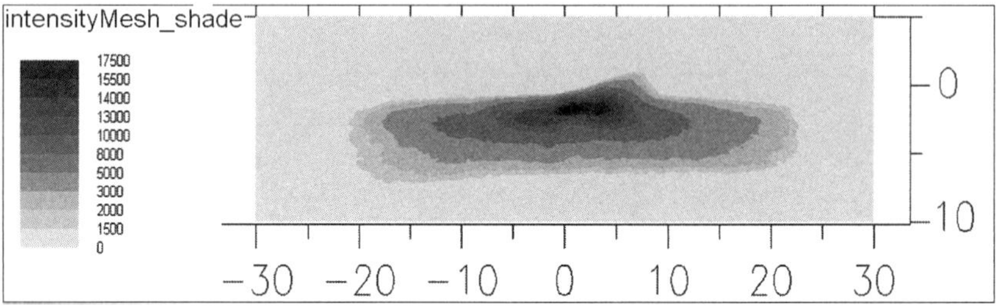

Fig. 6.18 Beam pattern of an LB reflector.

Above the cutoff line resides the glare-control zone, in which luminous intensity values are minimized for the safety of oncoming drivers. Below the cutoff line is the illuminated zone, in which luminous intensity values must observe appropriate values to ensure roadway visibility for the driver [6-3]. To help achieve this stringent beam pattern cutoff line, the headlamp design calls for precise alignment and dynamic adjustments for the headlamp reflector mechanism, which is achieved through an aiming system.

The trend toward large, clear PC headlamp lenses has further increased the surface requirements on the visible elements in a headlamp, such as the reflector and bezel. Packaging requirements have become more compressed as well, and today's reflectors (along with the materials are made of) must satisfy considerably stringent requirements regarding manufacturing tolerances, ability to process different designs, HDT, mechanical strength, and, in particular, surface quality of the parts.

This calls for a reflector surface that is free of sink marks, which can be achieved more easily if the mounting features and ribs can be removed from the back side of the reflector. A neat reflector (a reflector without B-side features) would also reduce the residual stresses in the part. A design that reduces the thermal stresses in the reflector is the most desired to assure minimal beam pattern distortion.

In general, a headlamp reflector consists of common features that enable or incorporate horizontal adjustment, vertical adjustment, pivot, bulb shield attachment, and bulb attachment. The aiming or adjustment features like horizontal, vertical, and pivot, which can be major contributors to high stress in the reflector, are removed from the back side of the reflector and incorporated in a separate reflector bracket, thereby making the headlamp reflector free of features on the rear surface. The new bracket with all the functional features for its aiming mechanism is shown in Fig. 6.19. The manufacturability can be further simplified by introducing plug-in snaps that can eliminate the side cores in the tool [6-19]. The reflector and bracket can be interlocked using a turn-and-lock concept. The turn-and-lock snap design comprises a hook on the reflector and locking feature on the bracket. The hook snaps into the lock by turning the bracket relative to the reflector and thereby locking the parts together as shown in Fig. 6.19.

Various nodes, as shown in Fig. 6.20, are analyzed to observe the angular deformation in the horizontal and the vertical directions. Node numbers of various nodes are also shown in Fig. 6.20.

The typical constraining conditions of a traditional reflector design and the reflector with the new bracket design are shown in Fig. 6.21.

Comparison of the angular deformations of a reflector with the new bracket versus a traditional mounting design of the reflector has been studied [6-19], and the results are shown in Table 6.5. From this table, the reduction of angular deformation of the massless 1-D beams (rays) can be seen at various locations on the reflector parabola. The angular deformations in both horizontal and vertical directions, along with percentage of

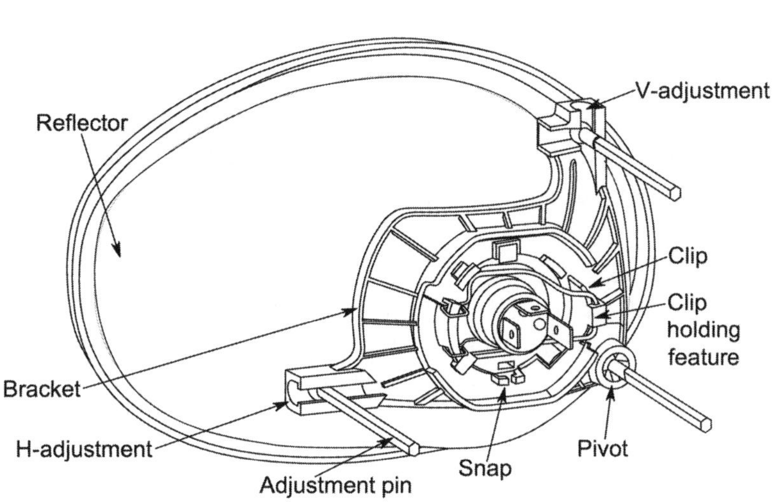

Fig. 6.19 Assembled view of headlamp reflector and bracket.

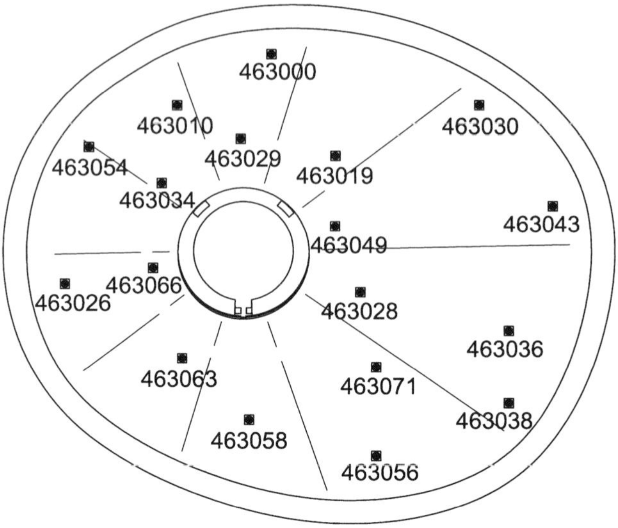

Fig. 6.20 Representative nodes considered for the study.

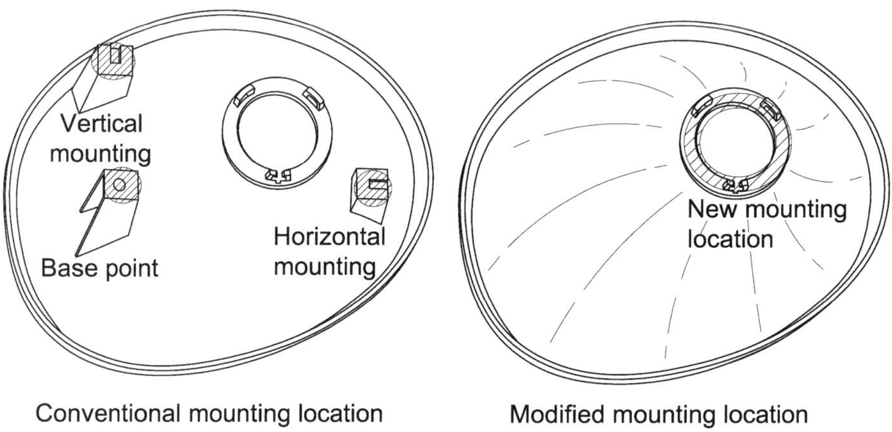

Fig. 6.21 Comparison of boundary condition details.

reduction, are shown. This study demonstrates that the reflector bracket contributes to a significant reduction of beam angular deformations in both the horizontal and vertical directions compared to the traditional reflector mounting design. The reduction in the vertical angle is expected to indicate the reduction in the vertical beam drift that would be measured by conventional vertical scan methods according to ECE regulations [6-3].

Table 6.5 Angular Deformation Results [6-19]

Location	Horizontal Angle Change (in milliradians)			Vertical Angle Change (in milliradians)		
	Conventional Mounting	Modified Mounting	% Reduction	Conventional Mounting	Modified Mounting	% Reduction
463058	0.403	0.105	74.0	0.260	0.045	82.6
463056	0.195	0.093	52.7	0.438	0.089	79.7
463034	0.412	0.089	78.4	0.380	0.009	97.7
463010	0.438	0.145	66.9	0.255	0.187	26.7
463054	0.534	0.274	48.7	0.185	0.009	95.3
463019	0.183	0.140	23.8	0.400	0.180	55.0
463029	0.291	0.031	89.2	0.342	0.001	99.7
463000	0.234	0.042	82.1	0.293	0.065	78.0
463049	0.199	0.141	28.9	0.194	0.009	95.5
463028	0.195	0.136	30.4	0.354	0.182	48.8
463071	0.175	0.143	18.0	0.400	0.056	86.0
463026	0.503	0.213	57.6	0.216	0.073	66.1
463066	0.454	0.143	68.5	0.316	0.061	80.7
463063	0.429	0.122	71.5	0.171	0.084	51.0
403030	0.080	-0.218	371.7	0.508	0.182	64.3

6.4.6 Tooling and Processing

While tool design for PEI reflectors is similar to any other thermoplastic injection-molding tooling, some additional precautions need to be taken.

6.4.7 Gate Design

Gate design plays an important role in ensuring uniform filling of the part with minimum pressure. Typically for this kind of part, uniform filling can be achieved by gating at the bulb holder area. Edge gates have also been effective in achieving uniform filling, but in few a cases, this may cause imbalance, which in turn causes higher pressure. Hence, selection of the gate at the center or the edge depends on the part geometry, flow length, and tool configuration (e.g., number of cavities and positioning of knit lines).

Figure 6.22 shows a recommended use for the diaphragm gate for reflectors [6-20].

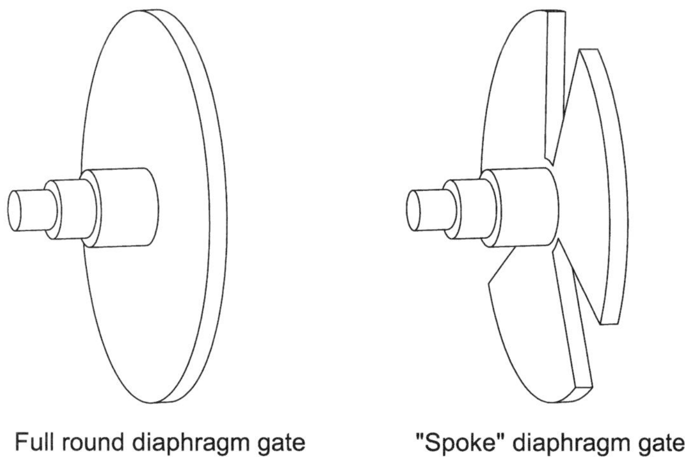

Fig. 6.22 Diaphragm gate design.

This type of gate provides good filling and packing control to minimize warpage. The disadvantage of using this gate is the difficulty in degating, as it requires proper fixture design for cutting the gate. Also, the gate dimensions play a critical role in both filling and degating using mechanical fixtures. Recommended dimensions can be as seen in Fig. 6.23.

The cold sprue can be replaced with a hot sprue to minimize scrap rate. The hot runner system can be successfully used to process PEI material.

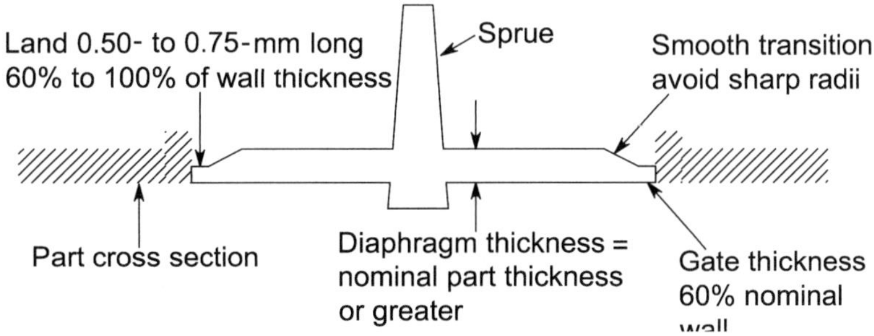

Fig. 6.23 Details of diaphragm gate design.

The system and controllers should be able to handle a temperature in the range of 350–400 °C (662–752 °F).

A typical edge gate design is depicted in Fig. 6.24.

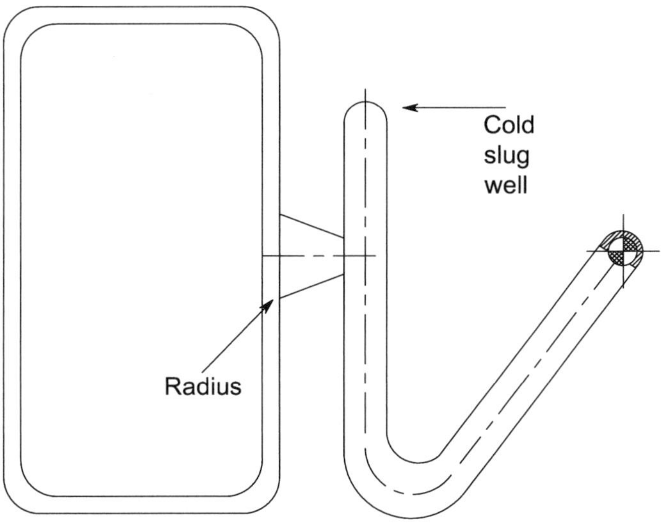

Fig. 6.24 Edge gate design configuration.

A cold slug well is recommended to avoid cold material entering the cavity. The dimensions may be similar to those in Fig. 6.23 representing a diaphragm gate. The radii at the gate location must be ample in order to minimize localized pressure drop and related imperfections. Typically, the edge gate can be placed in such a way as to achieve a uniform flow balance and to shift the knit line to the noncritical areas [6-20]. A mold-filling analysis may be recommended to evaluate gate design upfront for a proper tool design [6-21].

6.4.8 Venting

Venting of the tool helps in removing the air and other volatiles entrapped during the closing and filling of the part inside the cavity. Sufficient venting needs to be provided at the parting line area and at the end of fill regions. Also, vertical air vents in ejector pins or side cores will provide proper removal of air and avoid local degradation of material causing black spots. A typical air vent design that can be applied for PEI material [6-20] is as shown in Fig. 6.25. A tool designed with good air vents will help to reduce pressure, avoid air entrapment, and optimize the surface quality.

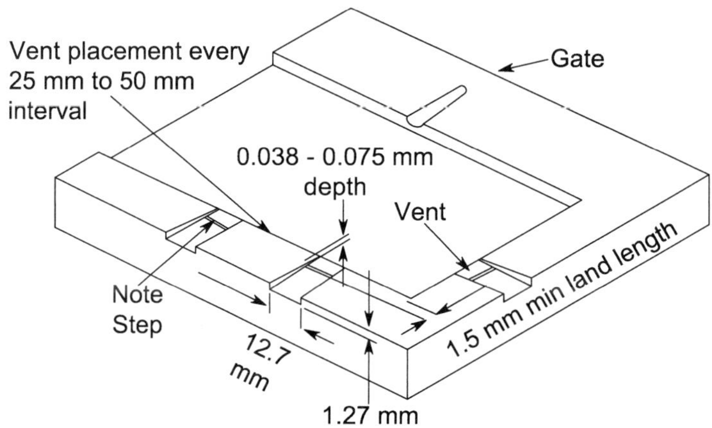

Fig. 6.25 Air vent design guide line.

6.4.9 Tool Thermal Management

PEI needs a higher tool temperature to be maintained, in the range of 135–165 °C (275–329 °F) [6-9], than the regular thermoplastic materials. Although this level of temperature is achieved through oil heating, the recommended way is to use high-pressure water-heating systems. With an oil heating system, there is a chance of oil leakage into the core/cavity that can get carried away with the molded part. It has been found that during metallization, the oil reacts and leaves patches on the surface, which may not be acceptable. Although oil leakage can be prevented by using high-temperature seals, it is better practice to use high-pressure water-heating systems to avoid long-term problems.

The design of a cooling circuit plays an important role in the uniform removal of heat. Uneven cooling causes differential cooling, resulting in poor surface defects and warpage. Hence, care should be taken to run cooling lines close to the part. If possible, use of conformal cooling, that is, a cooling circuit following the profile of the part, should be used [6-22]. Also, use of heat pipes or BeCu (beryllium copper) inserts is recommended for heat removal if the part region is inaccessible through cooling lines. Mold-insulating sheets are recommended to avoid heat loss to the surrounding areas, to improve heating efficiency, and to maintain uniform cycle time. The recommended areas where insulation can be used are shown in Fig. 6.26.

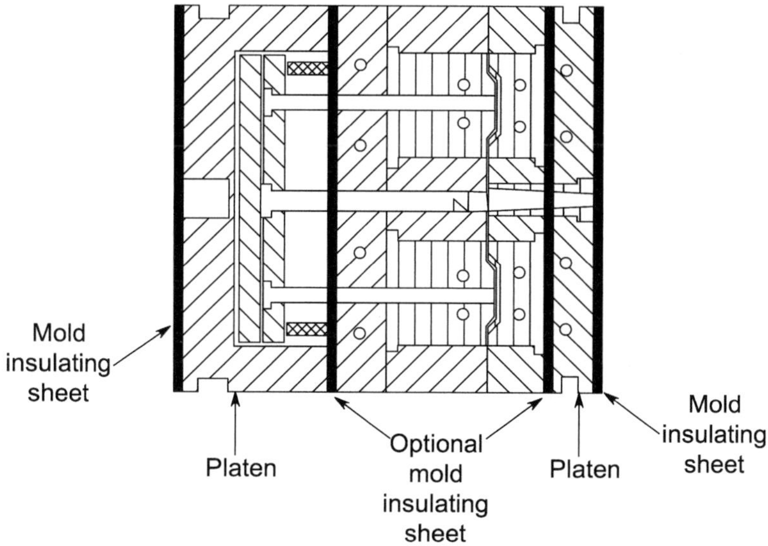

Fig. 6.26 Insulation of mold to minimize heat loss.

Another important factor to be considered is the possibility of moving parts getting mechanically bonded as a result of the high temperature of the tool. Parts such as the guide pillar and guide bush, sliders, and ejector pins need to be provided with adequate clearance to manage the expansion of the tool during heating and still maintain the smooth functioning of these moving parts. Also, use of high-temperature lubricants is recommended.

6.4.10 Tool Surface Treatment

Good tool steel is required to withstand high temperature and continuous cycle operation. Industry-recommended 420 stainless steel with minimum 46 HRc to use for a reflector tool offers good polish ability with minimal risk of corrosion. Standard P20 may not be advisable, as it is prone to corrosion as compared to 420 steel with PEI material. Double-vacuum remelt steel can also be used successfully, because the vacuum remelt process removes sulfides that can lead to surface pitting. It is strongly recommended not to use nitrided surfaces, as PEI has some affinity toward nitride.

6.4.11 Processing

PEI requires a higher processing temperature, in the range of 350–400 °C (662–752 °F). Before injection molding, the material has to be predried for 4 hours at 150 °C (302 °F). The dryer should be capable of removing moisture to less than 0.02% [6-9]. A low-shear screw design with an L/D ratio of 20 is recommended. While a general purpose screw design works well, nitriding for surface hardness is not recommended. Typically injection speed has to be faster to fill the low-wall thickness regions, and the injection time depends on the volume of part to be filled.

6.5 Headlamp Bezels

Headlamp bezels are the supporting structure between the housing and lens. The shape of the bezel becomes complex as the geometry of the headlamp becomes complex. Bezels offer many options for the styling of a headlight besides giving it a clean, finished look. It is important that bezels do not warp or shrink after long exposure to the high heat of the headlamps. It is equally important to have a smooth specular surface for direct metallization. With bezel designs getting more and more sophisticated, with features such as hot projector lamps and integrated fog lamps, the lighting industry is looking to plastics.

Temperature on the bezel is a very important parameter, influencing the choice of material for the bezel application. During the thermal simulation of a headlamp assembly, the temperature of the bezel is a critical output. Figure 6.27 shows the temperature profile of a bezel, where one can observe high temperature limited to a small area of the bezel.

With bezels requiring lower operating temperatures, direct metallizable PBT (polybutylene terephthalate) can be a good choice of material; while for higher operating temperatures, high-heat PC would fit the bill [6-23].

Typically, side edge gates are used for bezels, as shown in Fig. 6.28.

The tooling configuration would generally be a two-cavity tool with left and right side bezels. The gate position is selected based on the flow balance and location of knit lines. A bigger fan gate is suggested for uniform filling. On the other hand, disc gating at the bulb hole opening can be used if a hot runner system is employed [6-20].

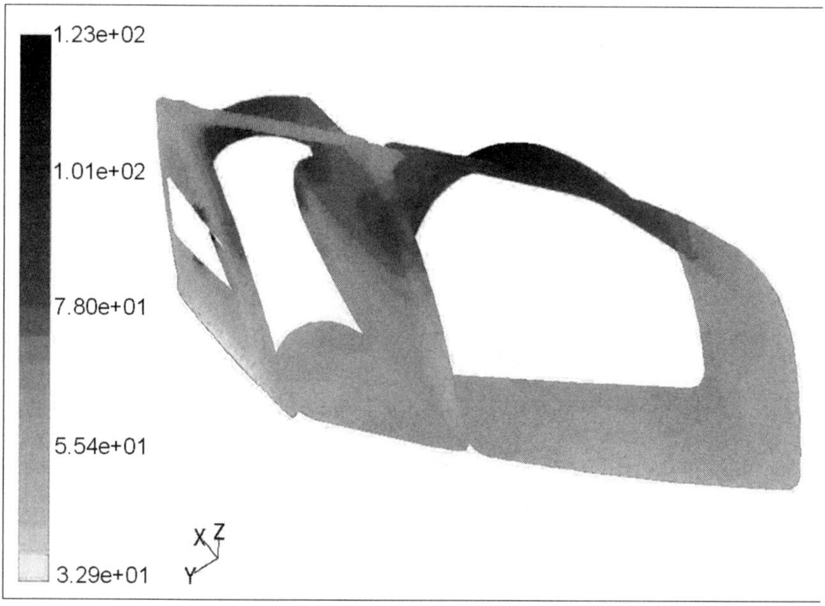

Fig. 6.27 Temperature profile of a bezel.

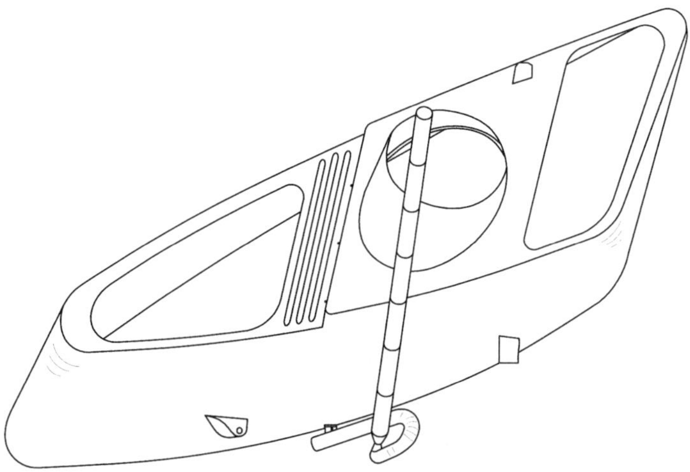

Fig. 6.28 Gating system for bezels.

6.6 Headlamp Lenses

Because of styling considerations, the lighting tiers are increasingly looking at complex headlamp designs with especially complex shapes for clear lens. PC is the popular choice as a lens material, given the feasibility of injection molding of complex shapes. The material grade for the lens is determined by the peak temperature, which has to be lower than the HDT of that material [6-24]. These PC resins are designed to resist high levels of heat and shattering, and they can be molded into almost any shape. The temperature profile of a lens is shown in Fig. 6.29.

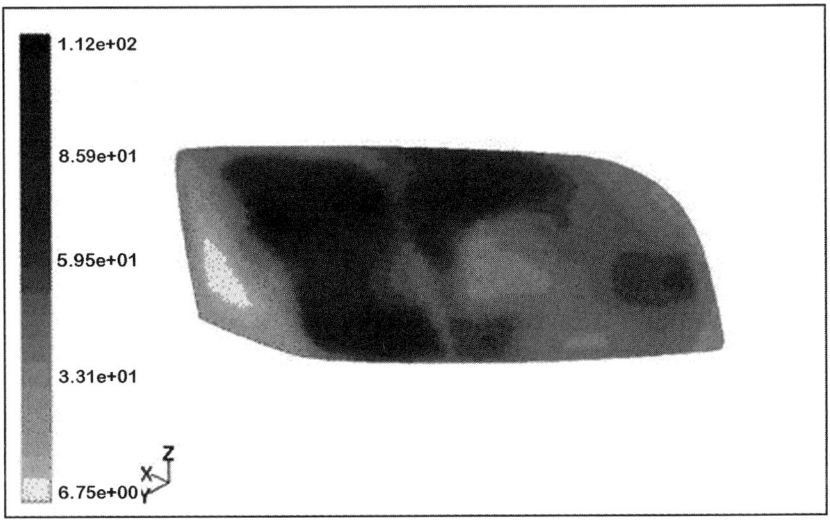

Fig. 6.29 Temperature profile of a lens.

The hot-spot temperature on the lens is primarily a result of the radiation heat transfer from the bulb surface close to the lens. The three hot regions close to the three bulbs of the headlamp assembly are shown in Fig. 6.29. The lens is also subjected to high ambient temperatures and solar radiation. Hence any lens material will have to withstand these loads.

Cold edge gates with either a hot runner system or cold sprue are used for lens parts. Lens parts require the highest level of tool polishing. Large fan gates with cold sprue are recommended for the hot sprue system to avoid black specks in the part, which is a common phenomenon while using a hot runner system. Typically, the gating is placed at the center edge of the part to balance the flow. However, modern lenses are of complex shape and larger in size, and hence the pressure required fill these parts is higher. Therefore, this factor needs to be considered when designing the feed system [6-10, 6-20].

The industry is moving toward thin wall lenses to reduce the system mass. This provides a challenge in the processing of lenses with the same material configuration. Hence, significant effort is being made to improve the flow ability of the material so that thin molding is possible [6-25].

6.7 Headlamp Assembly—Pedestrian Safety

According to a study on pedestrian protection in Europe [6-26], each year about 2,000 pedestrians are killed and 17,000 are injured from pedestrian–vehicle accidents. The percentage of pedestrians injured each year across the globe is around 2% of the 2.9 million (total persons injured in traffic crashes). On an average, every 7 minutes a pedestrian is injured in a vehicle impact. On the positive side, pedestrian fatalities have gone down by almost 16% over the last decade, primarily because of increased awareness toward pedestrian safety and impact regulations.

Generally, the lower-leg impact test is carried out by impacting the lower-leg model into the front of the vehicle, and the upper-leg impact test is carried out by impacting the upper-leg model into the front of the vehicle. Vehicle safety star ratings are assigned based on the cumulative impact performance of all the frontal components of a vehicle.

As pedestrian impact regulations are continuing to evolve [6-27, 6-28], there is a growing emphasis on each component of a vehicle front to be pedestrian friendly. Traditionally, pedestrian safety during an automotive impact has been ensured through appropriate design of bumper, grille, fender, and hood. Because the headlamp also participates in an overall frontal impact, a new concept, the energy absorber ring (EAR) around the headlamp housing, has been developed to make the headlamps pedestrian safe as well [6-29]. In this approach, the EAR will absorb the energy, when subjected to impact load and would crush and fail, if the impact energy crosses the threshold limit [6-30]. The arrangement of EAR in a headlamp assembly is shown in Fig. 6.30.

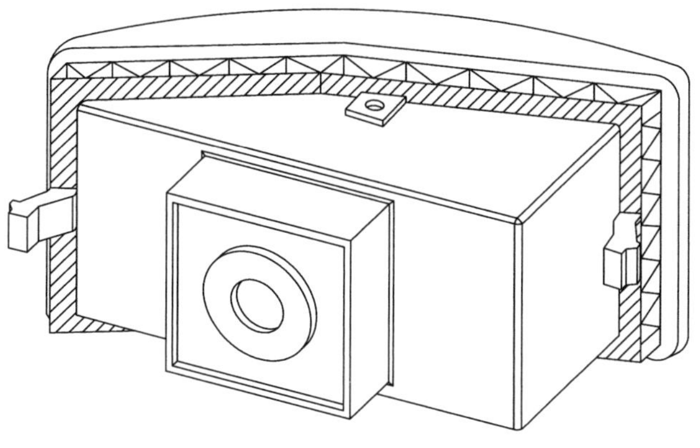

Fig. 6.30 EAR assembly location in headlamp housing.

The headlamp is allowed to move inside the vehicle to take the headlamp away from the impact area. Surrounding components such as the hood, bumper, and fender will come in contact with the impactor and thus absorb the remaining impact energy. This methodology will help in reducing damage to the headlamp, while reducing the reaction forces coming onto the leg impactor.

The EAR design is developed through a sequential approach. Helical spring elements are modeled connecting headlamp housing to its support structure. The spring geometry has to be chosen in such a way that during a pedestrian impact, the spring element would compress first and the headlamp would deform later. A DoE is performed to establish the required force-deflection behavior of the spring elements, for achieving the best performance. A bilinear spring behavior model has been adopted for the spring. Figure 6.31 describes specifications for the force-deflection curve.

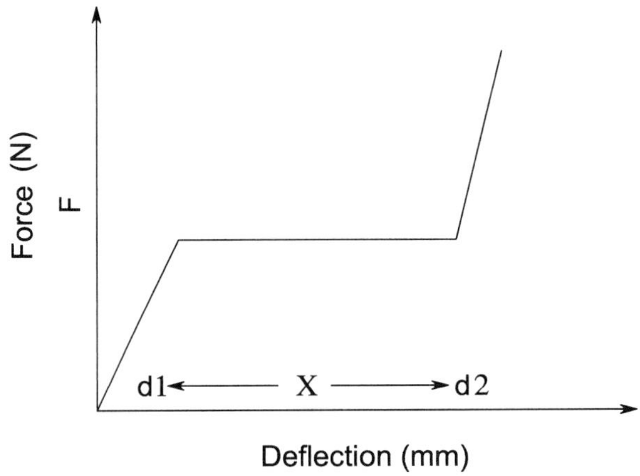

X:	Y:
250 < F < 750	g force
3 < d1 < 7	Force experienced by headlamp
7 < X < 13	

Fig. 6.31 Force-deflection response of the spring elements.

Maximum force can range from 250 to 750 N, and the deformation stroke varies from 10 to 20 mm. The first force peak experienced by the leg and headlamp was monitored. The DoE suggested that as the maximum spring force increases, the force experienced by the leg and headlamp increases.

However, if the time at which the peak-force attained decreases, then the leg and headlamp also experience lower forces. The effect of deformation length (*X*) was not found to be significant.

A component comprising three connectors was designed in the next step. The component geometry was modeled to envelop the headlamp housing and was connected with the housing at three points very close the actual connector locations. The connectors were constrained at the bolting locations, and the lower leg impact was carried out.

The EAR structure is depicted in Figs. 6.32 and 6.33 (Fig. 6.33 is the expanded view of the window shown in Fig. 6.32).

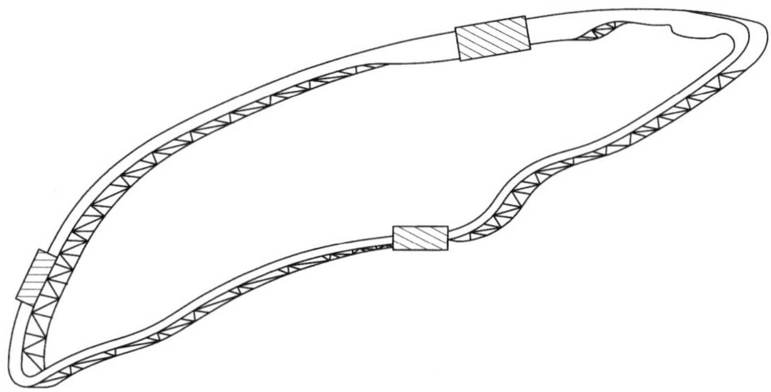

Fig. 6.32 Representative EAR element.

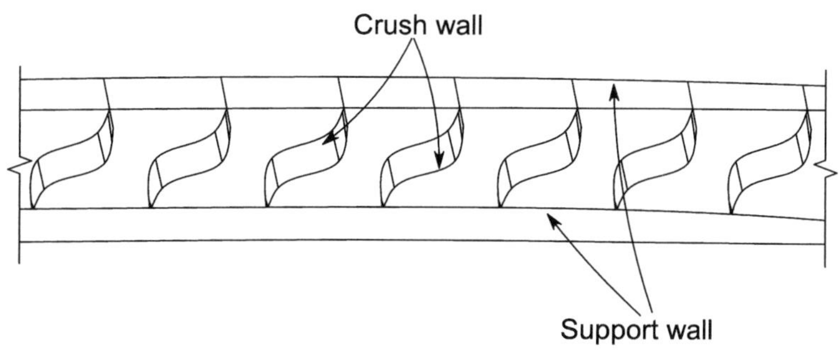

Fig. 6.33 Zoomed view of representative EAR element.

The EAR geometry has two main sections. The first one is the crush walls, which are made of shapes that would enable the crushing during an impact. The second is the support wall, which integrates all the crush walls into a single unit, holds the connectors, and envelops the entire headlamp housing. The EAR unit is attached to the headlamp housing using plug-in type connectors.

The headlamp model that was used to conduct lower-leg impact is shown in Fig. 6.34.

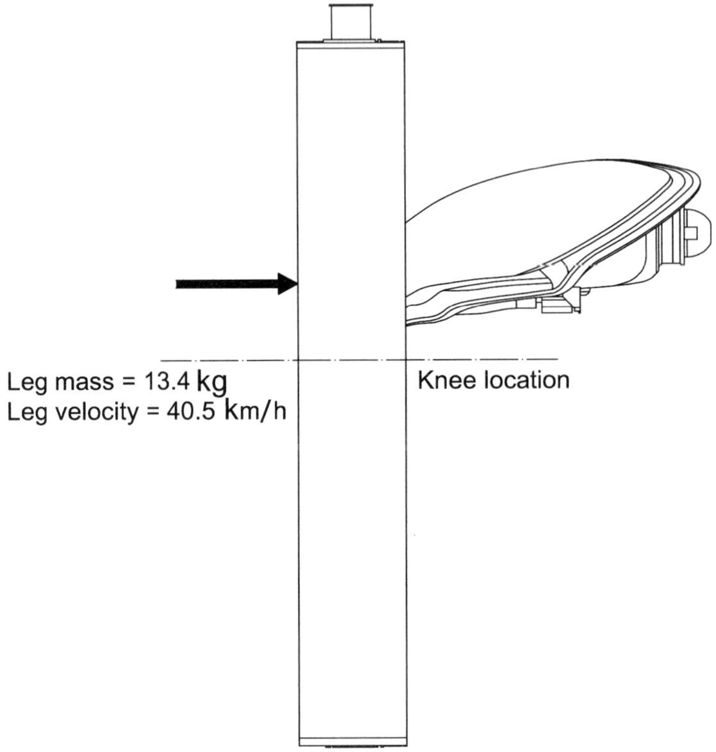

Fig. 6.34 Leg impactor model uses in the study.

A standard WG17 lower-leg impactor was used for this study [6-31]. The leg impactor mass was 13.4 kg, and it was impacted with a speed of 40.5 km/h to the headlamp. Only the headlamp geometry was used for this study, so as to avoid the effects of surrounding components on the impactor.

Table 6.6 Performance Comparison with EAR [6-27]				
		Without EAR	With EAR	Specification
Lower-leg impact	Peak acceleration	58.3	52.7	<150 g
	Plastic strain in lens (%)	33.6	14	
Upper-leg impact	max force (kN)	7.4	4.5	<5 kN
	Max bending moment (Nm)	403	234.8	<300 Nm
	Plastic strain in lens (%)	38.2	7.7	
Corner pendulum	max force (kN)	16	5.1	
	Plastic strain in lens (%)	56.7	26.6	

The headlamp used in the model has all its components such as lens, reflector, bezel, and housing. PC is used for the lens, PEI for the reflector and bezel, and PBT for the housing. Table 6.6 shows a performance comparison study between the impact with EAR and without EAR. This study showed that the g force decreases from 58.1 to 52.7 g when EAR was used. Plastic strain peak in the lens component decreased from 33.6 to 14%, which is from failure zone to no failure zone. When EAR was used, the maximum force experienced by the upper leg was reduced from 7.4 to 4.5 kN; the bending moment experienced by upper leg model was reduced from 403 to 234.8 Nm; and the plastic strains on the lens were reduced from 38.2 to 7.7%.

EAR not only converts the headlamp into a pedestrian-impact-safe component but also decreases the damage experienced by the lens and housing. EAR was observed to improve low-speed vehicle damageability performance. In addition to minimizing the damage to the headlamp, with its inclusion in a headlamp assembly, the replacement and maintenance cost will come down and the insurance rating of the vehicle can be improved depending upon the prevailing regulations. EAR manufacturability was also optimized for injection molding through crush walls alignment. Figure 6.35 depicts the crush wall orientation, which was designed for easy core removal from the cavity.

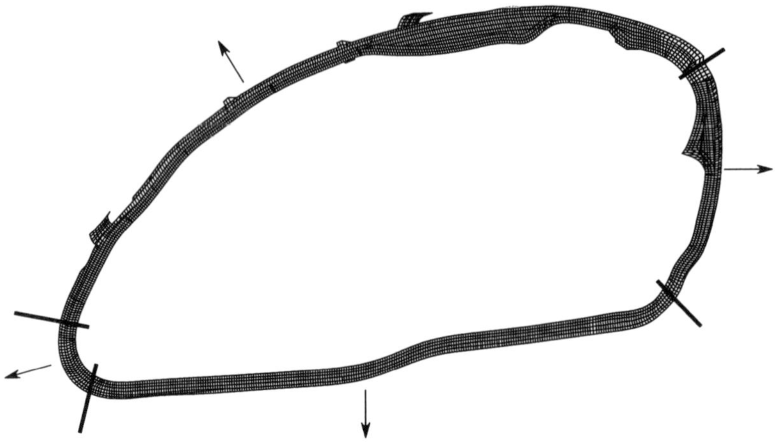

Fig. 6.35 Crush wall orientation for easy core removal.

6.8 Summary

Plastics are the materials of choice for a wide range of headlamp components such as lenses, housing, reflectors, and bezels. As the headlamp components are becoming more stylish and very low in mass, plastics provide the design flexibility to meet these challenging demands. Thermal, structural, and optical design of headlamp components is critical for a successful plastic design to meet all the design requirements.

6.9 Trends

Several studies revealed that approximately 40% of all serious road accidents occur during the night [6-32]. Hence, many automotive manufacturers and tiers are focusing on innovative lighting systems. One such system is dynamic bend lighting (DBL). These systems are just the beginning of future automotive lighting systems. Future trends will be variable lighting systems with fully automatic and situation-dependent light control. Halogen-based headlamps have been quite popular over the past few decades in the automotive lighting industry, while the recent shift has been toward LED-based advanced headlamps. LED lighting is generally powered by semiconductors, and it is very different from traditional light-source technologies that use electrical filaments. Entirely new possibilities of lighting using LED technology, such as dynamically changing light color temperature and flexible designs of headlamps, are challenging the standard interfaces that have been in place for decades. Figure 6.36 shows a headlamp with LED lights.

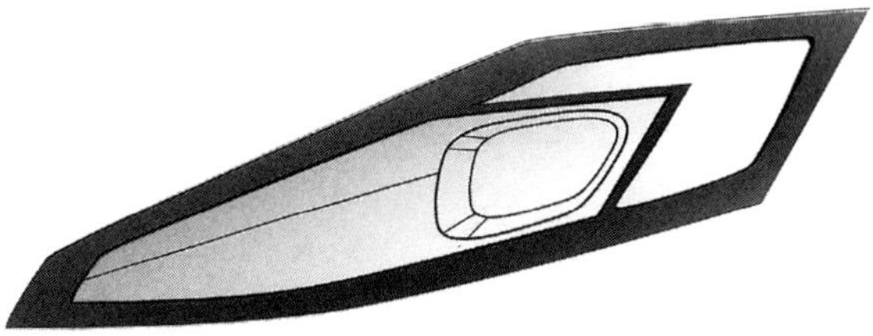

Fig. 6.36 Representative LED headlamp.

Through these LED lights, various combinations of signature styling can be achieved depending on the type of light. These functions can be LB, HB, highway lights, all-weather lights, city lights, indicator lights, and daytime running lights. Automotive designers are looking for new ways to use LEDs in automotive signal applications such as light guides.

6.10 References

6-1. Erjavec, J. 2010. *Automotive Technology: A Systems Approach* (5th edition). Delmar, New York: Cengage Learning.

6-2. Srinivasan, S. 2003. *Automotive Mechanics*. New Delhi: Tata McGraw-Hill.

6-3. "Uniform provisions concerning the approval of vehicles with regard to the installation of lighting and light-signaling devices 2008," http://www.unece.org/fileadmin/DAM/trans/main/wp29/wp29regs/r048r5e.pdf. Accessed January 31, 2012.

6-4. "Federal Motor Vehicle Safety Standard 108," http://www.gpo.gov/fdsys/pkg/CFR-2011-title49-vol6/pdf/CFR-2011-title49-vol6-sec571-108.pdf. Accessed January 31, 2012.

6-5. "ULTEM*—A Resin That Takes the Heat," Ultem* Resin, http://www.sabic-ip.com/gep/Plastics/en/ProductsAndServices/ProductLine/ultem.html. Accessed January 31, 2012.

6-6. Wordenweber, B., Wallaschek, J., Boyce, P. and Hoffman, D. 2007. *Automotive Lighting and Human Vision*. Berlin: Springer.

6-7. Zerda, A., Guggenheim, K., Grau, B. and Hooks, T. 2007. "Heat-aging performance of metalized thermoplastic for automotive lighting." SPE Paper number ANTEC-0396-2007.R1. Newtown, CT: Society of Plastic Engineers.

6-8. "BMC 100," http://www.bulkmolding.com/datasheets/informational/BMC_100.pdf. Accessed January 31, 2012.

6-9. "Ultem™ Resin 1000," Property Data Sheet, http://www.sabic-ip.com/gepapp/eng/weather/weatherhtml?sltRegionList=1002002000&sltPrd=1002003018&sltGrd=1002011252&sltUnit=SI&sltModule=DATASHEETS&sltVersion=Internet&sltType=Online. Accessed January 31, 2012.

6-10. Beck, R.D. 1980. *Plastic Product Design*. New York: Van Nostrand Reinhold.

6-11. Malloy, R.A. 1994. *Plastic Part Design for Injection Molding.* Munich: Carl Hanser.

6-12. "Deflection Temperature Testing of Plastics," MatWeb, http://www.matweb.com/reference/deflection-temperature.aspx. Accessed April 30, 2013.

6-13. "Automotive Lite* Design tool," SABIC Innovative Plastics, http://www.sabic-ip.com/gepapp/eng/automotive_tools.jsp. Accessed January 31, 2012.

6-14. Montgomery, D.C. 2004. *Design and Analysis of Experiments.* Singapore: Wiley.

6-15. Tankala, T., Marur, S. and Wilson, J. 2006. "Rapid Thermal Predictions for Automotive Headlamp Reflectors." SAE Paper No. 2006-01-1189. SAE International, Warrendale, PA.

6-16. Converse, T.O. and Wilson, J. 2008. "Headlamp assemblies having reduced operating temperatures." Patent Publication No. US 2008/0037267 A1.

6-17. *User Manual, FLUENT v12.1.* Canonsburg, PA: ANSYS Fluent.

6-18. *ABAQUS User Manual Version 6.11.* 2011. Providence, RI: Dassault Systèmes Simulia.

6-19. Tankala, T., Poovanna T.K., Arunachala P., Bloomfield, R. and Wilson, J. 2012. "Thermoplastic Reflector Design for Robust Optical Performance." SAE paper No. 2012-01-0266. SAE International: Warrendale, PA.

6-20. Pye, R.G.W. 1989. *Injection Mould Design: An Introduction & Design Manual for the Thermoplastics Industry*. Essex, England: Longman Scientific and Technical.

6-21. "Simulation Tools for Injection Molding," Autodesk Simulation, http://www.moldflow.com. Accessed April 30, 2013.

6-22. "Conformal Cooling: Why Use It Now?," *Plastics Today*, http://www.plasticstoday.com/imm/articles/tooling-conformal-cooling-1209. Accessed January 31, 2012.

6-23. "Headlamp Bezels—Direct Metalized," Lighting, http://www.sabic-ip.com/gep/Plastics/en/IndustryApplications/Applications/headlampbezelsdirectmetalized_lighting_automotive.html. Accessed January 31, 2012.

6-24. "Headlamp Lenses," Lighting, http://www.sabic-ip.com/gep/Plastics/en/IndustryApplications/Applications/headlamplenses_lighting_automotive.html. Accessed January 31, 2012.

6-25. "Study of Thin-Wall Injection Molding," http://etd.ohiolink.edu/send-pdf.cgi/Xu%20Guojung.pdf?osu1078788946. Accessed January 31, 2012.

6-26. "The Potential of Car Design and Impact Testing," http: //www.unece.org/fileadmin/DAM/trans/doc/2002/wp29grsp/inf-gr-ps-12e.pdf. Accessed January 31, 2012.

6-27. "European New Car Assessment Programme (Euro NCAP), Pedestrian Testing Protocol," http://www.euroncap.com/files/Euro-NCAP-Pedestrian-Protocol-Version-5.3.1---0-0560739b-595f-4a38-96b2-eeda5050dee8.pdf. Accessed January 21, 2012.

6-28. "Establishment of Regulations for Pedestrian Safety in Japan," http://www.unece.org/fileadmin/DAM/trans/doc/2002/wp29grsp/TRANS-WP29-GRSP-32-inf05e.pdf. Accessed January 22, 2012.

6-29. Garg, A., Marur, S., Tankala, T., and Wilson, J. 2010. "Pedestrian Safe and Impact Resistant Headlamp Design through a Novel Energy Absorber Ring Concept." SAE Paper No. 2010-01-0293. SAE International: Warrendale, PA.

6-30. Marur, S.R., Garg, A.K., Nayak, S., Tankala, T.C. and Wilson, J.R. 2011. "Energy Absorber Elements and Vehicle Systems." Patent Publication No. US 2011/0210579 A1

6-31. "Improved Test Methods to Evaluate Pedestrian Protection Afforded by Passenger Cars, EEVC, 1998, Working Group WG17 Report," http: //www.unece.org/fileadmin/DAM/trans/doc/2006/wp29grsp/ps-187r1e.pdf, Accessed January 31, 2012.

6-32. "Nighttime Car Accident Statistics," Auto Accidents, http://seriousaccidents.com/legal-advice/top-causes-of-car-accidents/nighttime-driving/. Accessed January 31, 2012.

Chapter 7
Body Panels

S.M. Kumar, Somasekhar Bobba, Narendra A. Hardikar, T.K. Poovanna, K.R.K. Ravikanth, and N. Venkatesha

7.1 Introduction

The global trend in automotive legislation is focused on enhancing fuel efficiency in cars and achieving reduced emissions. To achieve these objectives, lightweighting and enhanced aerodynamics have emerged as two key design trends of the automotive industry. These trends have also influenced the design and material usage for body panels—the external cladding surfaces of an automotive. Metals continue to dominate the body panels segment. However, advances in engineering thermoplastics (ETP), manufacturing processes, and predictive engineering tools have made ETP exterior body panels a reality, offering several unique advantages. Figure 7.1 depicts an ETP fender.

Use of ETP in body panels facilitates complex curved shapes and smooth curvatures to aid in the reduction of aerodynamic drag. For example, air vents can be incorporated in injection-molded ETP fenders. Also, exterior styling effects, a key differentiator for car design, can be achieved. Paint quality and surface finish equivalent to metals can be obtained by using specialized conductive high-heat and thermally stable thermoplastic.

ETP is six times less dense than steel, as shown in Fig. 7.2, and has lower stiffness as compared to metals. Despite that limitation, appropriate design of

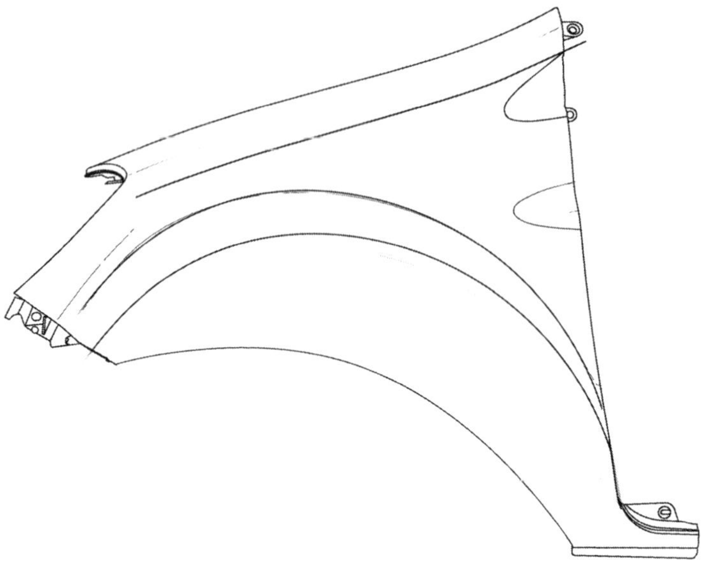

Fig. 7.1 Thermoplastic fender.

ETP can still deliver considerable savings on body panel applications, which aids reduced fuel usage and emissions during the use phase of the vehicle.

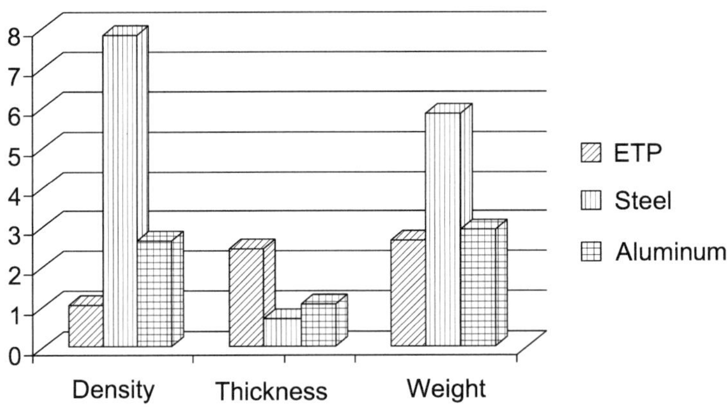

Fig. 7.2 Basic material property comparison.

Thermoplastics have the unique advantage of absorbing low-velocity impacts, such as a shopping cart hitting the body panel, but leaving no permanent deformation. Another advantage of using ETP is the possibility of integration of parts. For example, ETP fenders, unlike metals, can be manufactured in one operation with in-built energy-absorbing features for pedestrian impacts, as shown in Fig. 7.3. In addition, depending on the total number of cars built (<100,000 builds a year), ETP body panels can also offer a significant system-level cost savings over metallic ones.

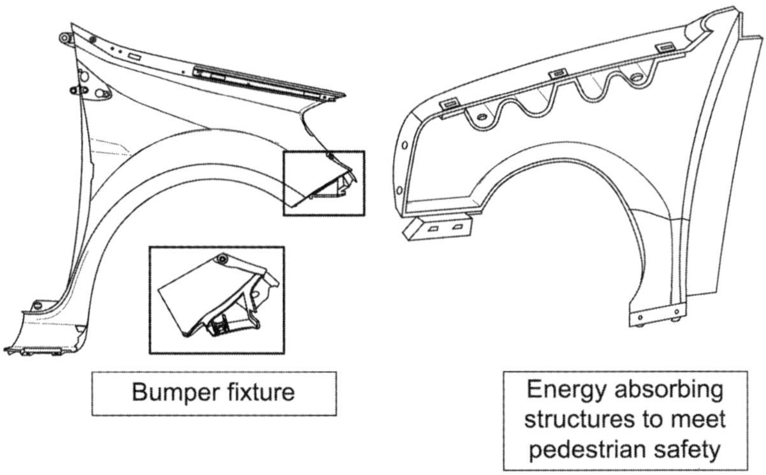

Fig. 7.3 Complex features in ETP fenders.

Given the importance of cost, in material selection for automotive components, ETP designs have to be optimized to meet varying functional and cost requirements. The design and optimization of ETP body panels rely extensively on predictive engineering to virtually ascertain various configurations for their effectiveness. Such studies require appropriate material models to mimic the ETP behavior and corresponding material characterization techniques to obtain the material constants required for the numerical model. This requirement is especially pronounced for fenders and is described in subsequent sections dedicated to fender design.

This chapter gives an overview of the body panel functional requirements, ETP materials used, and manufacturing methods. Among the various body panels successfully made in ETP, we focus on fenders, for which we provide a detailed design evolution process along with predictive engineering case studies.

7.2 Functional Requirements for Body Panels

Some of the typical automotive body panel applications using ETP are fenders, exterior door panels, tailgates, tank flaps, claddings, and spoilers, as shown in Fig. 7.4.

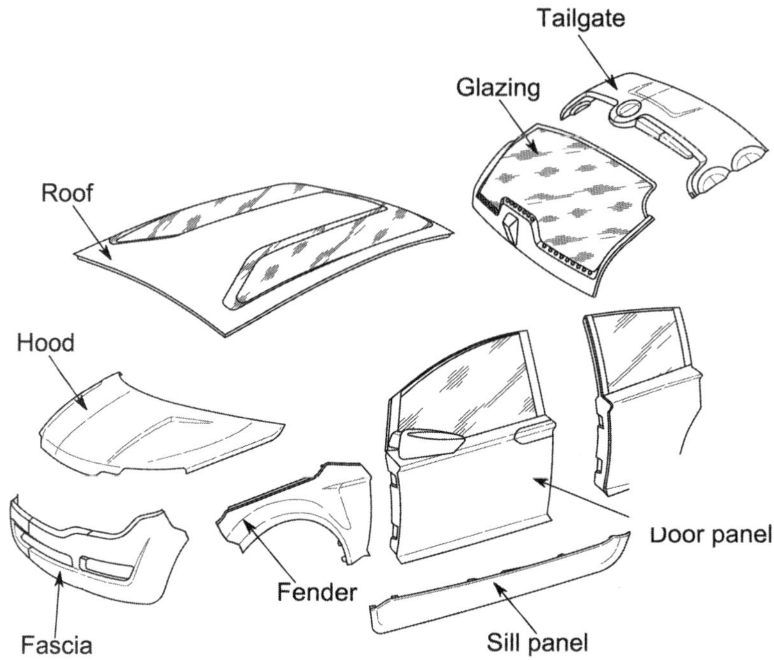

Fig. 7.4 Typical body panel applications.

In this section, we present the typical requirement for body panels, the factors that influence material selection, and the manufacturing techniques used to produce body panels. Body panels, especially the larger ones such as tailgates or fenders, require adequate surface rigidity to withstand minor abuse loadings such as a push from a hand, stone impacts, or a person leaning against it. Body panels like door panels and tailgates need sufficient toughness to absorb low-speed impacts arising from shutting of car door or backdoor. They should withstand other low-velocity impacts, such as a hit from a shopping cart. Depending on geographic location, some body panels such as the fenders may also need to meet pedestrian safety or insurance-related requirements.

Another key requirement of body panels is a high aesthetic surface finish. To meet this, exterior body panels, are often painted for Class A finish. The passage of an ETP body panel through the paint line imposes a requirement of high-heat resistance and the need for sufficient stiffness at elevated temperatures. In a typical paint cycle, a car gets two to three paint layers [7-1]. Every paint cycle is followed by a curing process in an oven before it goes to the next stage in the paint line. During the curing cycle, the peak temperatures may go as high as 200 °C (392 °F). ETP body panels have to be designed with an optimum support configuration to maintain the deformation within tolerances during their passage through the paint line. In addition to the heat load during the paint cycle, the body panels also need to withstand heat loads in the use phase of the car arising from the exposure to the sun.

7.2.1 Material Selection in Engineering Thermoplastics Body Panels

The choice of thermoplastic material primarily depends on the method used for painting the body panel. There are three types of painting: online, inline, and offline (Fig. 7.5). In the online paint processes, the body panel is assembled onto the body in white (BIW) before the e-coat bath, and it stays on the BIW right through the paint process as well as all the curing cycles.

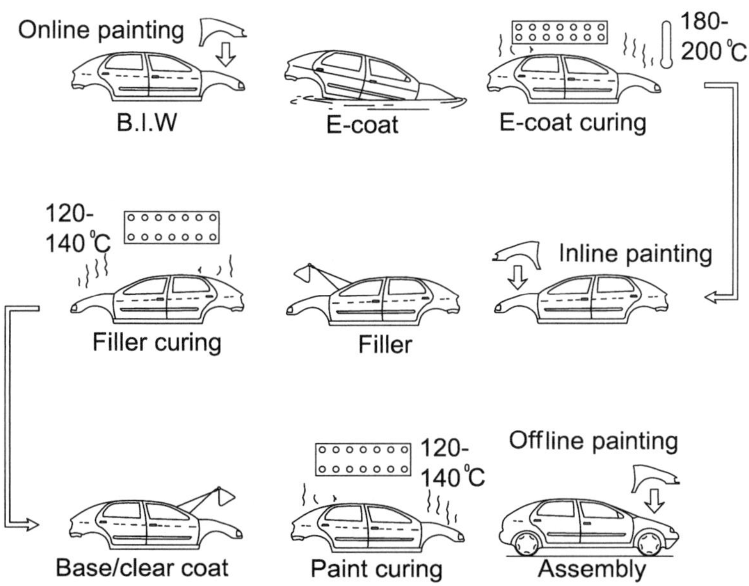

Fig. 7.5 Typical ETP fender painting types and processes.

In the inline paint process, the body panel is assembled after the e-coat stage and goes through the rest of the painting cycle. In the offline mode, the part is painted separately and then assembled on the BIW of the car. During the online painting process, ETP materials used in body panel applications need to withstand around 180 °C (356 °F), the peak curing temperature during the paint cycle. A specialized thermoplastic material blend of polyphenylene oxide (PPO) and polyamide (PA) is the ETP material used in the online application, and it has managed to replace metal in fender application [7-2]. In this blend, PPO, an amorphous polymer, is in a suspended phase in a matrix of semicrystalline PA. PPO with a glass transition temperature (Tg) of 215 °C (419 °F). As such, PPO contributes to heat resistance, dimensional stability, and low-density to the blend. PA, a semicrystalline material, contributes to good chemical resistance, processability, and paintability.

Inline painting, where the ETP sees curing temperatures lower than the e-coat cycle, requires a material with good temperature resistance, such as PC/ABS (polycarbonate and acrylonitrile butadiene styrene) blends. However, for offline painting where the temperature requirements are not high, a material with paint compatibility will serve the purpose. Materials such as PP (polypropylene), ABS, PBT (polybutylene terephthalate), and PC/ABS are the material options used for offline body panels. For an overview of the automotive paint process, interested readers may refer to Ref. [7-3].

Injection molding is the most common manufacturing process for the manufacture of thermoplastic body panels [7-4]. Injection molding has the advantage of producing complex parts within a short cycle time. As the size of the part increases, advanced processing techniques, such as injection compression molding, can be explored for molding body panels. In summary, a list of materials being used to produce different body panel applications and their processing methods are listed in Table 7.1.

7.3 Fenders

Fenders are one of the largest body panels in which ETP has successfully managed to make a mark for itself, meeting functional, regulatory, and original equipment manufacturer (OEM) requirements. The functional requirements aforementioned in this chapter are met through a combination of appropriate material selection and engineering design. A large optimization problem exists in the engineering design of fenders, in which the design variables relevant to the three key phases of the life of an ETP fender are optimized. The key phases are the following:

1. Manufacturing via the injection-molding process
2. Passage and residence through the paint line curing cycle
3. The loading condition during the use phase of the automotive

From a manufacturing perspective, the objective of the design is to identify the gate type, size, and location, which would help in achieving uniform

Table 7.1 Type of Materials Used for Body Panel Applications

Body Panel Application	Material Type	Thermoplastic Usage	Processing Method For Thermoplastics
Bumper (front and rear)	Thermoplastics/ steel	PP compounds	Injection molding
Grille	Thermoplastics	PP, PA, PPO+PA, ABS, PC+ABS, PC+PBT	Injection molding
Hood	Steel/composite/ SMC	Sandwich structure with thermoplastics	Thermoforming and plastic welding
Fender	Thermoplastics/ steel/aluminum	PPO+PA, PA, PP	Injection Molding
Door	Steel/ thermoplastics	PP with long glass fiber (LGF), PPO+PA, PA, PC+ABS	Injection Molding
Roof	Steel/ thermoplastics	PC	Thermoforming, 2-Shot Injection molding
Tank flap	Thermoplastics/ steel	PC+PBT, PPO+PA, PA	Injection molding
Spoilers	Thermoplastics/ steel	PC/ABS	Blow molding, gas-assist injection Molding
Trunk lid	Steel/ thermoplastics	PPO+PA, PA	Thermoforming, injection molding and plastic welding, 2-shot injection molding

flow, essential for high-quality surface finish in body panels. It is also essential to identify process parameters that would minimize the warpage and residual stresses in the part. Another priority is to avoid or relocate weld lines and sink marks from the exterior visible areas of the body panel. The various molding considerations are explored in the next section.

During its passage through the paint line, an ETP fender is subjected to temperatures in the range of 180 °C (356 °F) to 200 °C (392 °F) [7-5] in the curing oven. The deformation of the fender in the oven, caused by the interaction of several variables, is elaborated in detail in Section 7.3.2, "Design for Paintability." The design objective in this phase is to identify an optimum support configuration and thickness distribution that would maintain fender deformation within OEM-specified limits.

Fenders need to be designed for use phase conditions such as surface rigidity and energy-absorption features to meet regulatory impact criteria. They also need to withstand thermal deformations caused by environmental variations such as high temperature during summer and low temperature during winter. Typically, the support configuration optimized for the paint cycle is fine-tuned to account for thermal deformations during the use phase.

Predictive engineering is utilized heavily to analyze a variety of design options to meet requirements across all the three phases and identify the most promising one. As shown in Fig. 7.6, this process is iterative in nature and involves a *customer input—part design-predictive evaluation—customer input cycle,* until an optimized design in obtained. This process is prototyped to validate the predictions and fine-tune the design before starting production. Use of predictive engineering in the early stage of design would expedite the design process.

7.3.1 Manufacturing Considerations in Fender Design

An efficient injection molding process, for fenders, needs to ensure that the following conditions are met:

1. Balanced filling, which ensures that the thermoplastic reaches the extremities of the cavity at approximately that same time, so as to reduce overpacked regions and have a gradual pressure built up without any pressure spikes during filling
2. Injection pressure and clamp tonnage within the capability of the available machines to derive cost advantage

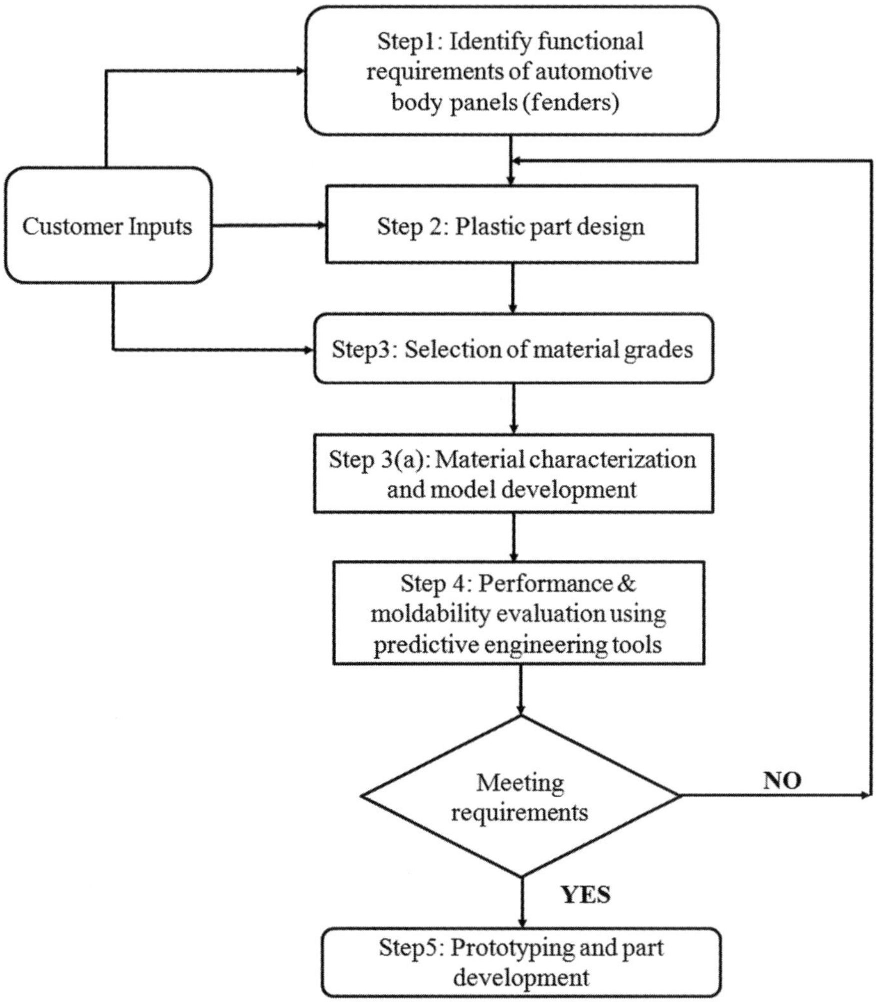

Fig. 7.6 Generalized thermoplastic fender development process.

3. And aesthetic surface, free of defects such as knit lines, flow marks, and burn marks.
4. Minimum part warpage

The type of gate and its location play an important role in achieving aesthetics and balanced filling, as well as ensuring that filling pressures are within the limits of available machines. Because of aesthetic requirements, gating cannot be on the visible surface or on the back side of the visible surface. Typically, the gating is provided on to the flange area of the wheel arch. There are several types of gates that are used in the molding of the ETP fenders. Table 7.2 gives an overview of the various types of gates used in manufacturing of fenders and the corresponding pros and cons for each of them.

Table 7.2 Types of Gating

Type of injection	Advantages	Disadvantages
1 to 4 pin points	Generally for small fenders Automatic removal of runners Low volume of wasted material	Possible flow line visibility
5 to 10 pin points	For large flat fenders Automatic removal of runners Possibility to improve flow balance	Possible flow lines visibility
Film gate along wheel arch	For easy to balance fenders	Need to cut the gate No in-mold modification possible
Film gate from sprue	Can be used where in there is large flow length variation from the center of wheel arch	Need to cut the film Increased clamping force High volume of waste material
Sequential cold runners	Automatic removal of runners	Patented; use is restricted.
Sequential hot runner system	Automatic removal of runners Possibility to use dynamic feed system	Expensive tooling; dynamic feed is still in validation stage for fenders.

Based on the gating at the wheel arch area, flow length can be calculated, and from the maximum part dimension and thickness of the part, the approximate pressure can be calculated. This pressure value should be within the injection-molding process window. Very high injection pressure leads to higher internal stresses and higher clamp tonnage. If the flow length is high, part thickness can be increased, whenever possible, to reduce the injection pressure. If the thickness increase is not possible, the number of gates or size of gates can be increased.

Warpage is an inevitable outcome of the injection-molding process. But its magnitude can be controlled through the use of optimum melt temperature, geometric design, and efficient cooling circuit design, which ensures that the optimum mold temperature is maintained uniformly throughout the core and cavity. A cooling circuit layout also needs to be optimized to achieve the cooling time of a large part such as a fender within manageable limits.

Influence of the gating system on the filling pattern, injection pressure, and part warpage can be predicted by using any injection-molding simulation software. The optimum feed system designed for mold-filling simulation is shown in Fig. 7.7. It consists of five tunnel gates, and the gates are positioned on the wheel arch in such a way as to achieve balanced filling, with all the flow paths filling at same time and under the same pressure.

The optimum gate locations are obtained after several iterations with different combinations of gate location and size evaluated through mold-filling simulation. The melt temperature of material is maintained at 305 °C (581 °F) at the tip of nozzle from which it is getting injected into the mold. A uniform mold temperature of 100 °C (212 °F) is maintained to eliminate the differential cooling effects. The part can be filled in 3 seconds of injection time, and injection velocity should be set in such a way that the flow front velocity is uniform to ensure good surface quality. The typical filling simulation results are shown in Figs. 7.8a and 7.8b.

A fill-time plot shows how the plastic melt fills in the mold cavity. The total time required to fill is 3.6 seconds. The material is flowing uniformly from the wheel arch and fills all the ends at the same time with uniform flow front speed and similar pressure. The in-cavity residual stress is calculated during the pack analysis and represents the stresses in the part before it is ejected. This result can be used as an input for viscoelastic stress analysis, to be carried out using the commercial finite element (FE) software. Reference [7-5] proposed using the molding history in the calculation of the gap and flush of the fender, during an e-coat paint cycle analysis. While several molding variables can be predicted with reasonably good

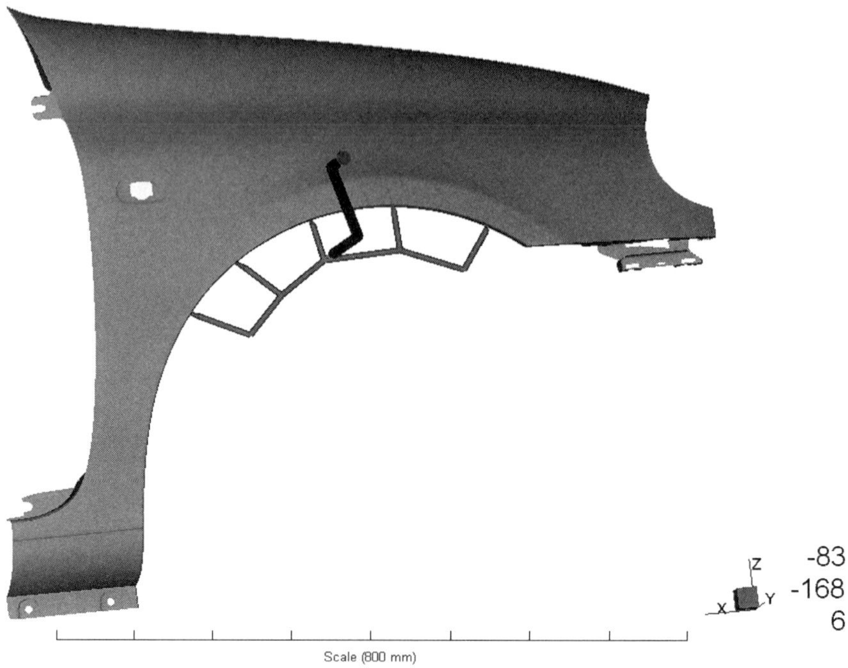

Fig. 7.7 Optimum feed system design.

accuracy, prediction of shrinkage of the part in an injection-molding cavity is still not possible with any accuracy through the use of predictive engineering tools. The exact sizing of the cavity is still an empirical process and depends on factors such as the use of extensive experimental shrinkage data on plaques, experience of the tool maker, and knowledge gained from prior fender designs. Hence, for a part like a fender, a prototype tool is cut to study various parameters and fine-tune the variables such as shrinkage before proceeding to production tooling for left–right components.

7.3.2 Design for Paintability

The periphery of a fender sits in close proximity to the door, hood, bumper, and reflector assembly. Given the importance of aesthetics and surface finish, it is essential to maintain the deformation of the fender edges during the paint cycle within prescribed limits. Typically, for a given body panel, tolerances on in-plane and out-of-plane deformations are defined by OEMs in the form of gap and flush values, respectively. Figure 7.9 depicts the gap and flush for a hood line edge (i.e., between hood and fender).

These gap and flush limits are prescribed for all the edges of the fender with respect to the working coordinate system, as shown in Fig. 7.10.

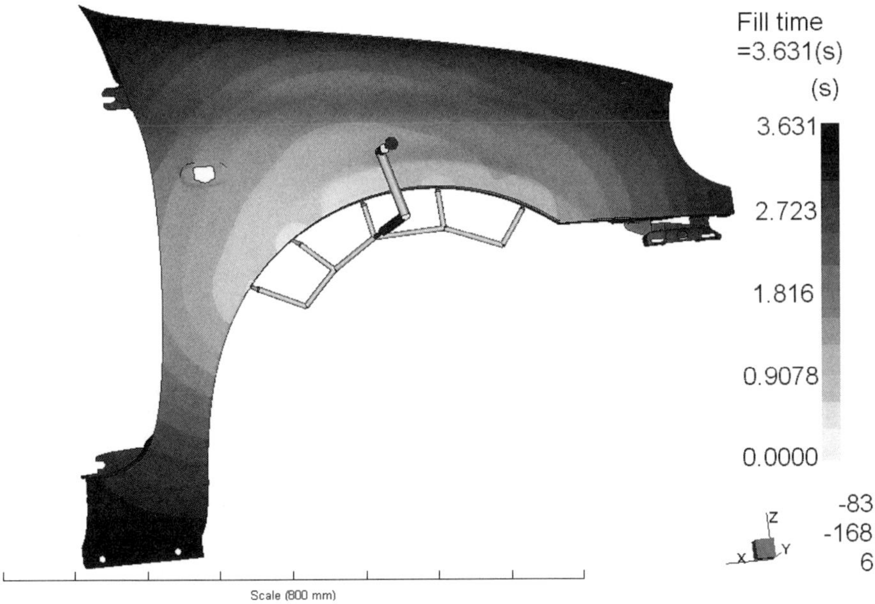

Fig. 7.8a Fill time for an ETP fender.

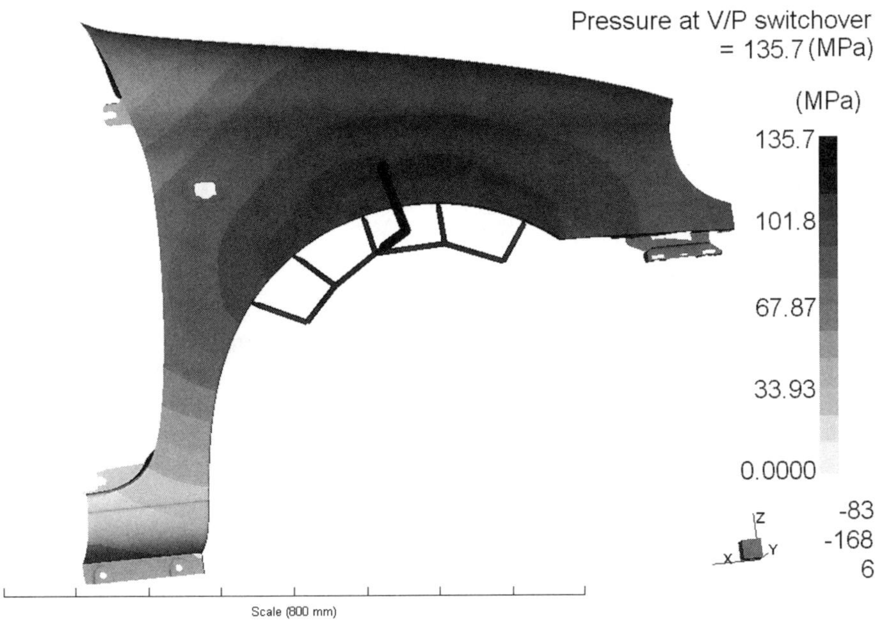

Fig. 7.8b Pressure distribution for an ETP fender.

Evaluation points	Evaluation point description	Potential problems
Gap	The horizontal distance between two measurement points	Inconsistent Tight Wide
Flush	The vertical distance between two measurement points	High Low

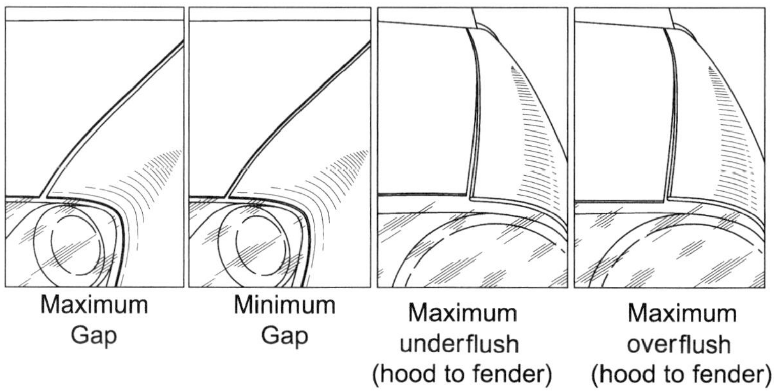

Fig. 7.9 Gap and flush for a fender hood line.

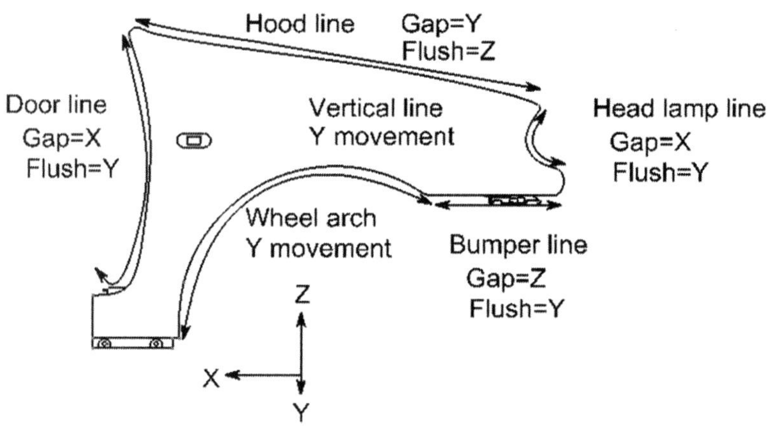

Fig. 7.10 In-plane (gap) and out-of-plane (flush) for various edges of the fender.

The deformations that arise during the paint curing process are an outcome of the interactions of several factors:

1. Transient temperature cycle and coefficient of thermal expansion (CTE) difference between ETP and metal BIW
2. Reduction of modulus at higher temperatures
3. Viscoelastic creep of the thermoplastic material
4. Shrinkage that occurs during the paint curing cycle at elevated temperatures

In an online paint cycle, the fender is assembled on steel BIW and subjected to a transient thermal cycle, a schematic of which is shown in Fig. 7.11.

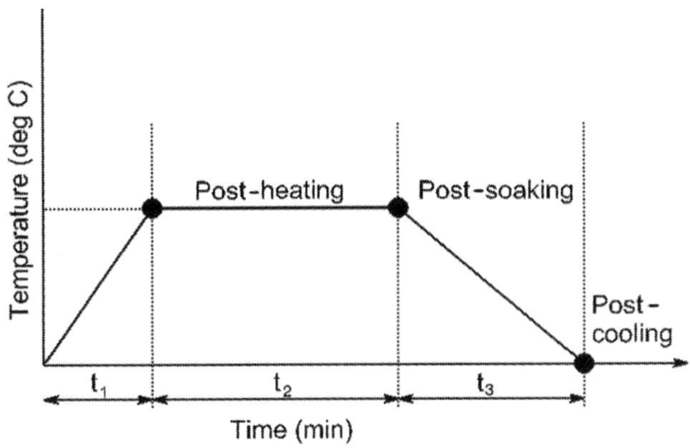

Fig. 7.11 Thermal cycle for e-coat paint processes.

The actual temperature cycle used in the e-coat oven and subsequent paint ovens is OEM specific and highly proprietary and confidential in nature. However, it has three segments: the heating, soaking, and cooling stages. In online painting, the peak temperature varies between 180 and 200 °C (356 and 392 °F). When an ETP fender is subjected to a temperature cycle, the CTE difference between BIW and ETP induces thermal expansions and stress that need to be managed. ETP materials are highly sensitive to temperature as compared to metal. During the paint cycle, the material stiffness decreases due to high temperature before it gets restored to a normal value at room temperature. This leads to an increase in deformation at constant self-weight during the heating and soaking phase of the temperature cycle. It is essential to ensure that the structure has sufficient rigidity and adequate support throughout the cycle so as to prevent the possibility of buckling or collapse under its self-weight.

Another factor to be considered is that ETP materials, depending on the temperature to which they are exposed and its relation with the glass transition temperature (Tg), can exhibit an elastic, viscoelastic, or hyper-elastic response [7-5]. The typical temperatures encountered in the e-coat curing oven are close to one of the Tg of the PPO+PA blend (which is 3 Tg), and hence the material response has to be considered viscoelastic. In a viscoelastic material, the deformation increases with time or temperature at the same load [7-6]. However, it is pertinent to point out that the viscoelastic effect is observed only at high temperatures and only during the residence of the ETP fender in the paint-curing cycle. One more phenomenon that contributes to the deformation process is ETP paint line shrinkage. This shrinkage is different from the shrinkage observed within the injection mold. Paint line shrinkage is observed typically only during the e-coat curing cycle in semicrystalline ETP material, where a part subjected to a heating-hold-cooling temperature cycle shrinks in dimensions at end of the cycle. Shrinkage in the ETP blend is primarily caused by the differential volumetric shrinkage between the amorphous and semicrystalline content of the PPO+PA blend used in an online painted fender. Thus the final dimensions of an assembled fender are obtained by subtracting the in-mold shrinkage and paint line shrinkage from the mold cavity dimensions.

A good fender design needs to manage the deformations that arise during its passage through the paint cycle. Usually, sliding joints are used to absorb thermal expansions and contractions during the paint cycle. A schematic of a sliding joint of a fender and typical locations along the hood line and door line are shown in Fig. 7.12.

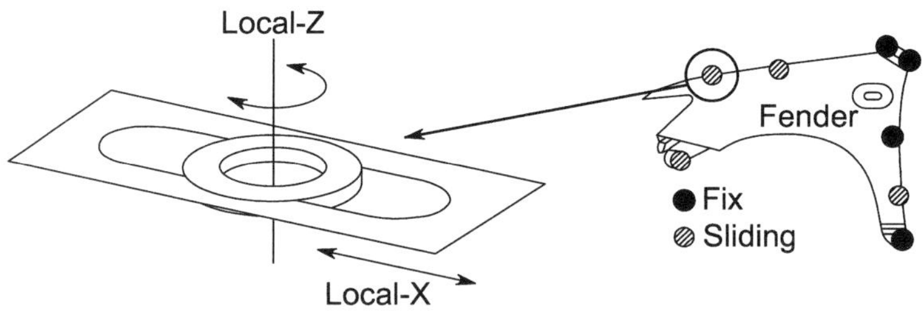

Fig. 7.12 Schematic of sliding joint and typical locations along the hood line and door line.

A typical fender has 8–10 support locations along the door line, hood line, and dog leg, as shown in Fig. 7.13.

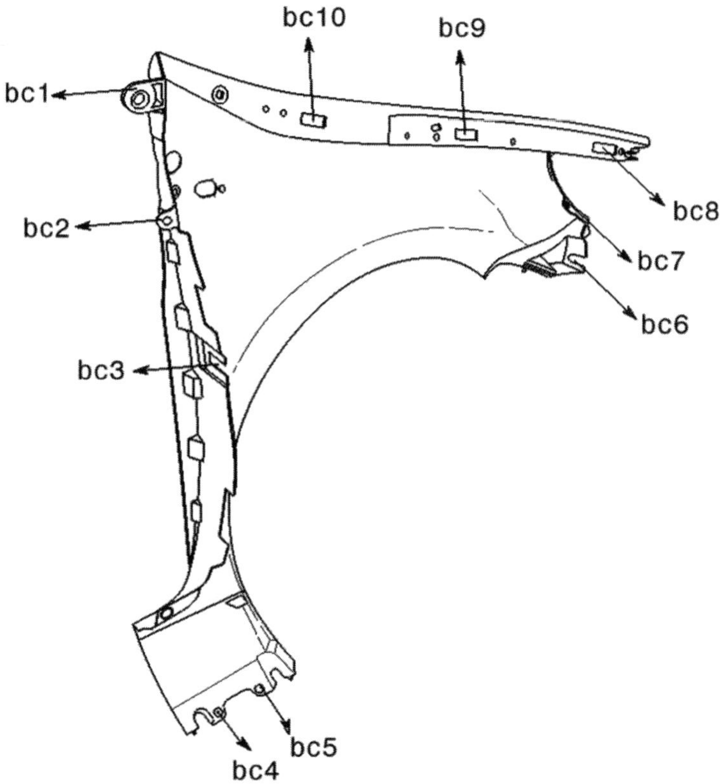

Fig. 7.13 ETP fender support locations.

Some of these may subsequently get fixed during the final assembly, while others may retain partial sliding degrees of freedom to account for environmental temperature fluctuations or sun load during the actual use stage. Thus the design problem is to identify the following:

- Fixing points where the body panel can be supported on BIW without interfering with other surrounding components
- The type of joint to be employed at each locations (sliding or fixed)
- The orientation of slots and degrees of freedom in which sliding is to be allowed

Given the high number of combinations, viscoelastic stress analysis is carried out using a standard finite element analysis (FEA) solver, which obtains comparative data for a few different support combinations and lists the constraint combination that minimizes the gap and flush deformation on most of, or at least a majority of, the fender periphery lines. A precursor for such an analysis is the material characterization required to get the relevant material data for the temperature and time range. The next section captures the details of the material characterization and material model used. The subsequent section gives a real-life case study of how the FEA results can be used by the designer to evaluate various combinations and identify the promising one.

7.3.3 Material Characterization and Material Model for Fender Predictive Studies

A viscoelastic model to capture the material response is the key in predicting fender deformations. Viscoelastic properties vary strongly with temperature as well as time [7-6, 7-7]. To mimic the response of the material, it is essential to capture the viscoelastic relaxation and effect of temperature on relaxation. In the case of fenders, viscoelastic relaxation over a temperature range of 20–200 °C (68–392 °F) is required for more than an hour of residence time, which makes the experimental characterization expensive. This can be resolved through the use of time temperature superposition (TTS) principle in polymer physics [7-8, 7-9, 7-10], which can extrapolate the limited rheology data extracted from dynamic mechanical analyzer (DMA) [7-11] and avoid testing material for a longer time.

TTS states that the viscoelastic effects such as creep or relaxation of materials are influenced in a similar way by temperature or time. Relaxation over a very long time can be obtained by testing for shorter intervals of time at higher temperatures. Relaxation or compliance master curves spanning over several decades of time or frequency can be obtained by TTS. The principle of TTS assumes that the relaxation curve at another temperature is obtained just by shifting the master curve on the time/frequency axis by the horizontal shift factor for that temperature [7-6]. Materials for which this assumption is valid are referred as thermorheologically simple (TRS) materials. The detailed procedure to validate the applicability of TTS to PPO+PA and creation of the relaxation master curve from temperature-frequency data extracted with DMA torsion test is explained in detail in Ref. [7-12].

In a commercially available FE tool—ABAQUS, the relaxation response is modeled by using Prony series coefficients [7-13], obtained by fitting a curve to the experimental relaxation data from DMA tests. The effect of temperature on the viscoelastic relaxation behavior and its inclusion in the viscoelastic stress analysis are obtained by the use of horizontal shift factor function and reduced time concept, which was first introduced by Leaderman [7-14]. For polymeric blends like PPO+PA, the horizontal shift factors do not follow the Williams-Landel-Ferry equation [7-7]. A user-defined thermorheologically simple (UTRS) function was developed to account for the influence of temperature variation in the viscoelastic stress analysis. Details on the need for horizontal shift factors and the concept of reduced time are elaborated in Ref. [7-14]. Use of the linear TRS viscoelastic model used for the e-coat cycle and the role of UTRS are explained in Ref. [7-5].

7.3.4 Case Study of Finite Element Analysis to Optimize Support Configuration

The main objective of this case study is to demonstrate the use of computer-aided engineering (CAE) tools to evaluate different fixing conditions and performance of fender subjected to e-coat cycle temperature. In the present case, a set of boundary (fixing) conditions and design geometry were considered as the baseline. We compared the baseline with two potential design modifications, referred to as DM-2 and DM-3, and studied their influence on hood line GAP_FLUSH performance [7-2].

The design changes to the base reference state, BASE, occur only in the vicinity of the sliding slots at bc2 and bc3 along the hood line. The common design changes in DM-2 and DM-3 compared to baseline are described in the following:

- Location of slot 2 moves backward by 10 mm along local 2 X (Fig. 7.14a).
- Location of slot 3 moves forward by 20 mm against local 3 X (Fig. 7.14a).
- Reinforcement in the form of supporting pillars at the front and back of modified slot 2 and slot 3 (Fig. 7.14b).

Specific change in DM-3 as compared to baseline is:

- Orientation of slot 3 is on the vertical wall instead of the horizontal wall (Fig. 7.14c).

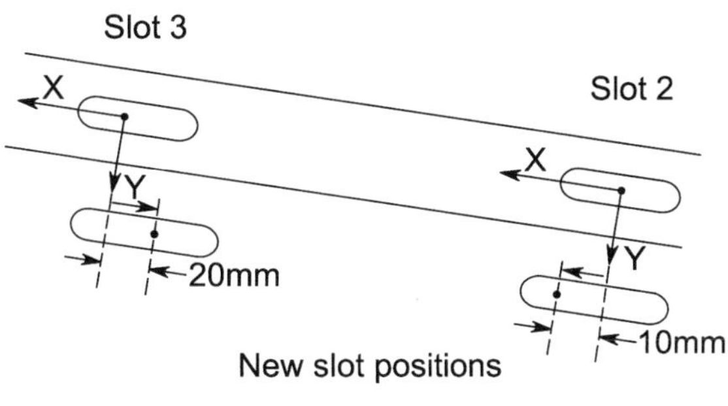

Fig. 7.14a Movement of slot as compared to baseline.

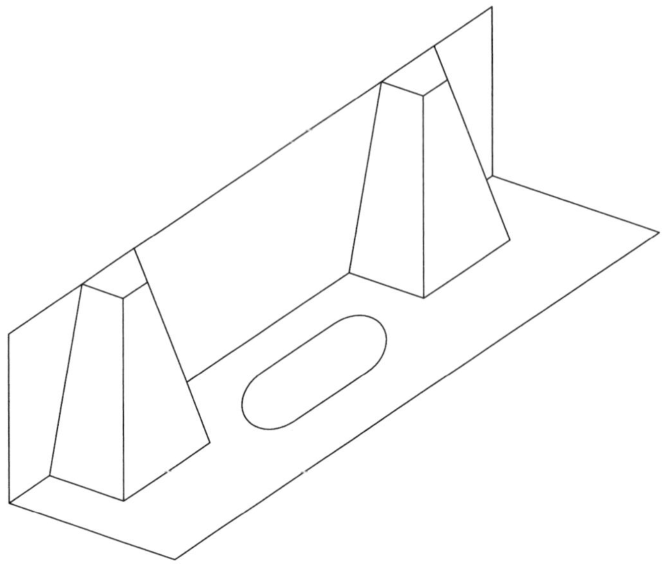

Reinforcements along each slot

Fig. 7.14b Reinforcement close to slot 2 and slot 3.

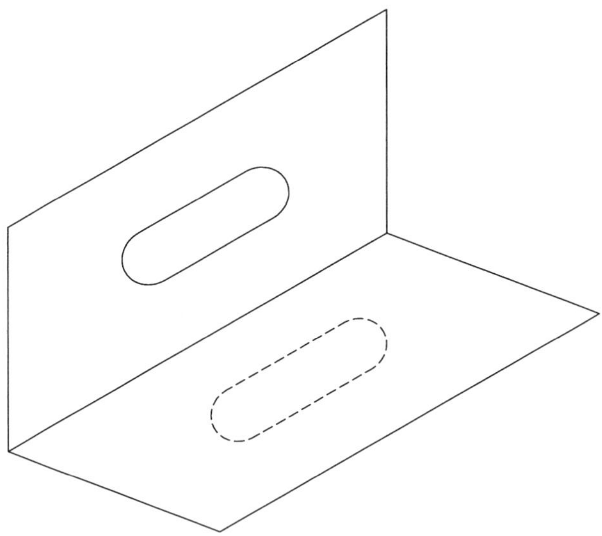

DM3 has vertical slot at SLOT3 position as compared to horizontal in DM2

Fig. 7.14c Slot on the vertical wall.

Comparative results of three simulations are presented in the form of gap and flush deformation graphs for the hood line in Figs. 7.15 and 7.16.

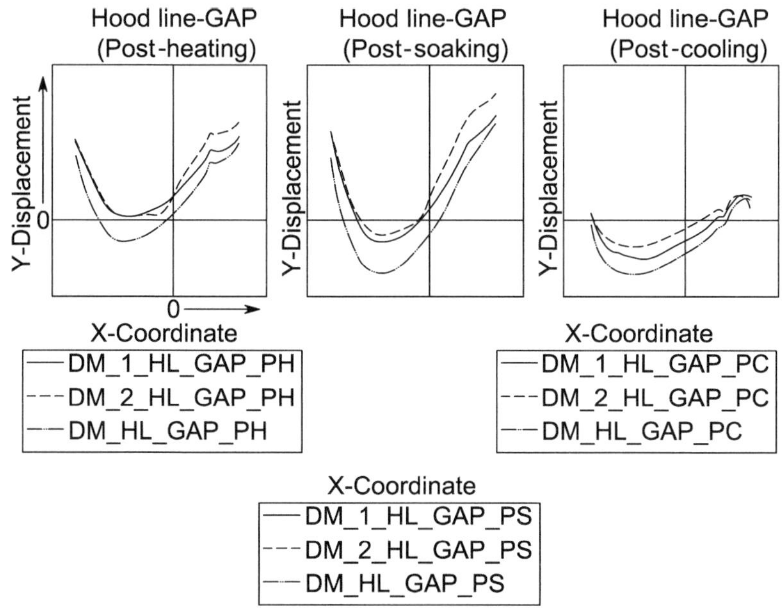

Fig. 7.15 Deformation plots for GAP along the hood line [7-2].

It can be seen that the DM-3 model reduces the hood line gap as compared to baseline, but hood line flush is higher. DM-2 model reduces the hood line gap as compared to baseline state, but flush remains almost same. Between the two modifications, DM-2 can be considered as optimum because it enhances gap performance as compared to baseline and maintains the flush at same level as baseline; DM-3 makes it worse than baseline. This comparative study gives the designer an insight into the influence of slot position, local reinforcements, and slot orientation on the hood line performance. A series of such iterations were performed to evaluate the efficacy of different real-life support conditions and identify the most promising ones among them. More details on modeling of assembly fit variations and design changes are presented in Ref. [7-2].

7.3.5 Fender Impact Resistance

Damages and injuries that happen during any automobile accident can generally be classified as three broad categories. The first one relates to the safety of pedestrians during vehicle-to-pedestrian impacts. The second and third relate to the damage of the vehicle components and the safety

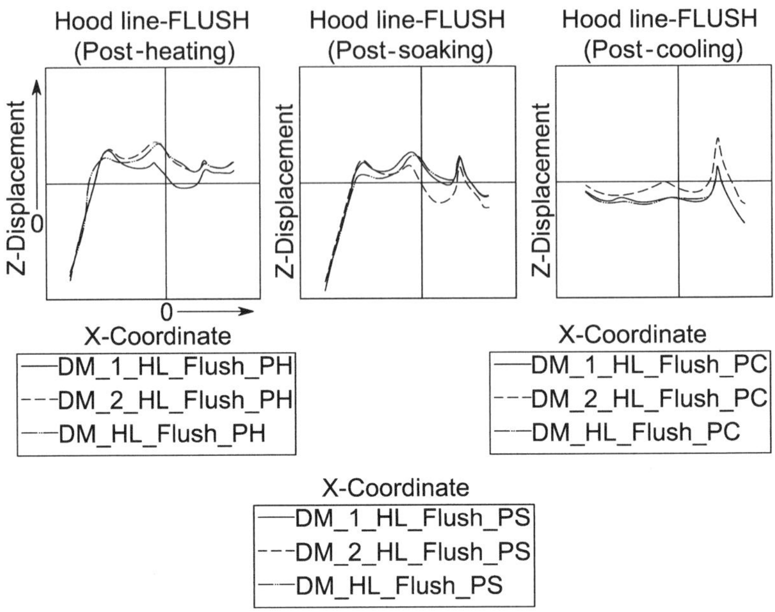

Fig. 7.16 Deformation plots for FLUSH along the hood line [7-2].

of the occupants, respectively, during automobile-to-automobile collisions. For pedestrian impacts, along with lower- and upper-leg impacts, the head (both child and adult) impact plays a major role in damages that lead to fatal injuries. In such cases, the head injury criteria (HIC) value has to be within the safe limit (<1000) [7-15, 7-16] as specified by European Enhanced Vehicle-Safety Committee (EEVC) [7-17].

The fender energy absorber plays an important role in minimizing damage to the pedestrian. A combination of materials, geometries, and manufacturing processes are being employed for the design of pedestrian-safe fenders [7-18]. Among the existing trends for pedestrian-safe fenders, the most commonly used method is the energy-absorber lobes arranged along the hood line beneath the fender surface [7-19].

7.4 Design and Development of the Thermoplastic Tailgates

The tailgate, also known as the fifth door, is arguably one of the largest body panel parts in a car. The tailgate is a hang-on part, which has to fulfill diverse requirements over its life span. The main function of the tailgate is to access the luggage/boot space or the last row seats. The frequent opening and closing of such a large door makes it one of the most demanding structural body panel parts. The tailgate is also one of the heaviest closures in a car. Tailgates come in various shapes and sizes, depending upon the type and style of the car. Some of the common tailgate variations are shown in Fig. 7.17.

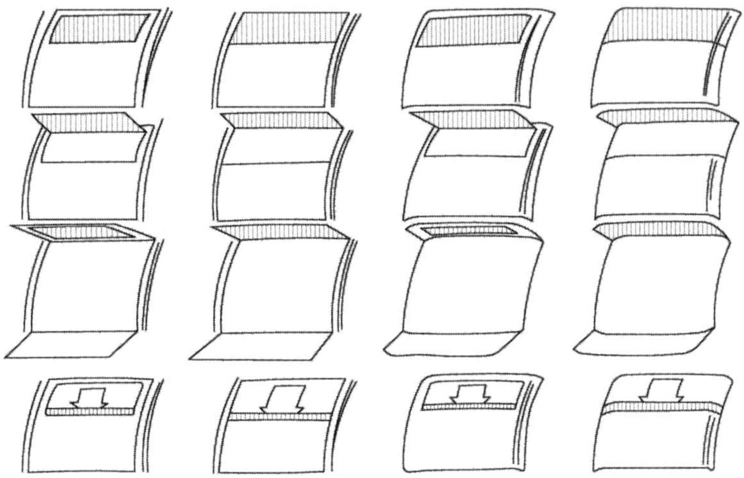

Fig. 7.17 Common tailgate variations.

When it comes to basic construction of tailgates, most of the advanced tailgates consist of a clamshell design (one single piece part that folds onto itself to close) to meet the structural performance requirements. Normally, the outer shell forms the A surface and inner shell forms the structural part. Most of the time, the inner panel is covered with plastic cladding to make it aesthetically appealing. Various materials are used for tailgate construction; mostly, aluminum and magnesium structures are used for weight reduction. The tailgate as a whole should meet all structural, safety, and aesthetic requirements, as warranted by the car maker. The materials used for tailgates should be capable of providing good torsional rigidity to the part, considering the way it is likely to be operated by the end user. The other key safety requirement is that it must not break into small parts while in a closed condition or during separation from the BIW during a rear collision. In addition to all these requirements, the outer shell needs to maintain Class A surface qualities after painting, and these parts need to meet the flush and gap requirements as well.

7.4.1 Functional Requirements of Thermoplastic Tailgates

The key challenges of designing a thermoplastic tailgate are (1) maintaining a close tolerance on flush and gap requirements and (2) the ability to withstand structural loads without sacrificing the cost advantage.

The outer panel of the tailgate can be designed either with online paintable or offline paintable materials. The inner shell is the main structural part, which gives a lot of rigidity to the tailgate along with glazing. Typically, filled composite material is the material choice for the inner structure. Based on the type of paint process the part has to go through, the base material for the outer panel is selected. Depending upon the severity of the structural load, the filler content is selected. There has to be a fine balance between the brittleness and stiffness of the material to meet both the static load and the impact load from rear collision. Normally, the plastic becomes brittle as the filler content is increased. Controlling the warpage of inner structure is a challenge considering the nature of the filler material and the dimensions of the part. Careful balancing of filling or the controlled filling of the cavity are some of the easily adoptable methods of controlling the warpage during molding. Another way of reducing warpage is through balanced ribbing, to give in-plane stiffness to the structure, as shown in Fig. 7.18.

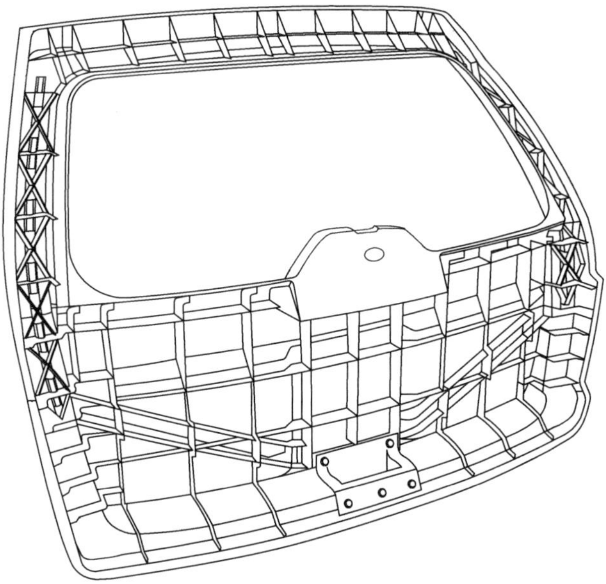

Fig. 7.18 ETP tailgate.

The other engineering challenge in tailgate application is connecting the outer panel to the inner panel. Traditional connecting methods such as adhesive, rivets, screws, and snaps are used to secure the outer panel to the inner one. If two different materials are used for the outer and inner panels, then the CTE mismatch needs to be accommodated, while designing the fixation elements. The other material properties that need to be considered while selecting the material include low moisture absorption, good chemical resistance, low CTE, good paintability, good impact strength, and good flowability. Hybrid concepts for the tailgate inner panel are also gaining popularity, because of the advancements of plastic metal hybrids and polymer composite hybrids. The strengths of both materials are utilized to the maximum, which brings a great cost advantage to the structure. The potential advantage of use of plastics in tailgates is realized when styling freedom, part integration, and lightweighting are fully achieved. Similar to thermoplastic fenders, tailgate parts can be evaluated using CAE tools to maintain gap and flush within the prescribed limits, with respect to the tailgate's surrounding components.

7.4.2 Tailgate Impact Resistance and Structural Rigidity

The design development processes adopted for a tailgate is similar to that of fender design processes, discussed in earlier sections. Along with the dent resistance requirements of the part, the tailgate structure has to withstand the rigidity requirements for repeated opening and closing cycles during its use phase. Predictive engineering simulations are normally carried out to optimize the part thickness and evaluate the plastic part performance. To simulate the opening and closing of a tailgate, static load is usually applied and its structural response is evaluated using commercial FE tools.

7.5 Tank Flap

The tank flap, one of the body panel applications, serves as a lid to the fuel tank and is a visible part blending perfectly with the car styling, as shown in Fig. 7.19.

Typically, tank flaps are opened remotely. They are in either circular or round-edged rectangular shape. Their inner features are designed according to the car's BIW features. Tank flaps are historically made with sheet metal. Just as in fenders, a suitable thermoplastic material is selected based on the painting process. In tank flap application, Class A surface finish and dimensional stability are the key challenges. Since the part has to match exactly with the car body surface, any dimensional variation costs a lot, in terms of the aesthetics of the car. Processing optimization is required to get a dimensionally stable part with very good surface finish. Since tank flaps are aesthetically important, gating on the flap surface or on the edges is not advisable: If gating is on the flap, it can cause some

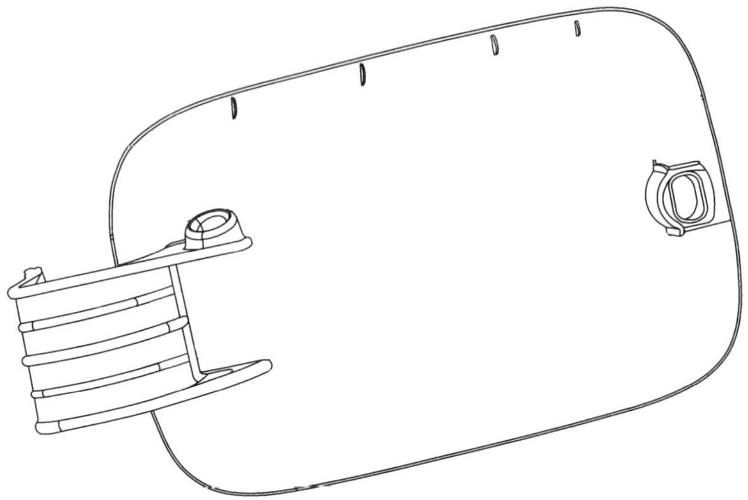

Fig. 7.19 Generic ETP tank flap design.

surface defects on the A surface, and if it is on the edge, it will leave a gate mark on the visible surface. The part is assembled to the BIW through the integrated bracket, which is a hinge mechanism for opening and closing the tank flap. The surface on the bracket is ideal for gating the part, as it will not cause any surface defect or leave a gate mark on visible surface. Very often submarine gates are used for molding the part. In the case of high shear rates, two subgates can also be used next to each other. While filling issues may not be high in tank flaps, warpage needs to be controlled within the tolerance limits.

7.6 Spoiler

The rear spoiler is an additional body panel application with similar requirements as those of other body panels of the vehicle. A typical rear spoiler used in a passenger car is shown in Fig. 7.20.

The main functions of rear spoiler are to improve the aerodynamic performance of the automobile and enhance its aesthetics. The spoiler will generally run across the entire width of the car. The part is typically molded with multiple sequential valve gates to make sure that there are no weld lines in the part. The gates should be located at invisible surface. Typically, this component is molded with four to six gates. In some of the designs, the rear spoiler will be of closed section, which is created by joining the upper and lower components by fasteners or adhesives.

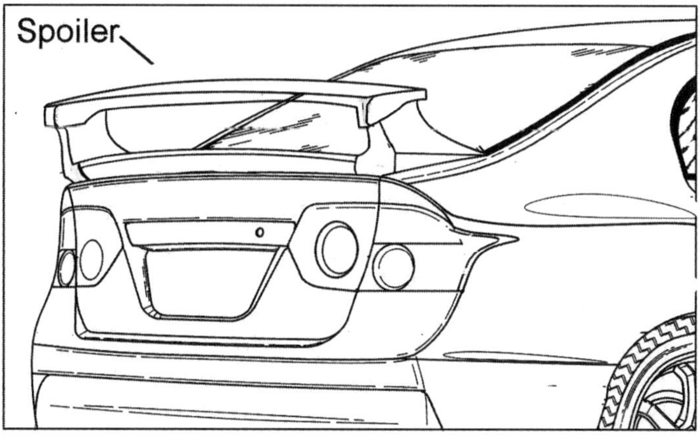

Fig. 7.20 ETP rear spoiler.

7.7 Summary

The automotive industry has adopted lightweighting and aerodynamic drag reduction to meet legislative requirements on fuel efficiency and emissions. This has led to the increased use of ETP in the body panel segment. This chapter elaborated upon several advantages of ETP in body panels, in addition to the key benefits of lightweight and aerodynamic friendly shapes. Various functional requirements and typical ETP materials used in the body panel segment have been summarized. The physics-rich thermo-viscoelastic problem associated with ETP fender design was described in detail. Also, the chapter brought out different engineering design objectives as well as the role of predictive engineering to optimize the design parameters, through a fender case study. A brief description of material characterization and material model for fenders was provided. Other body panel segments such as tailgates, tank flaps, and spoilers and their associated design challenges were briefly covered.

7.8 Trends

OEMs are looking toward weight reduction of the vehicle through the use of lightweight materials and design. The future trend is toward increased performance with higher mileage and lower CO_2 and greenhouse gas emissions. This translates into an aerodynamic design of exterior body panels along with integrated functional features such as air vents to enhance the performance of the vehicle. Further, as designs are becoming trendier, the size and shape of body panels are becoming geometrically more complex. Body panel parts, such as fenders, are getting larger in size, extending up to the headlamps, hood, and the B pillar.

Among the new material trends for exterior a body panel applications, CFRP (carbon-fiber-reinforced plastic) emerges as a very promising option that can be used to meet high-strength needs with a greater stiffness-to-weight ratio while retaining excellent impact and damping properties. One of the recent developments is a complete car made from CFRP [7-20]. Utilization of new materials such as hybrids and composites replacing traditional steel structures is an emerging trend for tailgate applications [7-21]. Another trend is the adoption of sustainable and recyclable materials, such as wood fibers for door applications, which has several advantages and is likely to increase in the coming years [7-22].

Also, the future will see further use of thermoplastics in other large body panels such as the door panels, roof, floor, and hood. Development of paintable hood scoops, without any need for additional cutting or drilling for easy installation, is the new trend in automotive hood applications. Design and development of sandwich structures that would minimize the weight with structural rigidity of hoods will be the future trend for this body panel segment [7-23]. In general, increased use of new materials to achieve lightweight with enhanced performance will be the path ahead for automotive body panels.

7.9 References

7-1. Pfanstiehl, J. 1998. *Automotive Paint Handbook*. New York: Berkley HP Books.

7-2. Hardikar, N., Bobba, S., Sugawara, M., Nemoto, T. and Shigeru, I. 2011. "Role of Predictive Engineering in the Design Evolution of a Thermoplastic Fender for a Compact SUV." *International Journal of Passenger Cars—Electronic and Electrical Systems* 4(1): 111–125.

7-3. Hess, F. and Gonzalez, U. 2003. "Automotive E-Coat Paint Process Simulation Using Finite Element Analysis," NAFEMS Ninth International Conference, Orlando, FL, http://www.algor.com/news_pub/tech_white_papers/e-coat_paint_process/default.asp. Accessed on January 19, 2012.

7-4. Malloy, R.A. 1994. *Plastic Part Design for Injection Molding: An Introduction*. Society of Plastics Engineering. Cincinnati, OH: Hanser Gardner.

7-5. Hardikar, N., Ratnesh, K., Venkatesha, N., and Doggen, J.G. 2010. "Prediction of Thermoplastic Fender Behavior During E-Coat Bake Cycle-Part 1 FEA Methodology and Problem Formulation." SAE Paper No. 2010-01-0232. SAE International, Warrendale, PA.

7-6. Ferry, J.D. 1980. *Viscoelastic Properties of Polymers*. Mississauga, Canada: Wiley.

7-7. Williams, M.L., Landel, R.F. and Ferry, J.D. 1955. "The Temperature Dependence of Relaxation Mechanisms in Amorphous Polymers and Other Glass-Forming Liquids." *Journal of the American Chemical Society* 77(1): 3701–3707.

7-8. Dealy, J. and Plazek, D. 2009. "Time-Temperature Superposition—A Users Guide." *Rheology Bulletin* 78(2): 17–21.

7-9. Brinson, L. 1990. "Time Temperature response of Multiphase Viscoelastic Solids through Numerical Analysis." PhD Dissertation, Division of Engineering and Applied science, California Institute of Technology, Pasadena.

7-10. Fesko, D.G. and Tschoegl, N. W. 2007. "Time-Temperature Superposition in Thermorheologically Complex Materials." *Journal of Polymer Science Part C: Polymer Symposia* 35(1): 51–69.

7-11. "TA Orchestrator Manuals, Rheometric—RDA III Dynamic Mechanical Analyzer," http://downloads.tainstruments.com/ftp/fetch/52757782-4bec-4ecf-96f4-d3411d860fd9/Readme72.pdf. Accessed January 19, 2012.

7-12. Hardikar, N.A., Bobba, S. and Jha, R. 2011. "Applicability of Time Temperature Superposition Principle to an Immiscible Blend of Polyphenyleneoxide and Polyamide," *Journal of Polymer Engineering* 31(2): 223–236.

7-13. *ABAQUS User Manual Version 6.11*. 2011. Providence, RI: Dassault Systèmes Simulia.

7-14. Leaderman, H. 1943. *Elastic and Creep Properties of Filamentous Materials and Other High Polymers*. Washington, DC: Textile Foundation, p. 175.

7-15. Farooq, S. and Schuster, P. 2003. "Body Concept Design for Pedestrian Head Impact." SAE Paper No. 2003-01-1300, SAE International, Warrendale, PA.

7-16. "Head Injury Criterion," Wikipedia, http://en.wikipedia.org/wiki/Head_injury_criterion. Accessed April 30, 2013.

7-17. "European Enhanced Vehicle-Safety Committee, EEVC Working Group 17 Report," http://www.unece.org/fileadmin/DAM/trans/doc/2006/wp29grsp/ps-187r1e.pdf. Accessed April 21, 2013.

7-18. de Melo Baleki, D. and Ferreira, A. 2009. "Development of Fender Attachment with Collapsible Brackets for Pedestrian Head Impact." SAE Paper No. 2009-36-0093. SAE International, Warrendale, PA.

7-19. Nagae, K., Oota, K., Nagumo, H., and Okabe, T. 2009. "Development of Fender Structure for Pedestrian Protection." SAE Paper No. 2009-01-1206. SAE International, Warrendale, PA.

7-20. "Carbon Fiber Reinforced Plastic in the Automotive Industry," http://www.doors-congress.com/uploadedFiles/EventRedesign/Germany/2012/March/12292004/Assets/AC3-Automotive-Composites_Carbon-fibre-reinforced-plastic-in-the-automotive-industry.pdf. Accessed on March 13, 2012.

7-21. "Tailgate Innovation Project," Instituto Superior Tecnico, http://in3.dem.ist.utl.pt/designstudio/proj05/innovation.htm#2. Accessed on March 13, 2012.

7-22. "Automotive New Technology and Trends,"http://www.doors-congress.com/uploadedFiles/EventRedesign/Germany/2012/March/12292004/Assets/Automotive%20Doors_NewTechnologyandTrends.pdf. Accessed on March 13, 2012.

7-23. "Design Analysis of a Sandwich Hood Structure for Pedestrian Protection," http://www-nrd.nhtsa.dot.gov/pdf/esv/esv21/09-0356.pdf. Accessed on September 24, 2012.

Chapter 8
Under-the-Hood Applications

T.K. Poovanna, Narendra A. Hardikar, K.R.K. Ravikanth, Triloka Tankala, and P. Arunachala

8.1 Introduction

Traditionally, using any material other than metal close to the engine block was unimaginable. That traditional mind-set is fast disappearing as automakers are forced to use lightweight materials to reduce the overall weight of the vehicle. The continuous performance enhancements and development of engineering plastics in areas such as dimensional stability and thermal and chemical resistance have made increased use of plastics in the under-the-hood (UTH) segment possible.

The engine compartment of any automobile would be the most nonconducive location for many traditional lightweighting materials and their applications. Constant pounding of the engine generates serious vibrations, which in turn create severe stress on most of the parts located in the UTH compartment. The heat radiated from the engine makes the world beneath the hood a very harsh environment for many of the parts made of engineering thermoplastics (ETP). Also, these parts come in contact with automotive fluids in UTH applications such as engine coolant, antifreeze in the coolant, engine oil, power steering fluid, transmission fluid, brake fluid, windscreen wiper fluid, gearbox oil, battery acid, A/C refrigerant, and other volatiles generated from various lubricants. Unlike many parts in cars, those UTH are directly exposed to extreme weather. The relentless thermal cycling of parts, from subzero temperature to in excess of 100 °C (212 °F), induces enormous thermal fatigue in individual parts.

Given the adaptability, moldability, versatility, and excellent chemical resistance of engineering plastics and the ongoing development of high-end materials that can withstand a wider temperature range, several ETP are being used for modern-day UTH applications. Government regulations on fuel economy, stringent emission standards, and the marketing needs of original equipment manufacturers (OEMs) and their tiers to reduce the system cost and minimize the overall weight of the car further pushed these lightweighting options into the UTH segment. Advancement in joining techniques, such as welding technology for plastics and development of compatible adhesives also created new application spaces, which were earlier either impossible or too expensive. Advancements in processing technology like gas-assisted molding and water-assisted molding have also created or redefined the way parts are being designed.

8.2 Material Requirements for Under-the-Hood Applications

Operating conditions in the UTH area of a motor vehicle are extremely challenging. The materials used in the engine compartment are highly stressed from many angles. Typically, they have to withstand temperatures of 160 °C (320 °F) or more (with the trend moving toward even higher temperatures) [8-1] and bear the punishing subzero temperatures. They are subjected to high mechanical and thermal stresses, while being exposed to extreme vibrations and G loads, reaching up to a maximum of 25 G. In addition, these parts are in direct contact with a variety of automotive fluids, and so chemical resistance to these fluids is a must. Some of the key tests that are performed to determine the suitability of a material for UTH applications are the heat-aging test and the chemical resistance test in addition to the standard tests such as, creep, flexural, and moisture intake test.

8.2.1 Heat Aging

When exposed to heat and oxygen for prolonged duration, the material properties of thermoplastics start to degrade. This phenomenon is known as heat aging, and it is a critical factor to be considered while assessing the expected service life of the component in the UTH area. To quantify the heat-aging behavior, a heat-aging test is performed. In this test, a specimen is exposed to heat for a prolonged period of time in an oven. Constant air change in the oven is ensured to simulate the effect of atmospheric oxygen. Specimens are taken out of the oven in regular intervals (e.g., at 500, 1000, 2000, and 3000 hours depending on the protocol and the OEM requirements) and subjected to mechanical tests. The mechanical properties of the material may decline depending upon the aging temperature and duration of exposure. Typically, the valuation is done at 85 °C (185 °F), 120 °C (248 °F) and 150 °C (302 °F), depending on the location of the part in UTH, for about 1000 hours [8-2]. The reduction in mechanical property is accounted for while the part is being designed. Sometimes the test is also conducted at higher humidity conditions to simulate the degradation of material properties resulting from moisture intake, depending on the material under consideration.

8.2.2 Chemical Resistance

Some of the applications require parts to be in constant contact with various automotive fluids. Applications that are in contact with, for example, engine oil transmission oil will also be at an elevated temperature and higher pressure. To check the compatibility of the material with motor oil or transmission oil, the specimen is cooked in the oil at an elevated temperature for about 1000 hours. Sometimes, the test is conducted at higher pressures as well to simulate real-life scenarios. Then, the reduction in the property after the testing versus the original property is recorded. Similar tests are performed using used engine oil, gasoline, and ethanol depending on the end use of the applications. The oxidative degradation of polymers is also significant since the service temperature is in excess of 100 °C (212 °F) in many cases. Because most of the parts are directly exposed to atmospheric oxygen, which can attack the polymers, oxidative degradation can occur, resulting in reduced mechanical and chemical properties. Some of the parts that are positioned very close to the road do come into contact with road salts typically used for deicing as well. Salts [8-3] such as $CaCl_2$ or $NaCl_2$ that are being used on the roads are easily solvable in water. These salts have an adverse effect on the mechanical properties of some of the materials that are used in UTH applications. The prolonged exposure to these chemicals can make some of the polymers brittle or degrade the mechanical properties drastically. Usually, materials are subjected to chemical resistance test for a prolonged period (up to 5000 hours) to study their behavior and quantify the property degradation.

8.2.3 Types of Engineering Plastics in Under-the-Hood Applications

The material choice for a given component depends on the environment in which the part operates and the types of load that the part is subjected to. In general, a typical UTH part can come into contact with air, oil, fuel, and water, and it may be near electric fields. While the part needs to withstand these fluids, it needs to be resistant to a several other solvents, acids, and chemicals as well. Mostly, the materials at the bottom of the pyramid are not preferred for applications close to engine; instead, engineering materials with good thermal properties and excellent chemical resistance make it to UTH applications, as shown in Fig. 8.1.

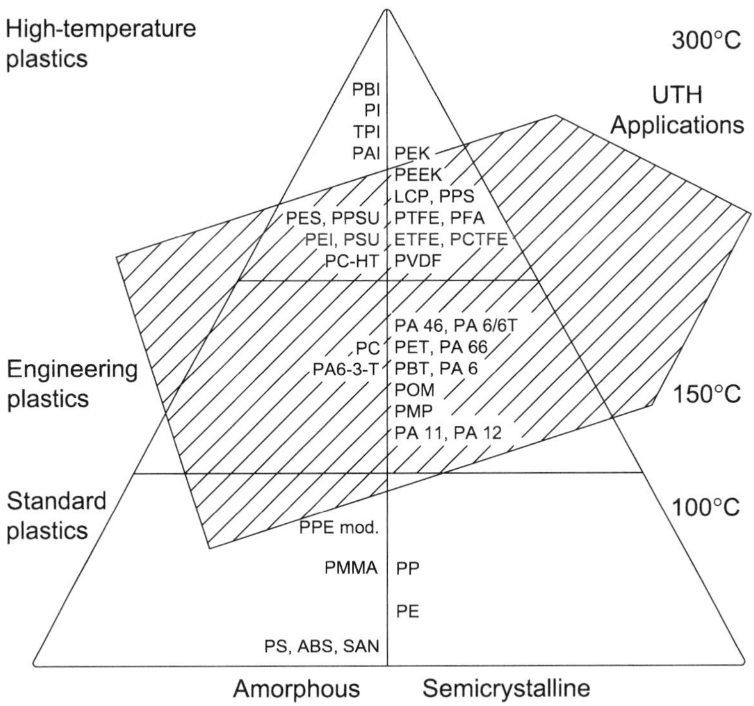

Fig. 8.1 Engineering thermoplastics in UTH applications.

Both amorphous and crystalline materials are employed for various applications depending on the end use and the medium with which the part makes primary contact. For example, if the part is in constant touch with gasoline, only materials that can withstand fuel can be used. Table 8.1 shows the more common plastics used in UTH applications.

Table 8.1 Common Engineering Thermoplastics in UTH Segments

Material	Abbreviation	Application Area	Inherent Property
Polyamide	PA	Air-intake manifold, fan shroud, fan, air inlet manifolds, engine cover, resonators, pulleys, water pump parts, oil sump, radiator end caps, chain tensioner	Excellent abrasion resistance, Good physical properties, including hygroscopic (moisture has significant effect on properties); good heat resistance; good chemical resistance; good flowability
Polybutylene terephthalate	PBT	Connectors, sensor encapsulations, brake booster valve, ignition coils	Creep resistance, dimensional stability, low moisture absorption, good dielectric strength, good chemical resistance
Polyphenylene sulfide	PPS	Crankshaft flange, throttle body, intercooler parts	High HDT (heat-deflection temperature), good chemical resistance, dimensional stability, creep resistance, low moisture absorption
Polyetherimide	PEI	Throttle body, connectors, engine head cover, pump housing, crankshaft gasket, sensor encapsulation, transmission housing, radar lens, fuel filter, Solenoid valve, Oil pump drive bearing, A/C compressor	Good dimensional stability, excellent electrical properties, long-term creep resistance, high strength retention at elevated temperatures, good resistance to fuels, good resistance to lubricants, very good chemical resistance

(Continues)

Table 8.1 Continued

Material	Abbreviation	Application Area	Inherent Property
Polyether ether ketone	PEEK	Butterfly valve, throttle body, distributor, turbo charger parts	Excellent mechanical properties, Excellent chemical resistance, Good resistance to thermal degradations
Polyphenylene oxide	PPO	Power distribution box, ignition coil bobbin, ignition coil	Low thermal expansion, low moisture absorption, excellent dimensional stability, good elongation

Also, nylon or polyamides dominate the UTH application space. Some of the popular nylon grades are shown in Table 8.2.

Table 8.2 Nylon Applications in UTH Segments

Family	Application
Polyamide 6	Air-intake manifolds, ventilation system, fans, water pump components, belt pulleys, tensioners, engine covers, electric air pumps
Polyamide 12	Fuel lines, fuel rails, fuel tanks, fuel float, connectors
Polyamide 66	Air-intake manifolds, resonators, engine mount, intercooler end caps
Polyamide 6/6	Fuel rails, fuel filter housings, fuel caps, fuel filler pipes, oil pan

8.3 Under-the-Hood Application Examples

Some of the prominent parts that have been converted successfully to thermoplastics include air-intake manifolds, valve covers, fuel rails, water pump housings, HVAC systems, brake fluid reservoirs, electronic throttle controls, engine covers, filtration systems, resonators, radiator heads, fuel rails, oil pans, and some of the powertrain parts.

8.3.1 Oil Pans

Traditionally, oil pans have been made from formed sheet metal or from die cast aluminum. Recently, thermoset polymers have been tried in pursuit of lighter, cost-effective alternatives. Thermoplastic oil pans [8-4] are making way into this segment, mainly because of weight savings over traditional materials, as shown in Fig. 8.2. However, the real benefit of moving into thermoplastic design is integration and parts consolidation.

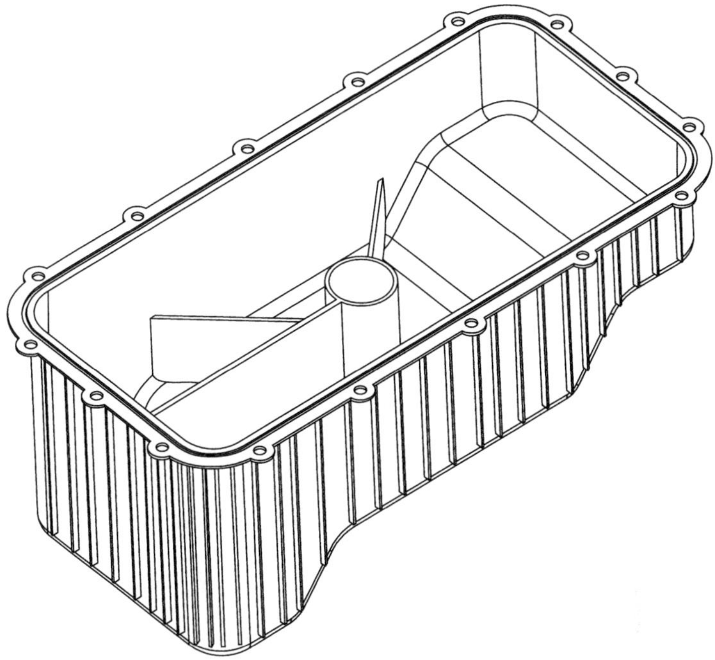

Fig. 8.2 ETP oil pan.

The main function of an oil pan is to hold the lubricant for the internal engine parts. Currently, most of the lubrication is enabled by forced pumping of lubricating oil into the required areas through an oil pump. There are various configurations of the pumping system in the marketplace. The oil filter sometimes becomes the part of the oil pan assembly, too. Most of the time, the oil sump is one of the physically low-hanging parts for metal replacement in an automobile engine, though it is quite difficult to convert into plastics, given the performance requirements of this part. Depending upon the region where the car is built or sold, the regulations governing the performance of the oil pan vary. Requirements such as stone-chipping resistance, curb impact, and chemical resistance vary from one OEM to another. These requirements need to be met, in addition to the performance requirements such as better acoustics performance, seal integrity, and the ability to withstand the vibrations transmitted from the engine block. Even the extreme load case of bearing the entire weight of the engine must be handled, for those times when the engine would be sitting on the oil pan during servicing of the engine.

The commercial success of the oil pan mainly depends on weight and cost reduction. Simply converting a die-cast or deep-drawn sheet metal to ETP would make it cost too much mainly because of the higher raw material cost. Sometimes, the weight reduction of 10–15% will be outweighed by the increase in cost. The recent trend in thermoplastic oil pans have been toward an integrated system, wherein parts such as seals, baffles, heaters, sensors, and oil suction tubes are being made as integral parts of the oil sump.

Depending upon the amount of integration, about 15–25% weight reduction for a simple configuration can be achieved. With complex oil pans, where multiple parts are involved, weight savings of more than 40% can be easily achieved.

8.3.2 Wire Coating

In modern-day cars, almost all the components are either electronically controlled or electronically monitored. In addition to this, many safety features and passenger comfort features has been added to the latest automobiles. The wire harness compactly bundles many wires together in a systematic way and routes these bundles into various components in the car. This has resulted in increased number of wires in the car, adding more weight to the existing wire harness [8-5]. A small- to medium-sized car can have about 20–25 kg of wire harness in its body; a typical luxury car can have up to 60 kg of wire harness. The wire harness is the main backbone of the vehicle control system and the safety system; hence, it is governed by high safety standards. At the same time, environmental awareness [8-6] and the drive for better fuel economy has created the need for reducing the automobile weight. In the past, the weight of wires was reduced by a using thinner [8-7] insulation layer and different conductors in the core, as shown in Fig. 8.3.

Polyvinyl chloride (PVC) and fluorinated ethylene propylene (FEP) dominate the cable insulation segment of the automotive market. Because of the health and safety issues related to some of the traditional materials used in wire coating, the industry is moving toward much safer thermoplastic alternatives [8-8]. Materials such as cross-linked polyethylene

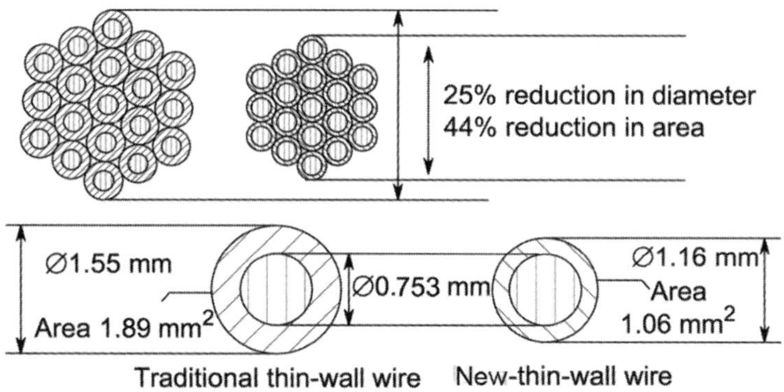

Fig. 8.3 Reduction in insulation thickness by ETP materials in wire coating.

(PE) and PPO are gaining popularity, because some countries have enacted regulations to restrict the use of certain chemicals in the wires. The thermoplastic alternative that does not require cross linking is PPO. Flame-retardant PPO resins with different properties have helped OEMs to opt for much thinner insulation layers. This has helped the OEMs reduce the thickness of insulation layer by 50%, resulting in insulation weight reduction of up to 25% and wire diameter reduction by an average of 25%. This, in turn, helps reduce the wire bundle size by about 40% [8-7]. These reductions enable OEMs to pack more cables into the same packaging space. The thinner bundle also gives additional flexibility in terms of routing the cable in confined spaces throughout the vehicle body. These materials also have twice the pinch resistance and up to four times the scrape abrasion resistance compared to PVC cable of similar insulation thickness. PPO resins offer many performance advantages, such as better abrasion resistance than cross-linked PE, better chemical resistance, lower specific gravity than either PVC or PE, and higher heat resistance than PVC. These materials meet the requirements of Class A and B wire heat aging. These materials offer a nonhalogenated, nonbrominated, lead-free flame resistance alternative to the existing materials. PPO resin is rated at 105 °C (221 °F), and PVC is rated at 85 °C (185 °F), which makes it feasible to be deployed near the engine areas without any additional insulation.

8.3.3 Engine Cover

The demand for comfortable and quiet vehicles has encouraged OEMs to develop methods to improve the comfort levels in their new car models. More particularly, the noise, vibration, and harshness (NVH) characteristics are being seriously taken into account while designing better noise isolation mechanisms.

The exhaust noise is the single loudest component of engine noise; however, it is not the only component. Sound originating from engine valves, piston heads, fuel injectors, timing chains or belts, pulleys, crankshaft bearings, and so forth, add to the engine combustion noise. Soundproofing the engine is a complex task, and completely isolating the engine in a way so as to bar the passage of sound to the passenger compartment and the external environment is impossible. Various attempts have been made to isolate or insulate the noise, to enhance passenger comfort. The plastic cladding on the top of the engine, known as engine cover or beauty cover [8-9], helps to some extent in reducing the noise levels.

The engine cover mainly serves two purposes: The first one is the beautification of engine compartment by hiding many components such as connectors, fasteners, wiring, cables, and the engine head. The second one is improving the acoustics performance by blocking some of the sound waves originating from the engine head. Over the years, many designs [8-10] on the engine cover have evolved. A typical thermoplastic engine cover is shown in Fig. 8.4.

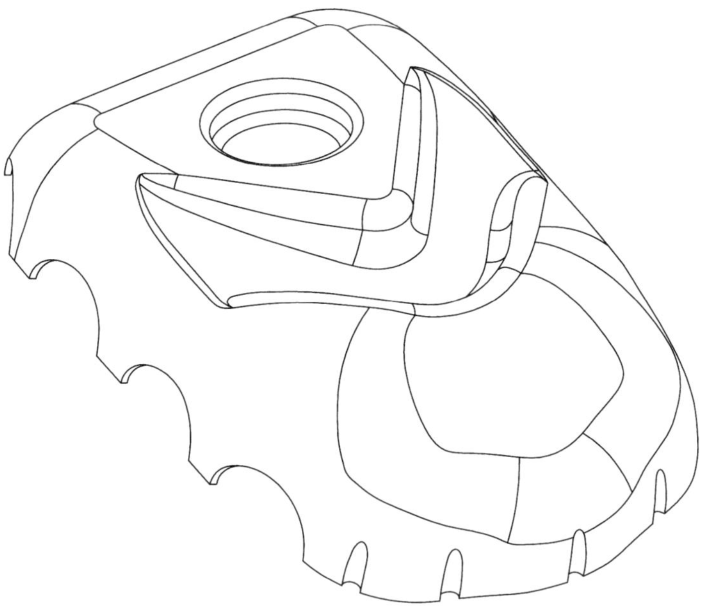

Fig. 8.4 Thermoplastic engine cover.

Many painted and molded-in color options with nylon have emerged to meet specific OEM demands. Multicomponent designs for the engine covers are also becoming increasingly evident in high-end cars. Integration of cladding materials and designing built-in resonators can also be seen in inexpensive models. Some designs with an integrated air filter assembly are also adapted to further increase the value proposition of the part or assembly. As the engine compartments are getting smaller, the UTH temperatures are on an upward trend. This requires new materials to meet these higher temperature needs. The blends of nylon and PPO are some of the options to meet these increased temperature requirements.

While designing the engine cover for acoustic performance, factors such as type of noise, noise levels, frequency distribution, noise sources, and noise propagation paths need to be considered. To meet the vibration loads, stiffening of the engine cover through the ribs or structure is generally adopted. Structurally foamed plastics are also used to get the benefit of stiffness and acoustics.

8.3.4 Fuel Lines

Fuel lines have one of the most complex form and shape of all UTH components. The fuel lines are also one of the safety-critical parts, as they tend to carry highly combustible auto fuels. Traditionally, these parts are made up of metal tubing, but recent emphasis on lightweighting and fuel economy standards are pushing automakers to search for lighter alternatives. When it comes to polymeric alternatives for fuel lines, extrudability and permeability of the materials are of primary importance. The chemical resistance to fuel and fuel additives is the other key factor for selecting the material. As the lines should withstand harsh conditions without rupture; notch resistance and environmental stress-cracking properties are also important. Some of the hoses handle hot fuel, such as the return hose, so they need to withstand temperature variations from –40 °C (–40 °F) to 110 °C (230 °F). To meet the standards on fuel emissions from the vehicle (not related to exhaust emissions), the permeability from the fuel systems must be reduced. To meet this standard, a barrier layer is introduced along with polyamide 12, which is widely used for fuel lines. The popular barrier material in the fuel system is polyvinylidene fluoride (PVDF). In most cases, a tie layer is used between nylon and PVDF to reduce delamination. In some cases, a conductive layer is also introduced to dissipate static charges in order to reduce the fire risks.

8.4 Designing of Under-the-Hood Components

Most of the time, different methodologies need to be adopted for designing various UTH components. While the components need to meet many of the load cases to realize a successful design, some of the load cases become critical and sometimes difficult to achieve. While the plastics give enormous freedom to tweak the design, inherent properties of the plastics sometimes become a limitation. A few case studies are discussed below to demonstrate the uniqueness and the challenges of designing UTH components with plastics.

8.4.1 Turbo Air Duct

To improve the efficiency of the engine, a turbo charger is commonly used. The turbo charger compresses the air and supplies it to the engine. This ensures that more oxygen is available for the combustion inside the cylinder. When the air is compressed in turbo charger, the temperature of air goes up drastically as a result of adiabatic compression. To improve the mass flow, the air needs to be cooled down, which is accomplished by the use of an intercooler. The duct [8-11] that connects the turbocharger to intercooler sees this high temperature. Traditionally, the duct is made up of cast metal. In recent times, high-heat thermoplastic is gaining momentum in this application. A typical air duct is shown in Fig. 8.5.

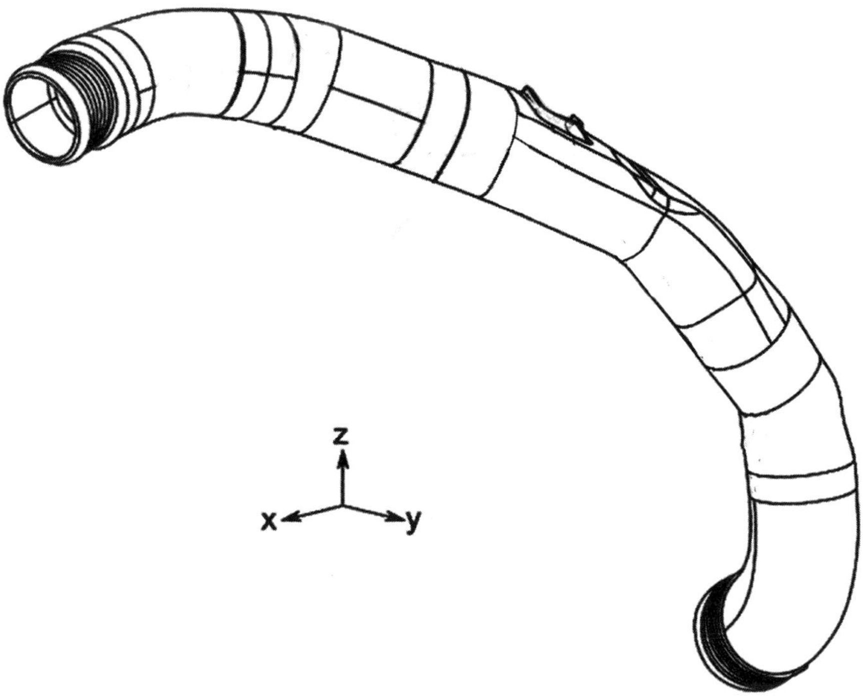

Fig. 8.5 Automotive air duct.

Some of the potential ETP materials are PPS, PEI/nylon and/or PEI/PPS glass-filled blends with higher melt strength for 3-D blow-moldability. The 3-D blow-moldability of the new blends makes it possible to produce products with complex geometry while meeting engineering performance requirements. The air duct needs to meet both temperature and stress loads, as the air intake is generally at high pressure and temperature and also has to sustain the inside air temperature of 230 °C (446 °F) and 0.25 MPa pressure load acting at a frequency of 1 Hz for 200,000 cycles. During the experimental investigation, the air duct is fixed at both the ends. The air is then passed through the duct at the operating temperature of 230 °C (446 °F), and the pressure of the air-intake system is 0.25 MPa, which is pulsated at 1 Hz. The experimental observation of the high stress is indicated in Fig. 8.6.

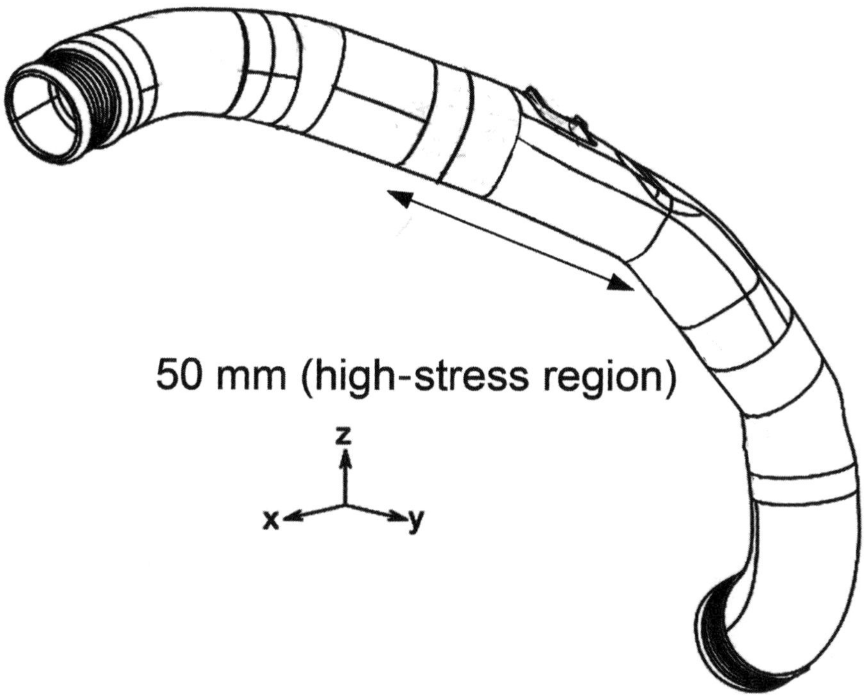

Fig. 8.6 High stress region during fatigue experiments on air duct.

The air duct in this study has failed during the fatigue load (cyclic load) cycle with some of the PEI material grades. To understand the behavior under fatigue load, one requires a curve between stress and number of cycles to failure (known as an S-N curve). As the S-N curves for these new materials are not available, a simple static analysis with the application of thermal and pressure loads is performed. The stresses experienced on the part were monitored to verify the use of PEI-based blends in this application. The part design is then evaluated through standard finite element analysis [8-12]. Once the magnitude and location of the stresses on the part are known, the performance and stress levels can be used for fatigue performance. The suitability of the material is determined on the basis of high-temperature strength and fatigue performance data.

8.4.1.1 Design Validation

Static structural analysis was performed on the air duct considering geometric and material nonlinearity with a pressure load at 230 °C (446 °F) for various material formulations. The air duct was modeled with various test grades of PEI with temperature-dependent elastic-plastic stress-strain and coefficient of thermal expansion (CTE) data.

The ends of the air duct are fixed in all degrees of freedom (DoF), as shown in Fig. 8.7a, simulating the experimental test setup.

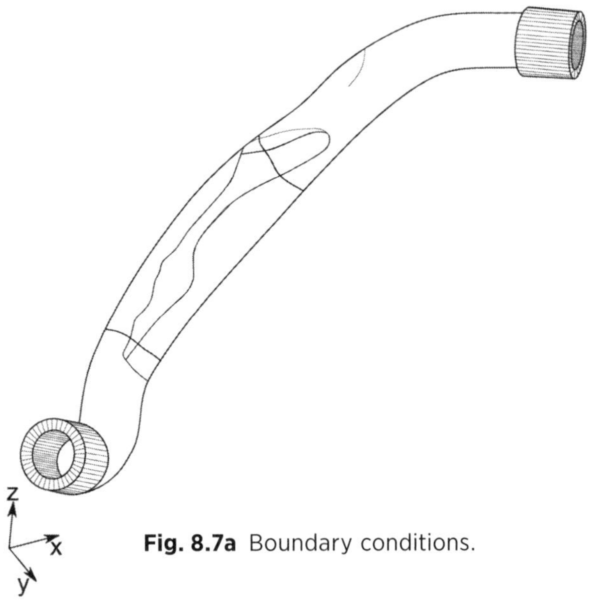

Fig. 8.7a Boundary conditions.

A uniform pressure of 0.25 MPa is applied to the inner surface of the air duct, as shown in Fig. 8.7b.

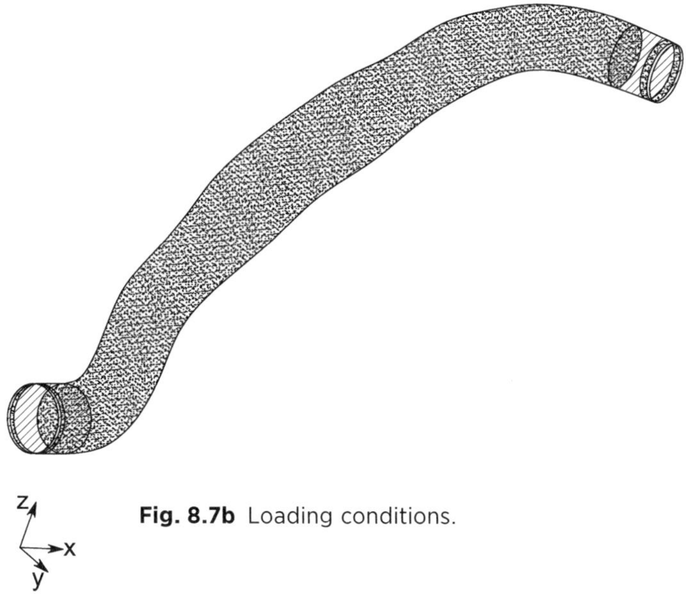

Fig. 8.7b Loading conditions.

Another loading on the air duct is the temperature, which is increased from 23 °C (73.4 °F) to 230 °C (446 °F), while maintaining the pressure of 0.25 MPa, for carrying out the numerical simulations [8-13]. The main challenge of the material is to sustain a temperature of 230 °C (446 °F) and 0.25 MPa load at high temperatures.

Thermomechanical analysis is carried out on the air duct, and a comparison is made between different PEI grades in terms of break plastic strains and stresses. The deformation contours for the air duct for one such material grade is shown in Fig. 8.8.

By carefully studying the stress contours and the plastic strain contours, the local thickening of the air duct can be seen at the center. All the material formulations were able to withstand the pressure and temperature loads with the increase in thickness to 4.5 mm. Through finite element (FE) modeling, good qualitative correlation between numerical and actual experiments could be established. The region of high stress is correlating well with the experimental high stress zone. By using a static study approach, qualitative predictions about the high-stress regions can be performed to simulate the fatigue loads.

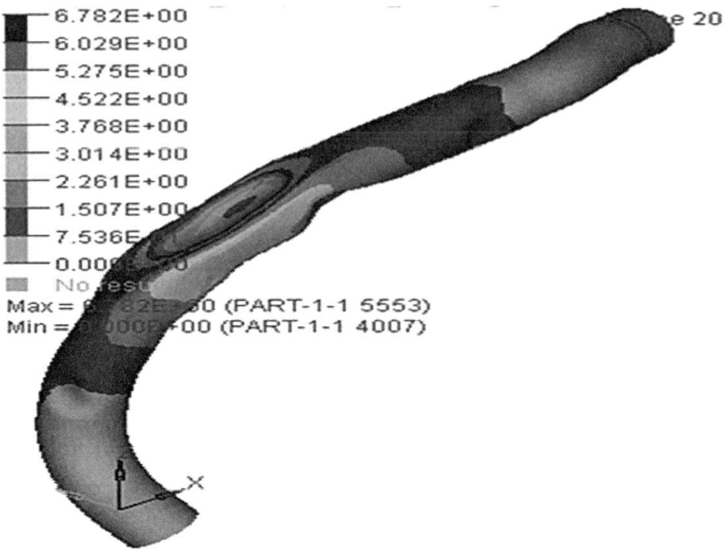

Fig. 8.8 Displacement contour of the air duct.

8.4.2 Throttle Body

In addition to the need for fuel efficiency enhancement through light-weighting, there is a global trend to reduce emissions as well as the harmful contents in them. The throttle body of the engine is a component that has significant influence on the quality of emissions. The primary function of the throttle body is to precisely meter the air supply to the engine. The air supply at idle conditions is often provided by maintaining the valve of the throttle body in the vicinity of the *closure angle*, which represents the geometric position of the valve where the annular clearance area between the throttle body bore and valve is at its minimum value. The *closure angle*, based on throttle body construction, can be from 0–5°. In this lower angle regime, the clearance between the throttle body bore and the valve is often a fraction of a millimeter. The quality of air metering depends on how well the clearances between the static bore and the movable valve are maintained over the operating temperature range.

8.4.2.1 Types of Throttle Body

Depending on the mechanism used to control the valve movement, throttle bodies can be classified into two main categories: mechanical throttle bodies (MTBs) and electronic throttle bodies (ETBs).

In MTBs, there is a direct cable linkage to the accelerator pedal, and thus manual inputs control the throttle valve movement and the air throttling as well. In contrast, the valve movement in ETBs is driven by a computer-controlled servomotor, which can be optimized for fuel efficiency and emission control. This key advantage of ETBs is enabling the transition from MTBs to ETBs in the auto industry.

ETBs, depending on the construction of their bores, can be further classified into conventional straight bore profile type and progressive bore type throttle bodies. A progressive ETB has a bore profile that is generated by extruding a circle about a centerline, defined by a combination of straight and curved segments. Geometric complexity, as compared to straight bores, allows higher degree of fine-tuning of airflow at lower valve angles. Such bores allow the valve to have more rotational movement with less effect on mass flow, resulting in a higher resolution at lower angles. The choice of bore profile is often OEM specific.

8.4.2.2 Materials for the Throttle Body

The throttle body, given its location in the hostile UTH environment, needs to withstand high temperature, vibrations, and chemical spills. Avoiding icing to prevent any blocking of the valve movement is also a key requirement.

Given these requirements, aluminum is the dominant incumbent material in the throttle body segment. However, glass- or mineral-filled thermoplastics such as PEI, one of the most dimensionally stable thermoplastics, has managed to replace aluminum as one of the materials of choice in the MTB segment. The CTE of some grades of glass-filled PEI in the flow direction (direction of fibers) is very similar to that of aluminum.

In addition to the usual advantages offered by thermoplastics such as weight saving and integration of parts, thermoplastic PEI throttle bodies reduce or eliminate the need to have water jackets, which are required in aluminum throttle bodies to prevent icing. There have been more than 10 throttle body applications with glass-reinforced PEI and, in some cases, with mineral fillers that have successfully entered production. Glass-filled PPS is another material that is considered for the development of throttle bodies. However, the dimensional stability of PPS is lower compared to that of glass-filled PEI.

While thermoplastics have managed to create a space in MTB segments, penetration in the ETB segment by replacing aluminum as the dominant material has proved to be a challenge and a slow process. OEMs generally require predictive engineering data before they make the transition from a traditional material to any new material systems.

8.4.2.3 Predictive Tools to Drive Thermoplastics Usage in Electronic Throttle Body

One distinct factor in an ETB as compared to MTB is the need for much tighter tolerance requirements on the bore-valve material. While the OEMs see the potential advantages that thermoplastics can offer, they also look for comparative data with metal on the following questions:

- How does a glass-filled thermoplastics bore valve deform under operational thermal or structural loads?
- How does the airflow resulting from the thermomechanical deformation of the throttle body vary?

The experimental validation, a potential option for answering these questions, requires manufacturing prototypes. Tooling for a thermoplastics throttle body is an expensive proposition, in terms of both cost and time. Moreover, in the case of filled thermoplastics, validation of results obtained on a prototype with a particular molding gate scheme will not be generally applicable to another gating scheme or to the use of another grade of thermoplastics, given the dependence of properties on flow-induced fiber orientation. Thus, the numerical modeling to compute the variation in material property resulting from flow, deformations under structural-thermal loads, and their effect on the clearance area and as a consequence on the mass flow variation is an essential precursor to an economical and optimal design and development of an ETP throttle body.

In this section a brief overview of the multiphysics-predictive methodology that was developed to evaluate the performance of thermoplastic throttle bodies is presented [8-14]. This methodology has three key steps:

1. Structural deformations that capture the effect of processing conditions and thermal loads
2. Mechanism to capture tolerance effects and structural deformations
3. Computational fluid dynamics (CFD) simulation on the deformed throttle body

Fig. 8.9 shows the simulation sequence capturing steps 1 to 3.

The initial phase consists of a processing simulation to capture the influence of the processing conditions, gate, and so forth, on the resulting orthotropic material properties (CTE, modulus) of the glass-filled

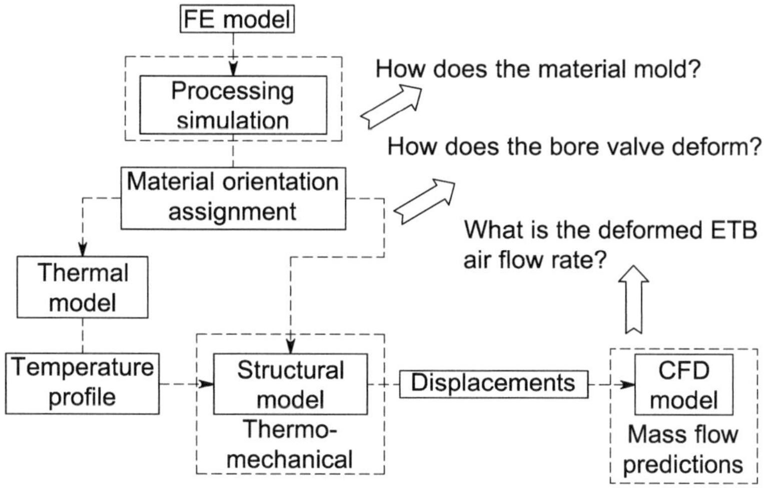

Fig. 8.9 Multiphysics approach for flow predictions.

thermoplastics. In the next step, the temperature field in the thermoplastic due to operational thermal loads is computed. The temperature field, fiber-orientation tensor, and orthotropic material properties are then fed as inputs to the structural model with the given structural loads, and corresponding deformations are predicted. The last step is the prediction of airflow through the deformed throttle body, using CFD.

Extensive work that was done to benchmark the CFD model and identify the right numerical options to simulate the airflow in a low-angle regime is reported in Ref. [8-15]. Numerical predictions of an undeformed 87-mm throttle body at the low-angle regime from 4° to 11°, with different flow models benchmarked with experimental data. The excellent correlation with experimental results and other details can be found in Ref. [8-16]. This work offers a guarantee that the CFD model has the fidelity to capture the flowfield accurately.

However, the key challenge in this methodology was the transfer of structural deformations details onto the flow grid. Figure 8.10 shows an overview of the patented procedure [8-17] developed to achieve the inter-linking between the structural and fluid solver.

This procedure is based on combining the analytical geometry, an FE interpolation algorithm, and CAD concepts in a unique way so as to link a commercially available FE solver with a CFD solver.

Figures 8.11a–8.11c depict diagrams based on the general equations for the clearance area of an n-segment progressive bore throttle body, which were

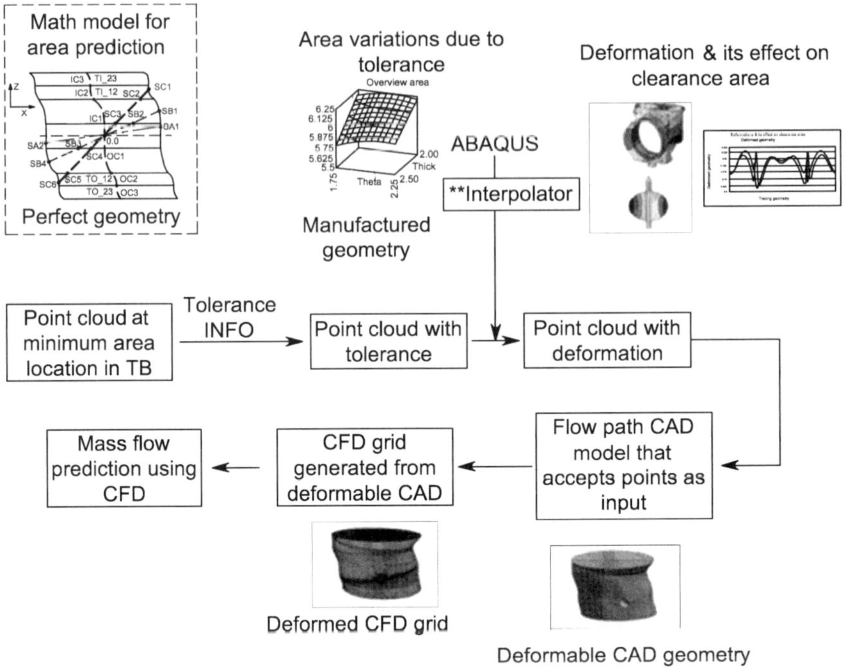

Fig. 8.10 Procedure developed to link FE and CFD solvers.

derived in Ref. [8-14], and with the final formulas in Ref. [8-16], they reduce to a straight bore throttle body under appropriate conditions.

These equations are essential to accurately estimate the clearance area as well as generate precise theoretical geometry, in the form of a point cloud (in X, Y coordinates). As the clearance area equations do not have a closed form solution, a Mathematica [8-18] program was developed for numerical integration. These equations were used to perform a six-sigma study on the influence of manufacturing tolerances on flow predictions of an ETB [8-16]

Since the airflow is dependent on the flow area, the CAD surfaces at the location of interest (minimum clearance area) were developed such that they accept the (analytically computed) point cloud as an input. Point-based CAD surfaces need a high number of points to capture the bore geometry precisely. The purpose of using a point-based surface was to update it with the deformations obtained from the FE solver. However, a practical FE mesh cannot have nodes that are equal to the number of points required for accurate CAD surface generation. Also, one cannot have a separate mesh made for every valve angle, with nodes available at the minimum clearance area. Hence, the geometry field was predicted

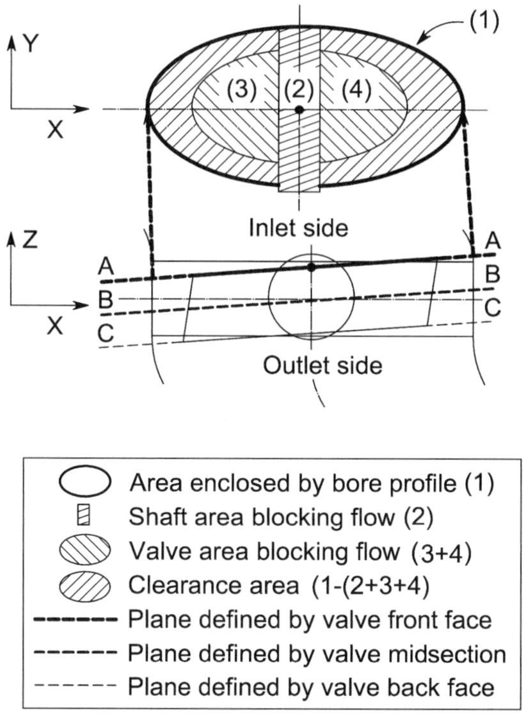

Fig. 8.11a Clearance area.

using analytical relations, while the FE calculations were used only for the displacement field calculation.

A search algorithm, utilizing ANSYS APDL (ANSYS Parametric Development language) [8-19] was developed to determine, for each geometry point in the point cloud, the corresponding element face and nodes in the FE mesh, which would be used for interpolating displacements at the geometry point. An interpolation routine using appropriate element shape functions was developed to interpolate the displacements obtained at nodes after the FE simulation. Structural displacements were then added to the point cloud to capture the structural deformations.

The deformed CAD surfaces were used to regenerate a CFD grid, and flow computations were performed using the new CFD grid. This tool was used to perform comparative predictions, considering various materials for the bore (aluminum, PEI) and valves (aluminum, brass, PEI).

8.4.2.4 Processing of Throttle Body

The plastic electronic throttle body is typically manufactured using an injection-molding process. Because of its complicated geometry, mold

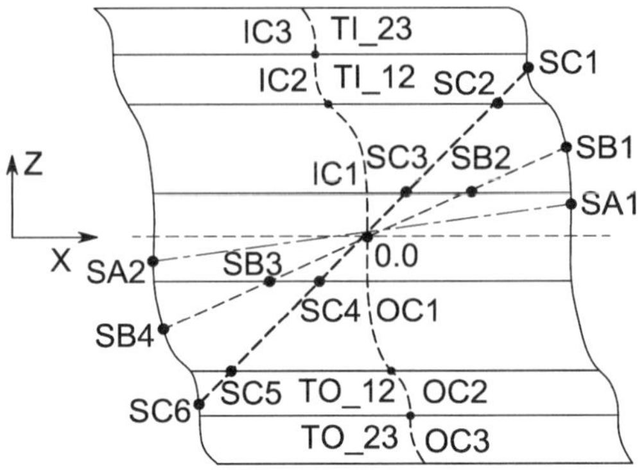

SA section cutting straight position

SB section cutting straight & first curved segment on inlet (IC1) and outlet (OC1)

SC section cutting straight & second curved segment on inlet (IC2) and outlet (OC2)

	Intersects with curvature no		
Section	Inlet	Outlet	Y variations
SA	0	0	1
SB	1	1	3
SC	2	2	5

Fig. 8.11b Segments in an *n*-segment progressive bore.

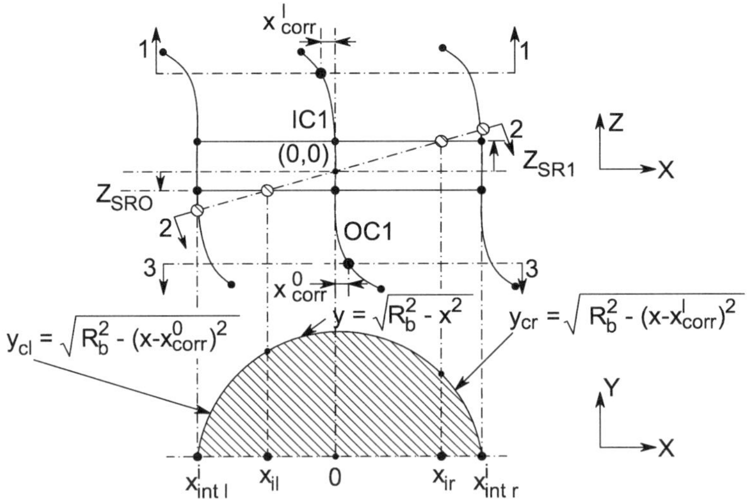

Fig. 8.11c Y ordinate variation in one section for a progressive bore.

construction has multiple moving cores. The dimensional stability in the injection-molded throttle body is extremely important, as its throughput is dependent on the rotation of the butterfly valve. Hence, the material selection, part design, feed design, processing of throttle body, and so forth, are quite important. Generally, throttle bodies are manufactured with aluminum die casting, and the inner surface is machined to the required accuracy. In aluminum design, there is a lot of variation in thickness; adopting the same design with plastic is not very practical. Therefore, the part should be redesigned suitably for plastics, from a molding perspective. Generally, the nominal thickness of an electronic throttle body will be 2.8–3.2 mm, to meet the various structural loading conditions.

The filler content influences the warpage through the fiber orientation in the part. The shape (cylindrical) and dimensions of the throttle body are important, because the butterfly valve therein must be assembled in such a way that when the butterfly valve is in the closed position, there should be no discharge of air; when the valve is rotated open, only the measured amount of air should be discharged though the gap. If the part is filled with one or two gates, the fiber orientation will not be uniform in the bore, and that causes the ovality in the cylindrical shape of the bore. So, it is better if the part is molded with more gates. The gating for the bore is more important than the servo drive housing or any other feature of the throttle body. To make sure that the filling is uniform in the bore, multiple gates in the form of spokes of a wheel are recommended for the bore. Optionally, one can design a gate for servo drive housing. In the past, submarine gates in the surface of the bore were also used. If feasible, the gates can be placed on the flange of the bore as well. If direct pin point gates are placed on the edge of the bore, then there will be a jetting problem, which causes disturbance in fiber orientation and can cause the warpage in the part. All the gates on inner bore of the throttle valve body can be fed with one hot nozzle or a cold sprue (in a three-plate mold construction). The second nozzle or cold sprue can fill the other gate on the servo drive housing. In some cases, the third gate on the housing body may also be required. It is important to optimize the feed system for better filling and fiber orientation.

8.4.2.5 Current Status of Thermoplastics in Electronic Throttle Body

Theoretical modeling predicts an acceptable performance for ETB as compared to aluminum. However, practical manufacturing tolerances, maintaining bore roundness, and, finally, the cost of the thermoplastics material as compared to aluminum will influence the successful entry of thermoplastics into this segment.

8.5 Summary

Thermoplastic applications in the UTH segment are growing steadily; particularly, non-nylon applications are increasing as a result of the recent advances in high-temperature ETP development. However, there are many challenges that must be overcome before successfully building plastics applications in this segment:

1. The traditional mind-set about plastics still does not embrace them as candidate materials for structural applications.
2. Acceptance of plastic parts requires long-term validation, a major challenge in application development.
3. The exact protocol for simulating or validating minimum life expectancy of 10 years in various environment conditions is still evolving.
4. The high material cost of plastics is a major deterrent for building new applications. The benefits of plastics must be shown to offset the current part cost by a huge margin in addition to substantial weight savings.

Many new applications are in the developmental stage and may take a few years to reach the market. The carbon-fiber-filled grades are being tried in many new structural applications, which could possibly be a cheaper alternative to machined aluminum. Also, various new parts are being designed exclusively with plastics during the initial phase of car or engine development cycle.

8.6 Trends

The stringent Corporate Average Fuel Economy (CAFE) standards [8-20] of 14.88 km/L (35 miles/gal) by 2020 and 23.17 km/L (54.5 miles/gal) by 2025 have put a lot of pressure on weight reduction. On an average, a 10% reduction in the vehicle mass yields about 6% increase in fuel economy [8-21]. OEMs are trying out new material options in pursuit of overall reduction of weight of cars. The recent trend in increase of plastics in UTH applications can be attributed to the developments in materials, processes, and secondary operations.

8.6.1 Material Advancements

New grades of nylon [8-22] that can withstand much higher temperatures than ever before are being developed by various resin suppliers. More recently, the continuous service temperature of nylon is being pushed beyond 180 °C (356 °F), thanks to the continuous improvements in material grades [8-23]. These new grades also have the ability to retain their mechanical properties, even after longer exposure to heat. These developments are pushing the designers to explore new areas in automotive UTH.

Several specialty compounds [8-24] have emerged to meet the specific needs of UTH, such as electromagnetic interference (EMI) shielding [8-25]. Electrically conductive, thermally conductive, and lubricated grades of nylon are replacing many of the traditional materials in the UTH segment. Many OEMs are trying out much larger parts such as engine front covers with some of the advanced materials such as carbon-fiber-filled PEI. The carbon-fiber-filled PEI materials exhibit exceptional strength-to-weight ratio while operating at higher temperatures. The tensile modulus can be increased significantly compared to glass-fiber-filled grades.

The long carbon-fiber-filled materials are also making inroads into applications in UTH segments. The metal-like strength and temperature-handling capability of carbon-fiber-reinforced PEI is finding new applications that were unthinkable or unsuccessful in the past. The amorphous PEI material also offers the additional advantage of dimensional stability compared to crystalline materials. The further development on copolymers of PEI is pushing the service temperature of these grades in excess of 250 °C (482 °F).

8.6.2 Processing Advancements

Embracing water injection technology to get hollow parts has taken the boundaries of hollow part molding to newer heights. With water-assisted molding, thin, slender, hollow parts [8-26] up to 500 mm have been achieved with ETP materials recently. These developments are opening up new areas in fuel rails and air ducts.

The ability to coinjection-mold along with water-assisted [8-27] techniques have helped OEMs to get smooth surface or multimaterial solutions with improved flow characteristics and enhanced barrier properties. Some of the nylon grades have also have been developed specially for this water-assisted molding to give a better surface finish inside the hollow part, which is helping OEMs to achieve better flow and hence improved fuel efficiency.

Adoption of induction-mold heating [8-28] and heat and cool technology has further improved the quality of the part, while reducing the overall cycle time. The ability to achieve higher mold temperatures through these innovations has pushed the flow length of the materials even further. The development of high flow grades [8-29, 8-30] is further reducing the thickness of parts, extending the weight savings. The long glass- and carbon-fiber material blends have started to take extreme structural metal parts head on.

8.6.3 Secondary Process Advancements

Advancements in joining techniques like laser welding and the development of advanced adhesives for engineering plastics are creating new spaces for plastic applications. Advancements in 3-D laser welding [8-31] are enabling some of the complex geometrical parts like integrated air-intake manifolds to be joined seamlessly with one another. The materials are also tweaked to give better weld strength at higher fiber loading, which was a limitation earlier. The additional freedom of part design is driving thermoplastics into extreme shapes, leading to further consolidation of parts and subassemblies.

According to recent trends, one may see more structural applications in the UTH segment, especially in powertrain and engine mountings. Advancement of fuel technology [8-21] will also push material development to make better materials that can withstand the green (environmentally friendly) fuels chemically. Use of laser direct structuring [8-32] on the molded polymer directly is also opening up space in sensors and connectors.

8.6.4 Design Trends

As the emission norms become stringent [8-18] and the fuel prices are increasing, smaller cars [8-33] are becoming more popular in Europe and Asia. To increase the space for the passenger cabin, the hood area is becoming more compact. Thus, components are getting closer to one another. This reduced space constrains the free flow of air into the engine area, leading to higher heat buildup inside. These constraints—along with the development of smaller engines, which operate at a much higher temperature compared to their predecessors—are pushing the thermal requirements of the materials to a new high. This new requirement of higher temperature is leading to the development of newer grades that are capable of withstanding much higher temperatures.

8.6.5 Green Trends

With today's emphasis on sustainability, the recyclability and green content of materials used in the cars are becoming the focus points of automakers. This trend is pushing traditional thermoset applications away from the car build, especially from the UTH area where thermoset material is used. The thermoplastics can be easily be recycled at the end of life of the parts or vehicle. The advanced polymers, which can withstand much higher temperatures and are stiffer than earlier versions of the materials, are helping automakers improve the sustainability of their manufactured products. The use of bio-based materials is an emerging focus for automakers. The use of natural fibers as reinforcing fillers, which often weigh less than glass and other traditional fillers, will potentially contribute to the design of lighter-weight applications.

8.7 References

8-1. Johnson, R.W., Evans, J. L., Jacobsen, P., Thompson, J.R., and Christopher, M. 2004. "The Changing Automotive Environment: High-Temperature Electronics" *IEEE Transactions on Electronics Packaging Manufacturing* 27(3): 164–175.

8-2. "ASTM D3045–92 (2010)—Standard Practice for Heat Aging of Plastics without Load," ASTM International, http://www.astm.org/Standards/D3045.htm. Accessed April 29, 2013.

8-3. "Road Salt and Water Quality," http://des.nh.gov/organization/commissioner/pip/factsheets/wmb/documents/wmb-4.pdf. Accessed February 14, 2012.

8-4. "ElringKlinger Increases Functionality of Oil Pan made of DuPont™ Zytel®," DuPont, http://www2.dupont.com/Plastics/en_US/News_Events/article20090520.html. Accessed February 14, 2012.

8-5. Muccio, E.A. 1999. *Plastic Processing Technology*. Novelty, OH: ASM International, pp. 101–102.

8-6. Harriman, L. 2002. "Environmental, Health and Safety Issues in the Coated Wire and Cable Industry." Technical Report No. 51, Massachusetts Toxics Use Reduction Institute, University of Massachusetts Lowell, MA.

8-7. "An Assessment of Mass Reduction Opportunities for a 2017–2020 Model Year Vehicle Program," http://www.theicct.org/sites/default/files/publications/Mass_reduction_final_2010.pdf. Accessed February 14, 2012.

8-8. "Green Wire," *Materials & Assembly*, http://www.designnews.com/document.asp?doc_id=226058. Accessed February 14, 2012.

8-9. Leaversuch, R.D. 2003. "Strength and Beauty—New Nylons Bring Both to Under-Hood Parts," *Plastics Technology*, January, http://www.ptonline.com/articles/strength-and-beauty-new-nylons-bring-both-to-under-hood-parts. Accessed January 20, 2012.

8-10. "Engine Covers," http://www.dsm.com/en_US/downloads/dep/enginecoverbrochure02US.pdf. Accessed January 20, 2012.

8-11. Schwaller, A.E. 1988. *Motor Automotive Mechanics*. Albany, NY: Delmar.

8-12. Zienkiewicz, O.C. and Taylor, R.L. 2000. *The Finite Element Method: Solid Mechanics Vol. 2*. Oxford, UK: Butterworth-Heinemann.

8-13. *ABAQUS User Manual Version 6.11*. 2011. Providence, RI: Dassault Systèmes Simulia.

8-14. Hardikar, N.A. 2004. "Predictive Methodology Development for Sequential Structure—Fluid Analysis of Throttle Body." GE Global Research TISCAT report 2004GRC255.

8-15. Vadari, H. and Hardikar, N.A. 2007. "Flow Prediction for Electronic Throttle Body-Study of Viscous Models and Variable Affecting Compressible Flow." GE Global Research TISCAT report 2007GRC187.

8-16. Hardikar, N.A., Chunduru, S., Rexius, K., Tankala, T., Yandooru, J. and Withey, C. 2005. "Six Sigma Study on the Effect of Geometric Tolerances at Low Airflow Rates in a Progressive Bore Throttle Body Using CFD," SAE Paper No. 2005-01-1916. SAE International, Warrendale, PA.

8-17. Dupont, K.A., Hardikar, N.A., Price, M.E., Chunduru, S. and Tankala, T.C. 2007. "Methods and Apparatus Predicting Variations in Material Properties." US Patent Application No: 2007/0139,444.

8-18. "Mathematica, Version 4.2." 2002. Champaign, IL: Wolfram Research.

8-19. *ANSYS™ APDL Programmer's Guide*. 2003. ANSYS USA.

8-20. "Automobile and Light Truck Fuel Economy: The CAFE Standards," http://nepinstitute.org/get/CRS_Reports/CRS_Energy/Energy_Efficiency_and_Conservation/CAFE_Standards_for_Light_Trucks_and_Autos.pdf. Accessed February 14, 2012.

8-21. Wenzel, T. and Ross. M. 2006. "Increasing the Fuel Economy and Safety of New Light-Duty Vehicles." U.S. Department of Energy Report LBNL-60449.

8-22. "Superior Aging Resistance for Easy to Process Specialty Nylons," http://www2.dupont.com/Plastics/en_US/assets/downloads/product/zytelplus/Zytel_PLUS_Tech_Paper_En.pdf. Accessed February 20, 2012.

8-23. "High Performance Polyamides Fulfill Demanding Requirements for Automotive Thermal Management Components," http://www2.dupont.com/Plastics/en_US/assets/downloads/technical/HTN_whitepaper_R8_06-2008.pdf. Accessed February 15, 2012.

8-24. "Compounds Offer Enhanced EMI/RFI Shielding," Materials News in *Plastics, Additives and Compounding* 10(2): 11, http://www.sciencedirect.com/science/article/pii/S1464391X08700473. Accessed February 15, 2012.

8-25. "A Thermoplastic Composite for Cost-Effective EMI Shielding," Sabic, http://kbam.geampod.com/KBAM/Reflection/Assets/Thumbnail/13588_15.pdf. Accessed February 15, 2012.

8-26. Knights, M. 2005. "Water Injection Molding: It's All Coming Together," *Plastics Technology*, September, http://www.ptonline.com/articles/water-injection-molding-it%27s-all-coming-together. Accessed February 28, 2012.

8-27. "Fluid Assisted Injection Molding," http://www.battenfeld-imt.com/fileadmin/user_pdf/aktualisierung110928/Airmould_Aquamould_en.pdf. Accessed February 12, 2012.

8-28. "Plastic Injection," ROCTOOL, http://www.roctool.com/plastic Injection.php. Accessed February 15, 2012.

8-29. Rosato, D. 2012. "Automotive Under-the-Hood Engineered Plastics Advances," Omnexus, http://www.omnexus.com/resources/articles/article.aspx?id=29527. Accessed February 15, 2012.

8-30. "Engineering Thermoplastics," 2000. *Plastics Technology*, http://www.ptonline.com/articles/engineering-thermoplastics. Accessed February 15, 2012.

8-31. "Polymers & Lasers," http://www.degussa-hpp.com/dl/brochure/polymerlaser_eng.pdf. Accessed February 15, 2012.

8-32. "3-Dimensional Circuitry Laser Direct Structuring Technology (LPKF-LDSTM) for Moulded Interconnect Devices," http://www.lpkf.com/_mediafiles/1797-lpkf-lds-process.pdf. Accessed February 15, 2012.

8-33. "Current Trends in Small Cars—Small is Big," Automobile, http://www.automobilemag.com/features/news/1004_current_trends_in_small_cars/viewall.html#ixzz1pYKhoGus. Accessed February 15, 2012.

Chapter 9
Sustainability in the Automotive Industry

Srinivasan Krishnaswami, Anju Baroth, and Neena Chandramathy

9.1 Introduction

In this chapter, we focus on the environmental sustainability of many of the polymer systems mentioned in the previous chapters. The purpose of environmental sustainability is to limit or control environmental damage such as (1) climate change and global warming caused by greenhouse gas (GHG) emissions [9-1], (2) water quality degradation caused by the release of toxic chemicals, (3) soil degradation, and (4) environmental and health hazards resulting from waste generation. For the automotive industry, GHG emissions are the predominant environmental risk and hence given greater coverage in this chapter. The primary constituent of GHG emissions is carbon dioxide (CO_2). It is well known that the transportation sector is already among the largest GHG emitters [9-2], necessitating a serious focus on automotive sustainability. The new field of *sustainable mobility* [9-3] deals with a wide variety of sustainability issues in the automotive and other transport sectors. Key regulations related to the sustainability of the automotive industry concern fuel efficiency, weight of cars, GHG emissions, and recycling, among others.

Two of the major sustainability challenges for the automotive industry are (1) to invent and employ appropriate technologies that enable greater control or mitigation of GHG emissions, thus minimizing the damage to the environment even in times of high economic and industrial growth; and (2) to develop technologies that will enable the industry to shift from nonrenewable to renewable resources for energy and materials in a significant way. These two challenges have given rise to many sustainability trends.

9.1.1 Sustainability Trends in the Automotive Industry

Some of the top sustainability trends of the automotive industry are the following:

1. Lightweighting to improve fuel efficiency and reduce carbon emissions
2. Use of bio-based or renewably sourced materials
3. Efficient recovery, reuse, and recycling at the end of life (EOL) of automobiles
4. Next-generation trends:
    a. Alternative propulsion systems
    b. Electrical/hybrid vehicles
    c. Efficient powertrain
    d. Use of biofuels in place of fossil fuels
    e. Use of renewable energy in operations

The first three trends are the most significant sustainability developments related to plastics products and therefore form the core of this chapter.

9.2 Lightweighting and Fuel Efficiency

The automotive industry considers vehicle weight reduction as one of the key elements for fuel economy improvement. Reduction of the weight of transport vehicles reduces energy consumption, because the energy required to move a vehicle is directly proportional to the weight of the vehicle (except for aerodynamic resistance) [9-4]. With increasing fuel price, unpredictable long-term availability of petroleum, and greater realization that auto emissions are hazardous to both the environment and public health, fuel economy improvement has become the top priority for vehicle manufacturers around the world.

Now legally binding targets are in place in the United States and the European Union (EU) to improve fuel economy and reduce tailpipe emissions, which will necessitate reduction of the overall weight of vehicles. In the United States the Corporate Average Fuel Economy (CAFE) standards will be administered for fuel efficiency of vehicles. Accordingly, the CAFE fuel efficiency targets for cars are 15.1 km/L (35.5 mpg) in 2016 and about 23.2 km/L (54.5 mpg) in 2025. This in turn translates to a carbon dioxide reduction target of 155 g/km (250 grams per mile [g/mile]) in 2016 and 101 g/km (163 g/mile) in 2025 [9-5].

The EU has set legally binding targets for new cars to emit 130 g of CO_2 per km by 2015 and 95 g/km by 2020. In terms of fuel consumption, the 2015 target is approximately equivalent to 5.6 L per 100 km of gasoline or 4.9 L/100 km of diesel. The 2020 target is approximately 4.1 L/100 km of gasoline or 3.6 L/100 km of diesel [9-6]. To comply with these stringent regulations, the original equipment manufacturers (OEMs) are exercising a variety of options such as engine modification; aerodynamic design; and lightweighting through material substitution, parts consolidation, and design optimization.

A general rule of thumb is that every 10% reduction in vehicle weight will reduce fuel consumption by about 7%. In absolute terms, every 100-kg weight reduction will yield a reduction in fuel consumption of 0.39 L/100 km for a current average midsize gasoline car [9-7]. The amount of weight savings resulting from using alternative materials in any vehicle component depends on the application and design intent. Over the years, plastics have been successful in the lightweighting of many automotive components.

9.2.1 Materials for Lightweighting

This section focuses on lightweighting plastics that replace metals such as steel and aluminum in many automotive parts. Plastics and polymer composites now make up about 10% of a vehicle by weight and 50% by volume, and these numbers are expected to increase gradually [9-8]. The main factors restricting the growth of polymer composites in vehicles today are long production-cycle times, the high cost of fibers, and the demanding strength requirements for structural applications. Table 9.1 gives an overview of plastic solutions used for lightweighting in the automotive industry.

Many of these polymer systems have lower densities compared to aluminum or high-performance steel, and hence they are the light-weighting materials of choice in many applications. The GFR plastic systems provide strong, hard materials that can replace metals in many automotive applications. Composites based on thermoset systems do

Table 9.1 Plastics Systems for Automotive Lightweighting Applications

Plastics Systems for Lightweighting	Matrix Type	Automotive Applications
Glass-fiber-reinforced (GFR) plastics composite	Polypropylene, polyamides, polyesters and polycarbonate/ polyester blends	Rear hatches, roofs, door inner structures, door surrounds, and brackets for the instrument panel
Carbon-fiber-reinforced plastic (CFRP) [9-9]	Polyesters, polyamides and thermosets such as epoxy, vinyl esters	Interior car parts such as sun shades, door panel trim, bumpers, parcel shelves and under-the-hood (UTH) structural parts
Glass mat thermoplastics	Polypropylene, polyamides, polyesters (PBT/PET), polyphenylene sulfide, polyetherimide, etc.	Headliners, load floors, sunshades, and trunk liners, UTH structural parts, Body panels
Polymers without glass reinforcement	PPO/polyamide blend, Polycarbonate/ polyester blends, PU foams, etc.	Fenders, bumpers, energy absorbers, seat cushions, etc.
Sheet molding compounds (SMCs)	Thermoset systems such as unsaturated polyesters and vinyl ester systems	Painted Class A exterior body panels, underbody shields, fenders, and semistructural components
Bulk molding compounds (BMCs) (prepreg + chopped glass strands)	Unsaturated polyesters and other thermoset resins	Instrument panels, body panels, underbody, front/rear modules, axles and suspensions, fuel and exhaust systems, and electrical and electronic components

Notes: PBT = polybutylene terephthalate; PET = polyethylene terephthalate; PPO = polyphenylene oxide; PU = polyurethane.

contribute to lightweighting of cars, but their recyclability is low. A discussion on natural-fiber-reinforced (NFR) polymer composites is in Section 9.3.3, "Bio-Based Materials for Plastics."

9.2.2 Quantifying Environmental Benefits of Lightweighting through Life Cycle Assessment

Life cycle assessment (LCA) is an analytical tool used to comprehensively quantify and interpret the environmental flows to and from the environment over the entire life cycle of a product or service. LCA studies cover various stages of a product's life cycle as given in one or the other of the following options.

1. **Gate-to-gate** focuses on one particular plant or operation.
2. **Cradle-to-gate** refers to the production phase where *cradle* refers to manufacturing of product by using all upstream providers (e.g., mining of raw materials, processing, and transportation), and *gate* refers to the end of production when the product leaves the factory gate.
3. **Cradle-to-grave** encompasses the entire life cycle of the product from extraction through disposal. Here *grave* refers to disposal options such as landfill and incineration.
4. **Cradle-to-cradle** includes the entire cradle-to-grave life cycle of the product with the additional step of recycling the product back to its original or an alternative application purpose.

LCA is the preferred tool to quantify the environmental benefits of lightweighting a vehicle through use of alternative materials such as plastics and composites [9-10]. It can be instrumental in the design of systems, product design optimization, material selection, and EOL considerations. Several LCA studies suggest that 85% of the automotive GHG emissions occur in the *use* phase of the product. About 10% of emissions occur in the *production* phase, and about 5% in the EOL or *recycling* phase [9-11]. It has been proven that lightweighting reduces carbon emissions during the use phase and hence represents the most impactful action to reduce carbon footprint of an automobile [9-12].

9.2.3 Life Cycle Assessment Case Studies for Lightweight Materials

Lightweighting with plastics has been quite successful in reducing the carbon footprint of automobiles. All the lightweight materials given in Table 9.1 have the potential to reduce the weight of a vehicle and associated environmental impacts in varying proportions. For example, fiber-reinforced polymer composites based on glass fiber and carbon fiber can reduce 25–30% and 50–60% of vehicle weight, respectively [9-13]. However, the current cost of carbon fiber and difficulty in processing the composites inhibit the growth of carbon-fiber-reinforced plastics. SMC usage has an advantage over steel in not only reducing weight but also lowering the tooling cost and parts consolidation. An example of parts consolidation can be found in the radiator supports in which SMC is used as a substitute for low-carbon steel [9-14].

An LCA study is normally carried out to assess the environmental impacts such as global warming, ozone layer depletion, human toxicity, freshwater aquatic ecotoxicity, marine aquatic ecotoxicity, terrestrial ecotoxicity, photochemical oxidation, acidification, eutrophication, and abiotic depletion [9-15]. Among all the impacts, GHG emission or global warming is the impact discussed most in the industry, consistent with the general concern about global climate change. Along with the global warming potential, cumulative energy demand (CED) for a product over all the stages of its life cycle is also of interest. Following are some of the automotive LCA case studies where traditional material (metal) was compared against alternative thermoplastic/composite material. As expected, the use phase is dominant among all the stages of life cycle in these case studies.

1. In a life cycle study, significant savings in global warming potential (GWP) and nonrenewable energy demand were observed when a steel fender was replaced with PPO/polyamide fender in a midsize car. The overall comparison of cradle-to-cradle phases revealed that the PPO/polyamide resin fenders have up to 47% lower GWP and up to 45% lesser CED as compared to steel fenders over the entire life [9-16, 9-17]. Another LCA study on a similar PPO/polyamide-based fender also pointed to a significant sustainability advantage from lightweighting [9-18]. Although steel fenders use less energy during production, higher overall environmental impacts over the entire life cycle are observed because of the heavier weight of the steel. As environmental impact varies among the life cycle stages, the full lightweighting benefit can be recognized only when LCA is performed for the entire life cycle.
2. Traditionally, glazing materials in a vehicle are the laminated glass for the windshield and tempered glass for side windows, rear window, and sunroof. Although the weight of the glazing material is only 2–3% of the total weight of a vehicle, one of the alternatives considered to reduce its weight is polycarbonate. An LCA study showed that if all of

a midsize car's glazing, except for windshield, were made of polycarbonate, a saving of CO_2 emissions by up to 330 kg per vehicle would result over a vehicle's service life of 150,000 km compared to cases where glass was used [9-19].

3. In another example, when a polypropylene-long glass-fiber-reinforced composite replaced a steel-polyamide hybrid system in a front-end module (FEM) of a midsize car, considerable weight saving was achieved, resulting in a 44% lower energy demand and an equivalent reduction in carbon footprint. The cradle-to-cradle assessment showed that the heavier weight of the steel-polyamide hybrid FEM resulted in overall higher environmental impacts [9-20].
4. Another life cycle study dealt with steering wheels made of magnesium alloy and polycarbonate. This study is a unique example, because the environmental footprint of the application was not driven by weight reduction (since there is only a 1–2% weight difference between the two products) but by the choice of material. It is well known that magnesium production is highly energy intensive, with high amounts of GHG emissions during manufacturing [9-21]. Replacing the Mg alloy with a cleaner technology material such as polycarbonate resulted in significant footprint reduction during the material manufacturing phase. An LCA study showed that there was an overall (entire life) reduction of up to 62% in GWP and up to 46% in CED [9-22].
5. The life cycle inventory (LCI) data for lightweighting a propeller shaft using a fiber-reinforced plastic over a 150,000 km use revealed that a 14.0-kg saving in part weight led to a saving in emissions of CO_2, NO_x, CO, and SO_x of 294 kg, 0.7 kg, 2.4 kg, and 0.364 kg, respectively [9-23]. LCI is the basic input data for any LCA study.
6. In a study of a bumper beam application, an SMC plastic offered less weight savings than either aluminum or advanced high-strength steel (AHSS) [9-24]. Hence it is imperative to choose the right plastic materials to replace metals so that weight savings can be achieved without sacrificing required performance. In meeting lightweighting goals, plastics will have tough competition from AHSS in the future.

The examples just provided illustrate that LCA as a tool provides a comprehensive view of the environmental impacts of a product or process and an accurate picture of the true environmental trade-offs in product selection.

To replace metals or traditional materials, plastics used for lightweighting need to have the required mechanical properties such as the modulus and toughness suitable for a given application. Many applications do not require the stiffness of fiber-reinforced composites. Given the significant benefits in to the environment as well as other domains such as freedom of design, lightweighting solutions for different parts of automotive vehicles will continue to evolve from the collaborative efforts of OEMs and plastic materials suppliers.

9.2.4 The Future of Lightweighting with Plastics

In the United States, the mileage requirement for passenger vehicles will increase to 15.1 km/L (35.5 mpg) in 2016 from the current mileage of 12.2 km/L (28.7 mpg) [9-25]. The currently known plastics lightweighting solutions such as polypropylene-based plastics are significant contributors to this goal [9-26]. However, the mileage requirement will become 23.2 km/L (54.5 mpg) in future, consistent with the industry's need to reduce carbon emissions to 50% by 2025 [9-27].

To meet such stiff targets, new lightweighting solutions based on novel materials and integrated designs are needed, which will raise the usage level of plastics to levels as high as 18–33%. Among the plastic-fiber composites used as lightweighting solutions, those containing carbon fiber can match that of steel used in many stiff and load-bearing parts, subject to lowering of their costs and improved processability.

Additionally, hybrid metal-polymer systems as lightweighting solutions are gaining prominence. In large-part applications, plastics will face significant competition from AHSS in achieving lightweighting targets, triggering development of new plastics-based solutions for lightweighting applications. The inherent advantages of plastics in fabrication of parts, such as lower tooling cost, low wear of the tools, and flexibility of design and production will continue to make plastics attractive to the automotive industry.

9.2.5 Design for Sustainability

In 1997 United Nations Environment Program (UNEP) in conjunction with Delft University of Technology published a ground breaking manual *Ecodesign: A Promising Approach to Sustainable Production and Consumption*. As a result and based on experience gained, ecodesign has evolved through Design for the Environment (DfE) to the broader concept of design for sustainability, more popularly called D4S—which encompasses issues such as the social component of sustainability and the need to develop new ways to meet consumer needs in a less resource intensive way [9-28].

The important elements for design for sustainability are [9-29] (a) design for manufacturing, (b) design for recyclability, (c) design for durability, (d) design for resource conservation (to minimize material usage), and (e) design for energy efficiency. Lightweighting as a concept addresses the last two of the elements, (d) and (e). In all these elements, the critical factor for design for sustainability is material selection. Guidelines for material selection [9-10] include criteria such as (1) use of abundant and nontoxic materials; (2) environmental indicators such as energy content, GHG emissions, and recyclability; (3) whether the material is bio-derived with less energy intensity during production phase; and (4) low energy intensity for processing and fabrication of a part to achieve appropriate design. Going forward, design for sustainability will become the primary customer requirement demanding greater degree of product innovation.

9.3 Renewable-Sourced or Bio-Based Materials for the Automotive Industry

Automotive sustainability strategy includes all phases of the value chain of products. Generally, compared to the use phase, the production phase of a product contributes less to global warming; yet it is substantial and cannot be ignored. Our focus in this chapter is on understanding how renewable resources or bio-based materials for polymers in automotive applications can reduce GHG emissions in the production or use phase of a vehicle's life cycle.

9.3.1 Why Renewable Resources?

As shown in Fig. 9.1, the basic difference between a renewable and a nonrenewable resource is the time scale of renewability.

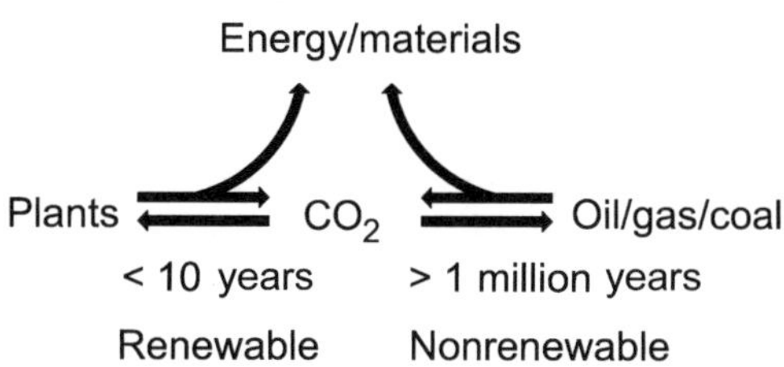

Fig. 9.1 Renewable and nonrenewable resources.

Renewable resources can be renewed in a fairly short time, say, less than 10 years. Nonrenewable resources such as petroleum crude oil cannot be renewed even in a million years. Hence sourcing a raw material from a renewable resource is a logical approach to sustainability.

The use of raw materials from renewable resources has the potential to address three sustainability aspects, often in combination: (1) to save fossil resources for future generations by providing an alternate non-fossil-based raw material resource, (2) to contribute to environment sustainability by enabling a smaller carbon footprint for a product, and (3) to reduce ecotoxicity.

Raw materials from renewable sources are obtained by methods such as direct extraction from plants and preparation by the action of microorganisms (such as bacteria) on plant-derived feedstock, such as carbohydrates, synthesis gas from biomass, and even CO_2. Therefore, the terms *renewable-sourced* and *bio-based* are equivalent and are used interchangeably in this chapter.

The use of bio-based materials is now becoming part of the sustainability strategy of several organizations [9-30], though there is no specific directive to use plastics made from renewable resources. The following are some examples: (1) Toyota is planning to have 20% of plastics used in its vehicles bio-sourced by 2015 as part of its efforts to reduce CO_2 emissions; (2) DuPont plans to double its revenues from nondepleting resources by 2015; (3) Ford vehicles in Europe use between 10 and 20 kg of renewable materials, depending on the vehicle size class [9-31]; and (4) PSA Peugeot Citroën plans to increase the proportion of green materials in its vehicles to 30% by 2015 [9-32].

9.3.2 Carbon Footprint of Bio-Based Raw Materials

The overall emissions generated by bio-based raw materials are generally considered low, according to the analysis in Table 9.2.

The analysis given in Table 9.2 also applies to biofuels such as biodiesel and bioethanol that have low carbon footprint. When these fuels derived from renewable resources are used in place of fossil fuels either in full or in part, the carbon footprint of automotive vehicles is reduced especially in the use phase. To reduce the carbon footprint of a product to a greater degree, lightweighting strategy is often combined with biofuels in the use phase.

Bio-based polyethylene derived from bioethanol, which in turn is obtained by fermentation of glucose, shows a lower carbon footprint when compared to polyethylene derived from fossil-based naphtha. The difference in the carbon footprint is substantial, that is, 3.2 metric tons (t)

Table 9.2 Carbon Footprint of a Raw Material from Renewable Sources

Stage	Carbon Footprint for a Raw Material (Cradle-to-Gate)	
	Renewable/Bio-Based Resource	Fossil Resource
Natural feedstock formation	Negative[1]	Zero[2]
Raw material production process steps	Positive[3]	Positive
Overall carbon footprint	Negative or small	Positive or large

Notes: [1] CO_2 was absorbed during photosynthesis as a recent event. [2] Since CO_2 was absorbed million years ago and not recently. [3] A desired assumption is that the conversion steps are not energy intensive.

of CO_2 per t of polyethylene produced and incinerated [9-33]. Similarly, bio-based polypropylene derived from sugarcane ethanol has a lower carbon footprint than fossil-based polypropylene [9-34]. The carbon footprints of many bio-derived materials indicate a significant saving of carbon emissions over fossil-based sources [9-35].

9.3.3 Bio-Based Materials for Plastics

Although the use of bio-based materials is increasingly becoming a part of many sustainability strategies [9-36], the field is still a fledgling one. The automotive industry has few concrete examples of using bio-based materials successfully. The types of bio-based materials fall in three categories, as shown in Figure 9.2.

9.3.3.1 Cellulosic Plant Fibers

Renewable plant fibers are used in place of heavier glass fibers in fiber-reinforced plastic composites [9-37] for lightweighting solutions. The plant fibers are cellulosic and are directly isolated from different parts of plants. Some of the examples are (1) bast fibers such as jute, kenaf, hemp, rami, china reed, and flax; (2) leaf fibers such as sisal and banana; and (3) fruit fibers such as coconut. Use of renewable plant fibers in place of glass fibers provides substantial benefits in the form CO_2 emission reduction, as shown in the Table 9.3.

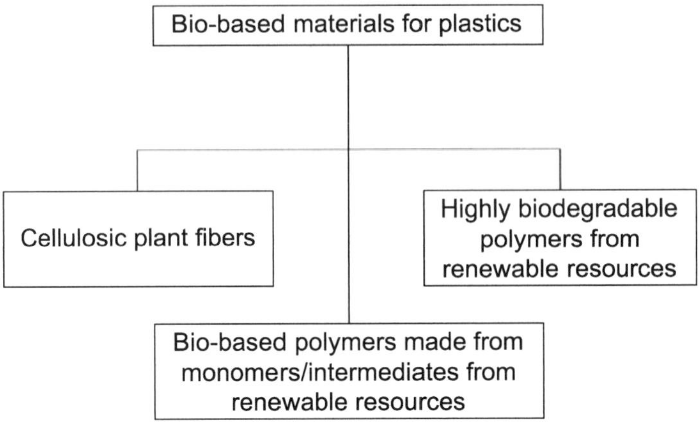

Fig. 9.2 Types of bio-based materials for plastics.

Table 9.3 Natural-Fiber-Reinforced (NFR) Versus Glass-Fiber-Reinforced (GFR) Plastics Composites

Natural Fiber	NFR Composite Matrix	GFR Composite Matrix	CO_2 Emissions Reduction (Cradle-to-Grave) kg/kg Composite	Reference
Hemp fibers	EPDM/PP	EPDM/PP	8.18	[9-38, 9-39]
China Reed	PP (53% China reed)	PP (42% Glass)	3.49	[9-38, 9-40]

Notes: EPDM = ethylene propylene diene monomer; PP = polypropylene.

The emission reduction occurs predominantly because of lightweighting in the use phase arising from the density difference between the composites based on natural fiber and glass fiber. To obtain the desired mechanical properties, in some applications natural fiber content in NFR composite needs to be higher than the glass fiber content in a GFR composite [9-40]. In that case, lightweighting would become less possible, and hence in some applications, the advantage of reduction of emissions could be lost.

Though the properties of plant-fiber-incorporated composites are different from those of glass-fiber composites, many applications are implemented in the automotive industry and are illustrated in Table 9.4.

Table 9.4 Selected Automotive Applications of Natural-Fiber-Reinforced (NFR) Composites

Class of Polymeric Product	Renewable Raw Material/Intermediate Used	Applications in Automotive
NFR plastics	Abaca/banana, flax, and other natural fibers in many media such as PP, PE, and PET	Seat backs, trim, floor area parts, acoustic insulators, under-body panels, spare-tire cover, ceiling liner, and exterior auto panels

Notes: PE = polyethylene; PET = polyethylene terephthalate; PP = polypropylene.

Some OEMs have developed commercial composites based on natural fibers [9-31, 9-41, 9-42, 9-43]. Another emerging bio-based lightweighting alternative for glass fibers is carbon fibers from biomass such as lignin [9-44, 9-45].

9.3.3.2 Bio-Based Polymers Made from Monomers or Intermediates from Renewable Resources

Some of the monomers such as bio-ethylene glycol, bio-ethylene, and sebacic acid can be made commercially using renewable resources. Bio processes for other monomers are being developed using known or genetically modified microorganisms as biocatalysts. Bio-based polymers (shown in Table 9.5) are reacted with other fossil-based monomers to make partially bio-based polymers (Table 9.6). The carbon emission savings for bio-based polymers over the fossil derived versions are justifiably significant [9-34, 9-46, 9-47].

If the manufacturers of those plastics listed in Table 9.5 seize the opportunity to source monomers from renewable sources, plastics with a lower carbon footprint will become available to users such as OEMs and hence the automotive industry in general.

The carbon footprints of the polymers mentioned in Table 9.6 are affected by the degree of bio-content in them. If one of the constituents of the polymer is fossil-based, then the polymer is said to be partially bio-based. The trend is moving from partially bio-based to fully bio-based content to maximize carbon footprint reduction. The technology barrier is that co-monomers such as terephthalic acid are as yet not available commercially from renewable resources.

Table 9.5 Fully Bio-Based Polymers Used in Automotive Applications

Bio-Based Polymers	Selected Automotive Applications Known	Bio-Based Monomer
PE	LDPE cover for dashboards, door panels, consoles, boot interiors, and handgrips	Bio-ethylene from bioethanol
PP	Front and rear bumper or fascia, instrument panels, seat panels, console trim parts, door trim, running boards, cowl grille, etc.	Bio-propylene from bioethanol
Polyamide 6	Carpeting and rugs; glass-filled systems in air intake manifolds, exterior automotive parts such as door and tailgate handles, exterior mirrors, front-end grilles, fuel caps and lids, and wheel covers and trim	Bio-caprolactam
PLA	Fiber for floor mats, seat cushion foam, cowl side trim, inner and outer scuff plates, and deck trim cover	Bio-lactic acid or lactides
Polyamide 11	Engine compartments and fuel lines, replacement of metal gears, radiator end-tank	Bio-Amino undecylinic acid from castor oil
Polyamide 10,10	Cable ties, wire insulation, flexible hosing, nozzles, damping cogwheels, flexible cover caps, sheet gaskets, sealing rings, and in other fuel-line applications	Sebacic acid and decanediamine both derived from castor oil

Notes: LDPE = low-density polyethylene; PE = polyethylene; PLA = polylactic acid; PP = polypropylene.

9.3.3.3 Highly Biodegradable Polymers from Renewable Resources

Starch, proteins, and certain aliphatic polyesters such as polyhydroxy alkanoates are highly biodegradable and are also considered carbon neutral. If these biodegradable polymers are used in cars, the challenge will be to prevent their degradation during the lifetime of the car.

Table 9.6 Partially Bio-Based Polymers in Automotive Applications

Partially Bio-Based Polymers	Selected Automotive Applications Known
PET	Nonwoven fabric applications for interior, trim and or other complex surfaces, wheel housing, dashboard insulation, hood liners and headliners, boot carpets, etc.
PBT	Windshield wiper covers, mirror housings, cowl vents, handles, fans, fuel system components, connectors, sensor housings; PBT alloys are used in bumper/fascia and safety applications such as airbag covers, fuel-line clips, and distribution boxes.
PUR	Instrument panel, A/C vent, seat cushion; headrests, head liner, etc.
PTT	Fibers in carpeting, headliners, and seat fabrics applications; injection molding of air ventilation louvers, electronic components, and headlight bezels; also in applications similar to those of PBT.
Polyamide 6,6	Automotive cooling systems, headlamp bezels, etc.
Polyamide 6,10	Engine compartments, fuel lines, and radiator end-tank.
PVC	Used widely in the past in many applications such as dashboard skins, carpets, etc. The trend is now moving away from PVC, as PVC contains toxic plasticizers and recyclability of PVC is limited.
Thermoplastic elastomers [9-48]	Hoses, tubing boots, energy dampers, airbag doors, etc.
Epoxy resin [9-49]	Coatings, castings, tooling, flooring, adhesives, and composites.
Unsaturated polyester resin [9-50, 9-51]	SMC panel composites with Class A surface-tractor hood and side panels, hood components.

Notes: PBT = polybutylene terephthalate; PET = polyethylene terephthalate; PTT = polytrimethylene terephthalate; PUR = polyurethane; PVC = polyvinyl chloride.

In contrast, cellulosic fibers used as plastic reinforcements are relatively less biodegradable, and their biodegradability is slowed down further by the enveloping nonbiodegradable matrix polymers. Compared to cellulosic fibers, the commodity and engineering polymers now used in plastics are even less biodegradable, irrespective of whether their source is renewable or nonrenewable.

Highly biodegradable polymers are not used widely in automotive applications primarily because of their high biodegradability and generally low values of glass transition temperatures (T_g). The higher is the value of T_g, the greater the heat resistance. These polymers are useful in different applications, however, such as packaging and can replace other low T_g polymers such as the nonbiodegradable polyethylene in those applications. However, some of these polymers possess $T_g > 50$ °C (112 °F) and are useful in blends with some engineering polymers. For example, PLA having a T_g of 58 °C (136.4 °F) is used in some blends with engineering polymers such as polycarbonate [9-52] potentially for automotive interior applications.

9.3.4 Limitations of Sourcing Raw Materials from Renewable Resources to Make Polymers

Following are some of the factors that limit the sourcing of raw materials from renewable resources:

1. Current availability of renewable resources for making polymers is much less than needed.
2. Development of genetically modified microorganisms to effect transformation of basic renewable feedstock into a desired monomer or intermediate is currently limited.
3. Currently, the cost of monomers from renewable materials is higher than that from fossil resources because of the low scale of manufacturing. However, as the cost of fossil resources rises, renewable materials will establish themselves as true alternatives.
4. The available bio-based materials from renewable sources cover some commodity polymers and certain engineering polymers of $T_g < 150$ °C [302 °F] [9-53] and virtually none for performance polymers of $T_g >$ 150 °C (302 °F).

9.3.5 Emerging Bio-Based Raw Materials

Many new bio-based raw materials are under investigation in various laboratories around the world. Some of these are listed in Table 9.7.

9.3.6 Bio-Based Plastics for the Future Automotive Industry

The use of bio-based plastics is a powerful way to reduce the carbon footprint of many automotive products. Although still in its infancy, the field of bio-based materials for automotive applications has been gaining in the last several years. A few of the bio-based materials, such as natural fibers, bio-polyethylene, castor oil–based polyamides and vegetable oil–based polyurethane foams are commercially available. A lot more bio-based raw materials are in the pipeline; some of them are processed in bench scale and some in pilot plant scale. Many bio-based materials are economically unattractive at this time. Using appropriate chemical methods to convert bio-derived intermediates into the raw materials of polymers is yet another active area of research. The following are some of the important factors that could shape the future of bio-based plastics in the automotive industry:

1. The use of cellulosic fibers with and without surface treatment in place of glass fibers for lightweighting will gain greater attention in meeting environmental regulations. The poor recyclability of cellulosic fibers and their propensity for biodegradation could limit their use to some extent. Carbon fibers developed from less expensive biomass to make carbon-fiber-reinforced plastics will provide fresh solutions for reducing carbon footprint.
2. Newer commercial processes with a lower carbon footprint for bio-propylene; bio-methyl methacrylate; bio-1,4-butanediol; bio-terephthalic acid; and bio-hexamethylene diamine will become available in a few years to enable large-scale production of bio-based polymers for various automotive applications.
3. The blends of common engineering polymers such as polycarbonate with bio-polymers such as polylactic acid provide additional scope to reduce the carbon footprint of plastics products for selected interior automotive applications.
4. The options for bio-based monomers for high heat ($T_g > 150$ °C [302 °F]) and high performance applications are meager. New, innovative approaches will be required to develop these bio-based monomers to meet the future requirements of the automotive industry.
5. Manufacturers will continue to maximize their carbon footprint reduction by combining bio-based matrix polymers and bio-based cellulosic fibers for their plastics products.
6. Opportunity exists for combining an EOL recycling option with the use of bio-based raw materials to further minimize carbon footprint.

9.4 End-of-Life Scenarios

Once a customer stops using a product, the product is said to have reached its EOL. To prevent an automobile at its EOL from becoming a waste product that could pollute the environment, there are several EOL options that can be exercised for treating the vehicle's parts: (a) recycling of parts in product assembly or recycling of material in production, (b) reuse, and (c) disposal via incineration, landfill, storage, or composting. Option (c) will not be discussed in this chapter, because incineration, landfill, and storage are nearly nonsustainable activities, and composting of polymers is applicable only for highly biodegradable materials, which are seldom used in automotive products.

The word *recycling* is often referred to as postconsumer recycling (PCR). It invariably refers to the process wherein a waste product or part is converted by physical or chemical means into some other product or broken down into raw materials, which in turn can be converted into some other product. That end product is very rarely the original product itself. The word *reuse* means that the part is used again for the same function, or it is used for a different function without any physical or chemical treatment except cleaning. An example of a reused part is a tire or bumper that can be used again directly without any physical or chemical modification.

Besides EOL, much of the waste produced in manufacturing plants is recycled in material production and product assembly. This recycling is generally called postindustrial recycling (PIR) or preconsumer recycling. PIR helps in improving material efficiency and reducing scrap in industrial processes.

In this section, the focus is on PCR, which is simply referred to as recycling. Automotive recycling serves a vital role in preserving natural resources and reducing the demand for scarce landfill space. In addition to conserving natural resources, automotive recycling plays an important role in reducing air and water pollution and solid waste generation.

9.4.1 Recycling in the Automotive Industry

The automobile is the most recycled consumer product in the world today [9-58]. Each year, more than 50 million vehicles reach the end of their service life throughout the world; more than 95% of these vehicles enter a comprehensive recycling infrastructure that includes auto parts recyclers/dismantlers, remanufacturers, and material recyclers (shredders). Today, about 75% of automotive materials are profitably recycled via part reuse

Table 9.7 Bio-Based Materials for the Future

Future Bio-based raw material	Status of development	Polymer Product potential	Reference
Terephthalic acid	Laboratory stage: many different routes being attempted	Polyesters such as PET, PBT, PETG, PCTG; liquid crystalline polyester, polyamides, polyaramides, PEI, etc.	[9-53]
Furan dicarboxylic acid	Microbial conversion of fructose from plants; commercially available	New polyesters and polyamides of high T_g	[9-53]
Isosorbide (a diol)	Obtained from sorbitol Commercially available	Polycarbonate, novel polyesters, isocyanates, etc.	[9-54, 9-55]
para, alpha -dimethyl styrene	Prepared from limonene isolated from plant oil; Commercially available	Partial use in ABS, SEBS type polymers	[9-53]
Acrylonitrile	From glutamic acid—laboratory stage.[1]	ABS, SAN, etc.	[9-56, 9-57]
Hydroquinone	Using quinic acid from carbohydrates—laboratory stage	High T_g PEEK polymer-engine parts, thrust washers and seals for power transmissions	[9-53]

(Continues)

Table 9.7 Continued

Future Bio-based raw material	Status of development	Polymer Product potential	Reference
4-hydroxybenzoic acid	Using shikimic acid from carbohydrates—laboratory stage	High T_g LC polyester-barrier film for automotive gas tanks	[9-53]

Notes: [1]Surprisingly, preliminary calculations have found that the emission savings are higher than that of fossil-derived acrylonitrile. Process improvements are expected.

ABS = acrylonitrile butadiene styrene; LC = liquid crystalline; PBT = polybutylene terephthalate; PCTG = polycyclohexylendimethylene terephthalate glycol; PET = polyethylene terephthalate; PETG = polyethylene terephthalate glycol; PEI = polyetherimide; PEEK = polyetheretherketone; SAN = styrene acrylonitrile; SEBS = styrene-ethylene/butylene-styrene.

and reprocessing [9-59]. Everything from floor mats and instrument panels to upholstery, glass, aluminum, steel, and plastic can be recycled for use in a new automobile or consumer product.

9.4.2 End-of-Life Options for Selected Polymer Families

Many types of polymers are used in more than 1000 different parts of all shapes and sizes in a vehicle. Although many different polymers may be used in a single car model [9-60], just three types of plastics make up 66% of the total plastics used in a car: polypropylene (PP, 32%), polyurethane (PU, 17%), and polyvinyl chloride (PVC, 16%) [9-60]. Out of the total plastics used, 50% by weight of the plastics goes to seating and upholstery, interior trim, and bumpers. The EOL options for different polymers are summarized in the Table 9.8.

Since the recyclability of thermoset systems is poor, thermoplastics systems for automotive vehicle parts are preferred by OEMs. Thermoplastics-based waste can be melt-processed, but thermoset composite wastes can only be reground and incorporated in certain compounds [9-60, 9-61].

9.4.3 Challenges and Limitations to Plastics Recycling

Recycling is a key element in developing the resource-conservation strategies of the automotive industry. However, recycling automotive

Table 9.8 EOL Recyclability of Selected Plastic Waste from the Automotive Industry

Polymer	Preferred EOL possibilities
PP	Recyclable. Most of the PP recovered is filled PP. It can be remolded into different plastics products. Waste availability is lower than those of PE or PET. It can be pyrolyzed into fuels and gaseous products such as ethylene.
PE	HDPE is the most recycled. PE can be remolded into different products such as pallets. Economic viability of recycling is uncertain since virgin PE is low cost. It can be pyrolyzed into fuels, aromatic products, and gaseous products such as ethylene.
PET	Technologies for chemical recycling of PET are well established. Automotive waste PET is heavily pigmented, and the pigments are hard to remove. Hence waste PET is broken down into monomers, and the monomers can be recycled to make other polyesters.
PBT	PBT can be broken down into monomers that can be repolymerized to make polyesters. Waste PBT is not available as organized waste.
PC	PC waste from automotive glazings can be used in other PC products. Integrity of the polymer depends on thermal history. Segregated parts can be recycled. Chemical recycling with methanol or ammonia can generate a BPA monomer that can be repolymerized into PC. Economic viability is uncertain because organized waste is not available.
PPO	PPO waste is available as blends such as PPO/polyamide. It can be recycled only if identified and segregated from other wastes. It can be remolded or worked off with virgin PPO-based compounds. Thermal stability of these blends is adequate for recycling.
Polyamide 6	It mostly ends up in landfill in the form of shredded fluff. If segregation is efficient through identification codes, it can be remolded into different parts. Chemical recycling is not feasible.

(Continues)

Table 9.8 Continued

Polymer	Preferred EOL possibilities
ABS	ABS ends up in landfill in the form of shredded fluff. It is difficult to separate from other substrates. Identified ABS parts can be worked off in some proportions with virgin ABS products. It is not economical to recycle. Combustion releases toxic vapors. However, identified PC/ABS waste can be recycled into virgin PC/ABS at different proportions.
Thermoset	Generally, thermoset parts are shredded and used as fillers in some composites or sent to landfill as fluff. Chemical recycling of unsaturated polyesters and polyurethanes is possible by glycolysis followed by conversion into polyester polyol for PU resin. However, recycling is not economical. New technologies are being developed to convert thermoset parts into fuel through catalytic pyrolysis [9-62].

Notes: ABS = acrylonitrile butadiene styrene; BPA = bisphenol A; HDPE = high-density polyethylene; PBT = polybutylene terephthalate; PC = polycarbonate; PE = polyethylene ; PET = polyethylene terephthalate; PP = polypropylene; PPO = polyphenylene oxide; PU = polyurethane.

plastics faces economic, technical, and infrastructural challenges as follows:

1. There are two major economic barriers to recycling: First, the labor cost to recover the materials in usable form is quite high, making it less likely that recycled stock can compete with the price of virgin stock. Second, recyclers cannot rely on a consistent and stable flow of plastic scrap, as retired automobiles vary greatly in the level and type of plastics. This makes it difficult, if not impossible, to establish end markets. Other economic barriers to successful recycling include the costs of transportation and recovery.
2. Technological challenges include effective segregation and separation of waste plastics from automotive waste with minimum contamination and making the adequately treated waste available for recycling in enough quantity. For example, a small amount of PVC contaminant present in a PET recycle stream degrades the recycled PET resin owing to evolution of hydrochloric acid gas at temperatures required to melt and reprocess PET. Many advances such as electrostatic separations

have already made significant improvement in this area to effect efficient separation of contaminants.

3. Economical methods for chemical recycling of many plastics such as unsaturated polyester or polyurethane systems, and many high-performance engineering polymer systems, are as yet not available. Availability of waste is critical to validate the economics of many current processes.
4. Infrastructural challenges include the availability of a proper facility for disassembly, storage, and sorting. In the absence of appropriate infrastructure, much of the potential recyclable waste ends up in landfill piles.
5. Apart from the above challenges, the recyclability of plastics products is also a limitation. For example, neat resins such as PE and PP can be recycled with minimal changes in their mechanical properties. Not all polymers can withstand repeated thermal cycles and exposure to ultraviolet (UV) light. Suppliers of resins generally add thermal stabilizers and UV stabilizers to resins to increase their recyclability. Filled polymers are more difficult to recycle. In the case of thermoplastics, they can be worked off at different usage levels in compounds for different applications. If a recovered polymer shows a change in property, one of the options is to combine it with a virgin resin in an appropriate proportion to make a new product of required performance. The other option is to chemically convert the recovered polymer into another useful polymer or monomer.

There are steps the industry can take to facilitate higher recycling rates for automotive plastics. First, plastic components and parts can be designed for easy disassembly and dismantling. Second, plastics can be clearly and consistently labeled, to avoid contamination in the recycle stock. Third, designers can try to limit the number and type of incompatible plastics in the vehicle and within any part or component. Fourth, governmental support on higher use of recycled material in automotive products is required. The United States and the EU have passed legislation that encourages the use of recycled material. The EU has set targets to increase the amount of recycled materials by 2015.

In the United States, organized efforts such as the Vehicle Recycling Partnership are targeting the practical recovery of plastic auto parts and recycling automotive vehicle waste. The EU has set high targets for vehicle recycling overall with its EOL Vehicles Directive and ELV Directive

(2000/53/EC) [9-63]. Accordingly, the industry has to increase reuse, recycling, and recovery as follows:

1. As of January 1, 2006, 80% for reuse and recycling and 85% for reuse and recovery
2. As of January 1, 2015, 85% for reuse and recycling and 95% for reuse and recovery

9.4.4 Effect of Recycling on Carbon Footprint Reduction

Recycling contributes to about 5% of the carbon footprint for the life of a product; the bulk of carbon footprint comes from the use phase. When a recycled material is used in a new product, this product's fossil-fuel dependence decreases in proportion to the amount of recycled material used, and accordingly a significant carbon emission saving is achieved. To illustrate, when waste PET is recycled into PBT, the carbon footprint of PBT for the cradle-to-gate phase decreases by 49%, from 5.3 to 2.7 kg CO_2/kg of PBT [9-64]. When the PBT obtained by recycling of postconsumer PET is used in PC/PBT blends for use in energy absorbers, the footprint of the energy absorber becomes lower by 15% than when it is made from fossil-resource-derived PBT. Hence, use of recycled material in a product is an important action toward sustainability. The carbon footprints of various recycled and fossil-resource-derived polymers have been reported by several authors [9-65].

9.4.5 Reuse

The carbon footprint of an automobile part that is reused is essentially zero because no new material or energy is spent in making the part again. Therefore, reuse as a strategy is theoretically attractive from a sustainability angle. However, there are only few examples of plastic parts that can be reused. One ubiquitous example is bumpers. Salvaged plastic bumpers become new bumper reinforcements in the recycling process for Ford Motor Company [9-66]. Toyota has recovered and recycled bumpers for new cars [9-67]. Mazda collects damaged bumpers at the point at which they were replaced with new ones and reuses the bumpers in the production of component parts for cars [9-68].

9.4.6 End-of-Life Scenario for the Future

Efficient recovery of waste plastics and their segregation through identification appears to be the key for efficient recycling, as chemical processes for recycling are known for their many polymer wastes. Regulatory mechanisms will become more stringent to ensure that recovery of waste gets streamlined and recycling will become routine.

Among the new technologies, pyrolysis of plastics waste is gaining momentum [9-62, 9-69]. Fuels similar to diesel or gasoline can be made through such pyrolysis. Pyrolysis also generates small molecules that can be reused as monomers to make polymers through known polymerization processes. Currently, the pyrolysis of nonoxygenated polymers such as polyethylene, polypropylene, and polystyrene appears more promising than that of oxygenated polymers such as polycarbonate.

9.5 Summary

Environmental sustainability efforts in the automotive industry are multifarious. The key observations are the following:

1. Use of plastics in lightweighting solutions is a powerful tool in automotive sustainability to reduce carbon footprint. It has the potential to enable OEMs and material suppliers to meet the exacting regulatory standards of the future on fuel efficiency. Many thermoplastics systems show promise in effectively replacing either metal or a thermoset plastics to achieve maximum weight savings.
2. Sourcing raw materials from renewable resources is becoming a reality to reduce the carbon footprint of automotive products. Only a few renewably sourced raw materials are now available, but many of these materials are in an advanced stage of development and commercialization.
3. Recycling of plastics is now a reality in practice, yet the efficiency of recycling depends on the availability of segregated waste plastics. It enables the lowering of the carbon footprint for products into which the waste has been recycled.

9.6 Trends

The following points summarize the trends discussed in this chapter:

1. With advances in the development of carbon fiber from biomass sources, carbon-fiber-reinforced plastics will find newer applications in future vehicles, especially in lightweighting applications of structural components. Advanced high-performance steel is fast becoming a competitive material to plastics in metal-replacement applications.
2. Use of cellulosic fibers in place of glass fibers in many products will increase. The availability of bio-based or renewable raw materials such as bio-propylene at an appropriate cost may become a reality because of the ongoing research and development in biochemical

process improvements. However, new chemical innovation is needed to develop bio-based materials for high-performance polymers. New building blocks from renewable sources are being explored for making bio-based polymers for automotive applications.

3. Designing for sustainability is a concept gaining popularity among automobile manufacturers and material suppliers. New technologies will be developed for making plastic products for improved recyclability, contributing to sustainability in the automotive industry.
4. Pyrolysis of waste products may become one of the viable EOL options for waste plastics, particularly for polyolefin wastes. Waste products made of thermoset composites are difficult to recycle and often get disposed of in landfills. Pyrolysis of thermoset composites could also become a sustainable EOL option.

9.7 References

9-1. (a) "Emissions Scenarios from Intergovernmental Panel for Climate Change," IPCC, http://www.ipcc.ch/pdf/special-reports/spm/sres-en.pdf. Accessed September 23, 2012.

(b) "Glossary," Ministry Environment, British Columbia, http://www.env.gov.bc.ca/cas/resources/glossary.html. Accessed September 23, 2012.

(c) Seinfeld, J.H. 2011. "Insights on Global Warming." *AIChE Journal 57* (12):3259–3284.

(d) Kamimoto, M., et al., 2011. "Renewable Energy and Climate Change," in *IPCC, 2011: IPCC Special Report on Renewable Energy Sources and Climate Change Mitigation. Prepared by Working Group III of the Intergovernmental Panel on Climate Change*, edited by O. Edenhofer, R. Pichs-Madruga, Y. Sokona, K. Seyboth, P. Matschoss, S. Kadner, T. Zwickel, P. Eickemeier, G. Hansen, S. Schlömer, and C. von Stechow. Cambridge, UK, and New York: Cambridge University Press, p. 164.

9-2. "Greenhouse Gas Emissions," EPA, United States Environmental Protection Agency, http://www.epa.gov/climatechange/ghgemissions/gases/co2.html. Accessed September 23, 2012.

9-3. (a) "Sustainable Mobility: Automotive Industry Challenges, Opportunities and the Role of PLM," Ora Research LLC and Sustainability Guild, LLC, http://www.plm.automation.siemens.com/zh_cn/Images/SMAI_PLM_091129_tcm78-92089.pdf. Accessed October 17, 2012.

b) "Mobility 2030," World Business Council for Sustainable Development (WBCSD), http://www.wbcsd.org/pages/edocument/edocumentdetails.aspx?id=69. Accessed September 23, 2012.

9-4. "Energy Savings by Light-Weighting—Final Report Commissioned by International Aluminium Institute (IAI)," Institute for Energy and Environmental Research (IFEU), http://www.world-aluminium.org/media/filer/2012/06/12/fl0000125.pdf. Accessed September 24, 2012.

9-5. "40 CFR Parts 85, 86 and 600," Department of Transportation, Environmental Protection Agency, Federal Register, http://www.nhtsa.gov/staticfiles/rulemaking/pdf/cafe/2017-25_CAFE_Final_Rule.pdf. Accessed November 9, 2012.

9-6. "Road Transport: Reducing CO2 Emissions from Vehicles," European Commission, http://ec.europa.eu/clima/policies/transport/vehicles/index_en.htm. Accessed November 9, 2012.

9-7. (a) Cheah, L., Evans. C., Bandivadekar, A. and Heywood, J. 2007. "Factor of Two: Halving the Fuel Consumption of New U.S. Automobiles by 2035," Publication No. LFEE 2007-04 RP. Cambridge, MA: Laboratory for Energy and Environment, Massachusetts Institute of Technology.

(b) Cheah, L. and Heywood, J. 2011. "Meeting U.S. Passenger Vehicle Fuel Economy Standards in 2016 And Beyond." *Energy Policy* 39:454–466.

9-8. "Plastics in Automotive Markets Technology Roadmap—A New Vision for the Road Ahead," American Chemical Council Plastics Division, http://www.energetics.com/resourcecenter/products/roadmaps/Documents/Plastics-Auto-Markets-Roadmap.pdf. Accessed October 17, 2012.

9-9. Rehkopf, J.D. 2011. *Automotive Carbon Fiber Composites: From Evolution to Implementation*. Warrendale, PA: SAE International, p. 40.

9-10. Tolinski, M. 2012. *Plastics and Sustainability towards a Peaceful Coexistence between Bio-Based and Fossil Fuel-Based Plastics*. Salem, MA: Wiley-Scrivener.

9-11. "2012 Automotive Sustainability Report," Society of Motor Manufacturers and Traders (SMMT) of the UK, http://www.smmt.co.uk/sustainability/#environmental-performance. Accessed September 23, 2012.

9-12. "Plastics the future of Automakers and Chemical Companies," A.T. Kearney Incorporated, Korea, http://www.atkearney.com/

documents/10192/28dcce52-affb-4c0b-9713-a2a57b9d753e. Accessed September 23, 2012.

9-13. Powers, W.F. 2000. "Automotive Materials in the 21st century." *Advanced Materials and Processes* 157(5): 38–41.

9-14. Mallick, P.K. 2007. *Fiber-Reinforced Composites Material, Manufacturing and Design* (3rd Edition; Dekker Mechanical Engineering). London: CRC Press, Taylor & Francis Group.

9-15. Ribeiro, C., Ferreira, J.V. and Partidário, P. 2007. "Life Cycle Assessment of a Multi-Material Car Component." *International Journal of Life Cycle Assessment* 12(5): 336–345.

9-16. Baroth, A., Karanam, S. and McKay, R. 2012. "Life Cycle Assessment of Lightweight Noryl* GTX* Resin Fender and Its Comparison with Steel Fender." SAE Paper No. 2012-01-0650. SAE International, Warrendale, PA. SAE 2012 World Congress and Exhibition, Detroit, MI.

9-17. "Sustainability Solutions," SABIC Innovative Plastics, http://www.pod-sabic-ip.com/KBAM/Reflection/Assets/Thumbnail/19924_5.pdf. Accessed on September 23, 2012.

9-18. ASM International Materials Life-Cycle Analysis Committee. 1997. "Life-Cycle Engineering and Design." *ASM Handbook* 20: 96–103. http://www.asminternational.org/content/ASM/StoreFiles/06481G_Sample.pdf. Accessed September 20, 2012.

9-19. "Greater Range Thanks to Lightweight Polycarbonate Glazing," Alliance for Sustainable Built Environments, http://www.greenerfacilities.org/Alliance_Member_Releases_Details.php?id=572. Accessed October 17, 2012.

9-20. "SABIC to Increase STAMAX* LGFPP Capacity at Genk, Belgium Facility to Meet Rising Automotive Customer Demand," SABIC, http://www.sabic-ip.com/gep/en/NewsRoom/PressReleaseDetail/july_24_2012_sabictoincreasestamax.html. Accessed September 23, 2012.

9-21. Dubreuil, A., Bushi, L., Das, S. and Tharumarajah, A. 2010. "A Comparative Life Cycle Assessment of Magnesium Front End Autoparts," SAE Paper No. 2010-01-0275. SAE International, Warrendale, PA. SAE 2010 World Congress and Exhibition. Detroit, MI.

9-22. Baroth, A., Karanam, S., and McKay, R. 2012. "Quantifying Environmental Benefits of Metal Replacement with Engineering Thermoplastic Resins—An Automotive Steering Wheel Case Study,"

Conference proceeding US LCA XII, Tacoma, WA." http://lcacenter.org/lcaxii/sessions/abstract-dynamic.php?id=628. Accessed October 17, 2012.

9-23. Kasai, J. 1999. "Life Cycle Assessment, Evaluation Method for Sustainable Development." *JSAE Review* 20: 387–393.

9-24. "Nature's Way to Mobility—Overview Report," WorldAutoSteel, pp. 55–56, http://c315221.r21.cf1.rackcdn.com/FSV_OverviewReport_Phase2_FINAL_20110430.pdf. Accessed October 18, 2012.

9-25. "New Fuel Economy Standards Would Boost Average Car to 54.5 mpg," *Los Angeles Times*, http://articles.latimes.com/2012/aug/28/nation/la-na-nn-obama-fuel-economy-mileage-20120828. Accessed October 17, 2012.

9-26. "Automotive Industry Seeks Light-Weight and High Performance Materials," ICIS.Com, http://www.icis.com/Articles/2012/07/27/9581541/automotive-industry-seeks-light-weight-and-high-performance.html. Accessed October 16, 2012.

9-27. Cackette, T., and California Air Resources Board. 2010. "Reducing Greenhouse Gases from Passenger Vehicles—2017 Models On," http://www.autosteel.org/sitecore/Content/Internal/Listers/NewsLister.aspx?targetItem=1b6e05b6-64b1-4d68-a879-ff032b9c35de&siteLocation=4971bb64-d151-4db4-9d28-e71c4338f3e6&contentSource=Primary. Accessed September 23, 2012.

9-28. "Design for Sustainability (D4S)," United Nations Environment Programme, Division of Technology, Industry, and Economics—Sustainable Consumption and Production Branch, http://www.unep.fr/scp/design/d4s.htm Accessed October 19, 2012.

9-29. Mayyas, A., Qattawi, A., Omar, M., and Shan, D. 2012. "Design for Sustainability in Automotive Industry: A Comprehensive Review." *Renewable and Sustainable Energy Reviews* 16: 1845–1862.

9-30. "The Knowledge Based Bio-Economy (KBBE) in Europe: Achievements and Challenges—Full Report," Flemish Government, Belgium, European Commission, p. 57, http://sectie.ewi-vlaanderen.be/sites/default/files/documents/KBBE_A4_1_Full%20report_final.pdf. Accessed September 23, 2012.

9-31. "Choosing More Sustainable Materials," Ford Company, http://corporate.ford.com/microsites/sustainability-report-2011-12/environment-products-materials-choosing. Accessed September 23, 2012.

9-32. "Clean Technologies: A Clean Car for Everyone," PSA Peugeot Citroen, http://www.psa-peugeot-citroen.com/en/inside-our-industrial-environment/innovation-and-research-development/clean-technologies-a-clean-car-for-everyone. Accessed September 23, 2012.

9-33. "Life Cycle Analysis Concludes Green Polyethylene Can Reduce GHG Emission Even with Half a World Away Transportation," Toyota Tsusho Corporation, http://www.toyota-tsusho.com/english/press/2010/11/20101118-3578.html. Accessed September 23, 2012.

9-34. "Bio-Plastics: 42% Reduction in Carbon Footprint," Bioplastics Innovation, http://bioplastic-innovation.com/2008/12/04/bio-plastics-42-reduction-in-carbon-footprint/#more-1565. Accessed September 23, 2012.

9-35. Hermann, B.G., Blok, K. and Patel, M.K. 2007. "Producing Bio-Based Bulk Chemicals Using Industrial Biotechnology Saves Energy and Combats Climate Change." *Environmental Science & Technology* 41: 7915–7921.

9-36. "Bio-Based Polymers Could Be Next Big Thing," Hydrocarbon Processing, http://www.hydrocarbonprocessing.com/Article/2966827/Bio-based-polymers-could-be-next-big-thing.html. Accessed September 23, 2012.

9-37. Sain, M., et al., 2004. "Green Fiber Thermoplastic Composites," in *Green Composites—Polymer Composites and the Environment*, edited by C. Baillie. Cambridge, UK: Woodhead, pp. 181–206.

9-38. Joshi, S.V., Drzal, L.T., Mohanty, A.K. and Arora, S. 2004. "Are Natural Fiber Composites Environmentally Superior to Glass Fiber Reinforced Composites?" *Composites: Part A* 35: 371–376.

9-39. Schmidt, W.P. and Beyer, H.M. 1998. "Life Cycle Study on a Natural Fiber Reinforced Component." SAE Technical Paper No. 982195. SAE International, Warrendale, PA. SAE Total Life-cycle Conference, Graz, Austria.

9-40. Corbiere-Nicollier, T., Gefeller, L.B., Lundquist, L., Leterrier, Y., Manson, J.A.E. and Jolliet, O. 2001. "Lifecycle Assessment of Bio Fibers Replacing Glass Fibers as Reinforcement in Plastics." *Resources Conservation and Recycling* 33: 267–287.

9-41. "Bio-Based Material Automotive Value Chain," Centre for Automotive Research (CAR), http://www.cargroup.org/?module=Publications&event=View&pubID=29. Accessed September 23, 2012.

9-42. "Renewable Materials for Automotive Applications," Interactive European Network for Industrial Crops and their Applications (IENICA), http://www.ienica.net/fibresseminar/schuh.pdf. Accessed September 23, 2012.

9-43. "Use of Renewable Resources," Toyota, http://www.toyota-global.com/sustainability/environmental_responsibility/automobile_recycling/design_for_recycling/use_of_renewable_resources.html. Accessed September 23, 2012.

9-44. "It Might Just Happen," in *REVIEW—The Next Generation of Biofuels*, Oak Ridge National Laboratory, http://www.ornl.gov/info/ornlreview/v43_3_10/article05.shtml. Accessed September 23, 2012.

9-45. Shimizu, K. and Sudo, K. 1987. "Production of Lignin Based Carbon Fiber." Patent Publication No. JP62110922A.

9-46. (a) "Application of Ecoprofile Methodology to Polyamide 11," Arkema, http://www.arkema.com/en/products/product-finder/range-viewer/Rilsan-Arkema/ Accessed September 23, 2012;

(b) "Castor Oil to the Auto Rescue," ICIS Green Chemicals, http://www.icis.com/blogs/green-chemicals/2012/04/castor-oil-to-the-auto-rescue.html. Accessed September 23, 2012.

9-47. (a) "Evonik Launches First Fully Vegetable Based Nylon-10, 10 Powder," Cosmetics Business, http://www.cosmeticsbusiness.com/news/article_page/Evonik_launches_first_fully_vegetable_based_Nylon1010_powder/75438. Accessed September 23, 2012;

(b) "Life Cycle Assessment of Bio Based Polyamides," Evonik, http://www.vestamid.com/sites/dc/Downloadcenter/Evonik/Product/VESTAMID/en/technical-literature/VESTAMID%20Terra%20-%20Life%20Cycle%20Analysis.pdf. Accessed September 23, 2012.

9-48. "Trends in Bio-renewable Thermoplastic Elastomers," Plastics Trends, http://www.plasticstrends.net/index.php/component/content/article/1-this-month/276-trends-in-bio-renewable-thermoplastics-elastomers. Accessed September 23, 2012.

9-49. "Epoxy Resins New Reports Alert," Nexant's Chem Systems, http://www.chemsystems.com/reports/search/docs/abstracts/0405S11_abs.pdf. Accessed September 23, 2012.

9-50. Loza, R. and McDaniel, P. 2001. "Process of Preparing Polyester Resins." Patent Publication No. US6222005B1.

9-51. "Bio-Composites Update: Bio-Based Resins Begin to Grow," Composites World, http://www.compositesworld.com/articles/

bio-composites-update-bio-based-resins-begin-to-grow. Accessed September 23, 2012.

9-52. Chiao, W., Hung, T.F. and Yang S.Y. 2012. "Improvement in Toughness and Heat Resistance of Poly(Lactic Acid)/Polycarbonate Blend through Twin-Screw Blending: Influence of Compatibilizer Type." *Journal of Applied Polymer Science* 125(S2): E402–412.

9-53. Fuessl, A., Yamamoto, M. and Schneller, A. 2012. "5.03—Opportunities in Bio Based Building Blocks for Polycondensates and Vinyl Polymers." *Polymer Science—A Comprehensive Reference* 5: 49–70.

9-54. "Isosorbide as a New Building Block for Polymers and Plasticizers," Roquette Group, http://www.vinyl.org.au/Assets/314/1/1103Fuertes-BioHubRoquette.pdf. Accessed September 23, 2012.

9-55. Dingerdissen, U., Haas, T., Korek, U., Ortelt, M., Pfeffer, J., Schmidt, H. and Spyrou, E. 2011. "Use of Isocyanates Based on Renewable Raw Materials." Patent Publication No. WO2011000586A1.

9-56. "A review on Bio-based Chemicals—Value Added Products from Bio-refineries," IEA Bio Energy, http://www.ieabioenergy.com/LibItem.aspx?id=7311. Accessed September 5, 2012.

9-57. (a) "Big Environmental Footprints for Bio-Based Chemicals," Polymer Solutions Incorporated, http://www.polymersolutions.com/blog/big-environmental-footprints-for-bio-based-chemicals. Accessed September 23, 2012;

(b) "Bio-based Chemicals under Scrutiny," RSC, http://www.rsc.org/chemistryworld/News/2011/September/23091102.asp. Accessed September 23, 2012.

9-58. "Steel: Driving Auto Recycling Success," Steel Recycling Institute, http://www.recycle-steel.org/en/Steel%20Markets/Automotive.aspx. Accessed October 16, 2012.

9-59. Jody, B.J. and Daniels, E.J. 2006. "End-of-Life Vehicle Recycling: The State of the Art of Resource Recovery from Shredder Residue." Argonne National Laboratory. Publication No. ANL/ESD/07-8.

9-60. (a) "The European Market for Plastics Automotive Components," EuPC, http://www.plasticsconverters.eu/organisation/division/automotive. Accessed September 23, 2012;

(b) http://www.plasticsconverters.eu/markets/automotive. Accessed September 23, 2012.

9-61. Thomas, R., et al., 2011. "Recycling of Thermosetting Polymers: Their Blends and Composites," in *Recent Developments in Polymer Recycling,* edited by A. Fainleib and O. Grigoryeva. Kerala, India: Transworld Research Network, pp. 121–153.

9-62. Allred, R.E. and Busselle, L.D. 2000. "Tertiary Recycling of Automotive Plastics and Composites." *Journal of Thermoplastic Composite Materials* 13(2): 92–101.

9-63. "A Study to Examine the Benefits of the End-of-Life Vehicles Directive and the Costs and Benefits of a Revision of the 2015 Targets for Recycling, Re-use and Recovery Under the ELV Directive," GHK, http://ec.europa.eu/environment/waste/pdf/study/final_report.pdf. Accessed September 23, 2012.

9-64. (a) "Environmental Benefits of Using Post-Consumer Recycled PET/PC as Feedstock in Manufacturing Engineering Thermoplastic Resins," USLCAXI http://lcacenter.org/lcaxi/final/363.pdf. Accessed September 23, 2012

(b) "Environmental Product Datasheet—Xenoy iQ," SABIC, http://kbam.geampod.com/KBAM/Reflection/Assets/20334_2.pdf. Accessed September 23, 2012.

9-65. Vogtländer, J.G. 2011. *A Quick Reference Guide to LCA DATA and Eco-based Materials Selection*. VSSD Science and Technology, Delft University, The Netherlands.

9-66. "Ford's Bumper Recycling Scheme Uses Thousands of Parts," European Motor News, http://europeanmotornews.com/2010/02/09/fords-bumper-recycling-scheme-reuses-thousands-of-parts/. Accessed September 24, 2012.

9-67. "Expanding the Reuse of Parts and Promoting Their Recycling," Toyota, http://www.toyota-global.com/sustainability/environmental_responsibility/automobile_recycling/recycling_at_dealers/expanding_the_use_of_used_parts.html. Accessed September 24, 2012.

9-68. "Bumper to Bumper Recycling—Reusing Recycled Bumpers in New Cars," Mazda, http://www.mazda.com/csr/environment/special_features/2010_02_01.html. Accessed September 24, 2012.

9-69. Scheris, J. 2006. "Overview of Commercial Pyrolysis Processes for Waste Plastics," in *Feedstock Recycling and Pyrolysis of Waste Plastics: Converting Waste Plastics into Diesel and Other Fuels*, edited by W. Kaminsky, and J. Scheris. Sussex, UK: Wiley.

Abbreviations

2K-ICM	two-shot injection compression molding
ACEA	Association of European Car Manufacturers
AHSS	advanced high-strength steel
BIW	body in white
BMC	bulk molding compound
CAD	computer-aided design
CAE	computer-aided engineering
CAFE	Corporate Average Fuel Economy
CCFC	continuous carbon-fiber composite
CCHS	closed-channel hybrid structure
CED	cumulative energy demand
CFD	computational fluid dynamics
CGFC	continuous glass-fiber composite
CHMS	Lcenter high-mount stop lamp
CLTE	coefficient of linear thermal expansion
CTE	coefficient of thermal expansion
DfE	Design for the Environment
DMA	dynamic mechanical analyzer
DoE	design of experiments
DSC	differential scanning calorimetry
EA	energy absorber
ECE	Economic Commission for Europe
EEVC	European Enhanced Vehicle-Safety Committee
EMI	electromagnetic interference
EOL	end-of-life
EPP	expanded polypropylene
ETB	electronic throttle body
ETP	engineering thermoplastics
EU	European Union

Euro NCAP	European New Car Assessment Program
EV	electric vehicle
FE	finite element
FEA	finite element analysis
FEC	front-end carrier
FEM	front-end module
FEP	fluorinated ethylene propylene
FMVSS	Federal Motor Vehicle Safety Standards
GHG	greenhouse gas
GoR	grille opening reinforcement
GVWR	gross vehicle weight rating
H	horizontal
HB	high beam
HDT	heat-deflection temperature
HIC	head injury criteria
HID	high-intensity discharge
HPS	hybrid plastic structure
HVAC	heating, ventilation, and air conditioning
ICM	injection compression molding
IHRA	International Harmonized Research Activities
IIHS	Insurance Institute for Highway Safety
IP	instrument panel
IR	infrared
ISO	International Standards Organization
LB	low beam
LCA	life cycle assessment
LDS	laser direct structuring
LED	light-emitting diode
MLIT	Ministry of Land, Infrastructure, Transport and Tourism
MTB	mechanical throttle body
NCAP	New Car Assessment Program

NVH	noise, vibration, and harshness
NHTSA	National Highway Traffic Safety Administration
OCHS	open-channel hybrid structure
OEM	original equipment manufacturer
PA	polyamide
PBT	polybutylene terephthalate
PC	polycarbonate
PC-Si	polycarbonate-siloxane
PCR	postconsumer recycling
PEI	polyetherimide
PIR	postindustrial recycling
PMH	plastic-metal hybrid
PPO	polyphenylene oxide
PPS	polyphenylene sulfide
PSH	pressure sequence hydroforming
PVC	polyvinyl chloride
PVDF	polyvinylidene fluoride
RCAR	Research Council for Automobile Repairs
REACH	Registration, Restriction, Evaluation, and Authorization of Chemicals
SUV	sports utility vehicle
TRL	Transport Research Laboratory
TRS	thermorheologically simple
TTS	time temperature superposition
UL	Underwriters Laboratories
UNEP	United Nations Environment Program
UTH	under the hood
V	vertical
VoC	voice of the customer

Index

About the Editor

Dr. Sudhakar R. Marur has more than 23 years of technical and leadership experience in industrial R&D. His areas of expertise include computational mechanics, mechanical design, and CAE for engineering thermoplastics and thermoset composite materials. He received a PhD from the Indian Institute of Technology (IIT), Bombay, in computational nonlinear structural dynamics, and he later continued his postdoctoral work on nonlinear vibrations and elementology at National Aerospace Labs. A certified six-sigma Black Belt, he has several scientific journal publications, patents, and corporate awards to his credit. Under his leadership, the plastics application technology team has successfully developed several applications for various OEMs across the globe.

Contributors

Anju Baroth: Anju obtained her PhD from Rajasthan University for her research work on environmental toxicology. She has over seven years of experience in industry in areas such as environment management, toxicity assessment, life cycle assessment (LCA) and sustainability. She has been contributing to the LCA studies of materials and various plastic applications for the automotive sector.

P. Arunachala: Arunachala holds a master's degree in tool engineering from NTTF, Bangalore. He has over six years of industrial experience in automotive plastics part design, sheet metal, and tool design. Many of his CAD designs have been successfully implemented in India and abroad.

Dhanendra K. Nagwanshi: Dhanendra is an alumnus of the Indian Institute of Technology (IIT), Delhi, majoring in mechanical design. He has over eight years of industrial experience, primarily in design and development of energy management solutions using plastics and hybrids. His areas of interests include design for crashworthiness, concurrent mechanical engineering design, and dynamic design. He has several successful energy absorber translations, papers, patents, and awards to his credit.

Dinesh Mana: Dinesh received his MS in mechanical design from the Indian Institute of Science (IISc), Bangalore. He has over five years of industrial experience in CAE and plastics design. He specializes in automotive crashworthiness, topology optimization, compliant mechanisms, and MEMS. His energy-absorber designs are part of many automobile models on the road.

N. Krishnamoorthy: Krish is an alumnus of the Central Institute of Plastics Engineering and Technology (CIPET), Chennai, through a master's degree program in plastics engineering. He has over 13 years of industrial experience and specializes in plastics product design, material selection, processing, CAE, and application development. He has successfully developed many applications for various automotive companies around the globe.

S.M. Kumar: Kumar holds a master's degree in tool engineering from NTTF, Bangalore. He has over 17 years of experience in plastics industry. He specializes in plastics product design, injection molding solutions, tool design, and CAE and has developed many successful automotive applications for the global market.

Narendra A. Hardikar: Naren graduated from the Indian Institute of Technology (IIT), Bombay, with a master's degree in aerospace engineering and has more than 13 years of industrial experience. His specialization includes mechanical design for plastics, CAE, material modeling, and multiphysics problems. He has developed many plastic part designs, which have been adopted by automotive companies.

Neena Chandramathy: Neena graduated from Kerala University with a master's degree in environmental engineering and has been working in the industry for over nine years in the area of environmental engineering and management. Her current areas of focus include sustainability and LCA.

T.K. Poovanna: Poovanna received his master's degree in tool engineering from NTTF, Bangalore. He has worked in plastics R&D for more than a decade. His areas of expertise include plastics product design, plastic processing design, sheet metal formability, and tool design. As a technologist, he has developed many successful applications, and as a prolific inventor, he has several patents and awards to his credit.

K.R.K. Ravikanth: Ravikanth completed his master's in plastics engineering from the University Institute of Chemical Technology (UICT), Mumbai. He has over 15 years of experience in the plastics industry and specializes in injection-molding solutions for automotive applications with various materials, CAE, and tool design. His plastic designs and applications have been part of many automotive models across the globe.

Sandeep C. Kulkarni: Sandeep obtained his PhD from Indian Institute of Technology (IIT), Bombay, in structural engineering. He has over six years of experience in plastic application development. His areas of interests include crash and safety analysis, computational mechanics, anisotropic material, and failure modeling. He has many successful application translations to his credit.

Somasekhar Bobba: Soma is an alumnus of Indian Institute of Technology (IIT), Chennai, through its doctoral program in machine design. He has over five years of industrial experience in mechanical design and CAE for automotive safety and crash energy management. He has contributed to many successful energy-absorber designs for various automotive companies.

Srinivasan Krishnaswami: Srini holds a PhD in organic chemistry from Vanderbilt University, Tennessee. He did his postdoctoral research in physical organic chemistry from the University of Houston. He has over 26 years of industrial experience in rubber chemicals, surfactants, surface coatings and engineering thermoplastics. Currently, he is specializing in intellectual property work products, technology, and market intelligence, particularly with reference to sustainability platforms.

Triloka Tankala: Triloka got his M.Tech. degree in thermal and fluids engineering from the Indian Institute of Technology (IIT), Bombay. He has a decade of experience in the area of computational fluid dynamics and heat transfer applications with plastics such as headlamp assembly, electronic throttle body, LED, steering wheel, and hard disk drive (HDD). He has many papers and patents in this area and awards for successful translations.

N. Venkatesha: Venkatesha graduated from NTTF, Bangalore, with specialization in tool engineering. During his 11-plus years of experience in the industry, he has specialized in plastics application development, processing technologies, metal-to-plastic replacement applications, product design, CAD, and CAE. He has several presentations and patents based on his successful plastic automotive applications.